Undergraduate Topics in Computer Science

Undergraduate Topics in Computer Science (UTiCS) delivers high-quality instructional content for undergraduates studying in all areas of computing and information science. From core foundational and theoretical material to final-year topics and applications, UTiCS books take a fresh, concise, and modern approach and are ideal for self-study or for a one- or two-semester course. The texts are all authored by established experts in their fields, reviewed by an international advisory board, and contain numerous examples and problems. Many include fully worked solutions.

Faron Moller · Georg Struth

Modelling Computing Systems

Mathematics for Computer Science

 Springer

Faron Moller
Department of Computer Science
Swansea University
Swansea, UK

Georg Struth
Dept. Computer Science
University of Sheffield
Sheffield, UK

ISSN 1863-7310   Undergraduate Topics in Computer Science
ISBN 978-1-84800-321-7   ISBN 978-1-84800-322-4 (eBook)
DOI 10.1007/978-1-84800-322-4
Springer London Heidelberg New York Dordrecht

Library of Congress Control Number: 2013943907

Contents

Starred sections are optional and often represent advanced material.

List of Figures

Preface

The good news about computers is that they do what you tell them to do. The bad news is that they do what you tell them to do.

- Ted Nelson.

Computer Science is a relatively young discipline. University Computer Science Departments are rarely more than a few decades old. They will typically have emerged either from a Mathematics Department or an Engineering Department, and until recently a Computer Science degree was predominantly about writing computer programs (the mathematical software) and building computers (the engineering hardware). Textbooks typically referred to programming as an "art" or a "craft" with little scientific basis compared to traditional engineering subjects, and many computer programmers still like to see themselves as part of a pop culture of geeks and hackers rather than as academically-trained professionals.

However, the nature of Computer Science is changing rapidly, reflecting the increasing ubiquity and importance of its subject matter. In the last decades, computational methods and tools have revolutionised the sciences, engineering and technology. Computational concepts and techniques are starting to influence the way we think, reason and tackle problems; and computing systems have become an integral part of our professional, economic and social lives. The more we depend on these systems – particularly for safety-critical or economically-critical applications – the more we must ensure that they are safe, reliable and well designed, and the less forgiving we can be of failures, delays or inconveniences caused by the notorious "computer glitch."

Unlike traditional engineering disciplines which are solidly rooted on centuries-old mathematical theories, the mathematical foundations underlying Computer Science are younger, and Computer Scientists have yet to agree on how best to approach the fundamental concepts and tasks in the design of computing systems. The Civil Engineer knows exactly how to define and analyse a mathematical model of the components of a bridge design so that it can be relied on not to fall down, and the Aeronautical Engineer knows exactly how to define and analyse a mathematical model of an aeroplane wing for the same purpose. However, Software Engineers have few universally-accepted mathematical modelling tools at their disposal. In the words of the eminent Computer Scientist Alan Kay, "most undergraduate degrees in computer science these days are basically Java vocational training." But computing systems can be at least as complex as bridges or aeroplanes, and a canon of mathematical methods for modelling computing systems is therefore very much needed. "Software's Chronic Crisis" was the title of a popular and widely-cited Scientific American article from 1994, and, unfortunately, its message remains valid today.

University Computer Science Departments face a sociological challenge posed by the fact that computers have become everyday, deceptively easy-to-use objects. A single generation ago, new Computer Science students typically had teenage backgrounds spent writing Basic and/or Assembly Language programs for their early hobbyist computers. A passion for this activity is what drove these students into University Computer Science programmes, and they were not disappointed with the education they received. Their modern-day successors on the other hand – born directly into the heart of the computer era – have grown up with the internet, a billion dollar computer games industry, and mobile phones with more computing power than the space shuttle. They often choose to study Computer Science on the basis of having a passion for using computing devices throughout their everyday lives, for everything from socialising with their friends to downloading the latest films, and they often have less regard than they might to the considerations of what a University Computer Science programme entails, that it is far more than just using computers.

There is a universal trend of large numbers of first-year students transferring out of Computer Science programmes and into related programmes such as Media Studies or Information Studies. This trend, we feel, is often unjustified, and can be reversed by a more considered approach to modelling and the mathematical foundations of system design, one which the students can connect and feel at home with right from the beginning of their University education. This has been our motivation in writing this textbook aimed at teaching first-year undergraduate students the essential mathematics and modelling techniques for computing systems in a novel and relatively lightweight way.

The book is divided into two parts. Part I, subtitled *Mathematics for Computer Science*, introduces concepts from Discrete Mathematics which are in the curriculum of any University Computer Science programme, as well as much which often is not. This material is typically taught in service modules by mathematicians, and new Computer Science students often find it difficult to engage with the material presented in a purely mathematical context. We attempt here to present the material in an engaging and motivating fashion as the basis of computational thinking.

Part II of the book – *Modelling Computing Systems* – develops a particular approach to modelling based on state transition systems. State transition systems have always featured in the Computer Science curriculum, but traditionally (and increasingly historically) only within the study of formal languages. Here we introduce them as general modelling devices, and explore languages and techniques for expressing and reasoning about system specifications and (concurrent) implementations. Although Part I covers twice as many pages as Part II, the title of the book is nonetheless justified: much of the Mathematics presented in Part I itself is used directly for modelling systems, and forms the basis on which the approach developed in Part II is based.

The main benefit of mathematical formalisation is that systems can be modelled and analysed in precise and unambiguous ways; but formal precision can also be a major pitfall in modelling since it can compromise simplicity and intuition. In this book, therefore, we always try to start from intuition and examples, and we aim at developing precise concepts from that basis. How and when to be precise is certainly not less important to learn than precision itself: the ability to give mathematical proofs often does not depend on knowing precise formal definitions and foundations. One can, for example, write down recursive functions without having a precise formal concept in mind.

There is a long standing tradition in disciplines like Physics to teach modelling through little artifacts. The fundamental ideas of computational modelling and thinking as well can better be learned from idealised examples and exercises than from many real world computer applications. This book builds on a large collection of logical puzzles and mathematical games that require no prior knowledge about computers and computing systems; these can be much more fun and sometimes much more challenging than analysing a device driver or a criminal record database. Also, computational modelling and thinking is about much more than just computers!

In fact, games play a far more important role in the book: they provide a novel approach to understanding computer software and systems which is proving to be very successful both in theory and practice. When a computer runs a program, for example, it is in a sense playing a game against the user who is providing the input to the program. The program represents a strategy which the computer is using in this game, and the computer wins the game if it correctly computes the result. In this game, the user is the adversary of the computer and is naturally trying to confound the computer, which itself is attempting to defend its claim that it is computing correctly, that is, that the program it is running is a winning strategy. (In Software Engineering, this game appears in the guise of *testing*.) Similarly, the controller of a software system that interacts with its environment plays a game against the environment: the controller tries to maintain the system's properties, while the environment tries to confound them.

This view suggests an approach to addressing three basic problems in the design of computing systems:

1. *Specification* refers to the problem of precisely identifying the task to be solved, as well as what exactly constitutes a solution. This problem corresponds to the problem of defining a winning strategy.

2. *Implementation* or *Synthesis* refers to the problem of devising a solution to the task which respects the specification. This problem corresponds to the problem of implementing a winning strategy.

3. *Verification* refers to the problem of demonstrating that the devised solution does indeed respect the specification. This problem corresponds to the problem of proving that a given strategy is in fact a winning strategy.

This analogy between the fundamental concepts in Software Engineering on the one hand, and games and strategies on the other, provides a mode of computational thinking which comes naturally to the human mind, and can be readily exploited to explain and understand Software Engineering concepts and their applications. It also motivates our thesis that Game Theory provides a paradigm for understanding the nature of computation.

There are over 200 exercises presented throughout each chapter, all of which have complete solutions at the back of the book, as well as over 200 futher exercises at the end of each chapter whose solutions are not provided. The exercises within the chapters are often used to explore subtleties or side-issues, or simply to put lengthy arguments into an appendix, and as such should all be attempted; their solutions at the back of the book should be looked at as well, as they often explain the issues which the exercises are attempting to highlight.

Most of the material in this book has been used successfully for over a decade in first-year Discrete Mathematics and Systems Modelling modules. Countless eyes have passed over the text, and a thousand students have solved its exercises. Nonetheless there will inevitably be a (hopefully small) flurry of errors in the text for which we accept full responsibility and offer our sincere apologies.

Faron Moller   Georg Struth
Swansea   Sheffield

Chapter 0

Introduction

... for by the error of some calculator the vessel often splits upon a rock that should have reached a friendly pier ...

— Henry David Thoreau.

We all know from personal experience that computers do not work correctly all the time. For most of us this realisation manifests itself with nothing more serious than delays and frustrations as we encounter automatic bank tellers which are out-of-order or Web sites which are faulty, or face long waits at airports as glitches in the booking, check-in, or even the flight control system are being catered for.

However, the problems of systems failures become more serious (costly, deadly) as automatic control systems find their way into almost every aspect of our daily lives. It is recognised – and accepted – that complete reliability of any major software system is beyond expectation. While, for example, civil and mechanical engineers can build impressive bridges which are guaranteed to remain standing, and aeronautical engineers can design aeroplane wings which behave in precise and predicatable ways, Software Engineers are almost never so successful. Computers carry viruses, hang, crash or die; and their software is full of security leaks and bugs. Designing dependable high quality computing systems remains a challenge for Software Engineers.

Quality expectations, of course, have much to do with culture and context. Many of us are willing to accept as a mere inconvenience that a train can be delayed, a cash machine can be out of order, or a mail server can temporarily be down. But we don't tolerate bridges that fall down, nuclear meltdowns in power stations, or security leaks in public data bases. Engineers speak about *safety critical* applications when system failure cannot be tolerated, but we should expect software systems to be user friendly, safe and dependable in any application context. Why is this so difficult to achieve?

There are several answers to this question. One of them is that software systems can be extremely complex – more complex even than most other systems that Engineers can design and build. They may consist of large numbers of heterogeneous components that can change over time and interact

F. Moller, G. Struth, *Modelling Computing Systems*,
Undergraduate Topics in Computer Science,
DOI 10.1007/978-1-84800-322-4_1, © Springer-Verlag London 2013

in intricate and sophisticated ways. Another answer is that our knowledge and experience in designing such systems, and our expertise in organising their design process, are still in their infancy compared to building bridges, chemical plants or railway networks. A third answer – and perhaps the most important one – is that rigorous mathematical tools and methods are much less employed in Software Engineering than in other engineering disciplines. While traditional engineers have always been academically trained to use tools and techniques from mathematics and physics to guarantee the quality of their products, software designers and programmers are still often self-taught and have traditionally relied very much on their intuition and intelligence. Many of them seem to have a rather fatalist attitude towards bugs.

This attitude is more and more difficult to defend. Programs and software systems are themselves mathematical structures; many software systems rely on sophisticated mathematical mechanisms such as audio compression, public key cryptography, or Web search ranking; and there are powerful domain-specific mathematical tools and techniques that can help us to understand, design, implement and analyse them in a better, more structured, and more scientific way.

The aim of this book is to introduce some of the techniques which, when applied, can help to reduce the number of errors present in a system. Errors can arise at many points in the software development process, from understanding exactly the requirements and behaviour of the system being built, to ensuring that these requirements are correctly captured in the design and implementation of the system. By working within the confines of a precise structured method, the occurrence of such errors can be drastically curtailed.

0.1 Examples of System Failures

To understand and appreciate the role of mathematics in modelling computing systems, it is helpful to look at a variety of examples of system failure. Some of these failures are of an entirely technological nature, others have to do with the ways in which humans and machines interact or in which rules of communication between different agents have been designed. In every case, they arise from errors in information processing, which is at the very core of computational modelling.

0.1.1 Clayton Tunnel Accident

Up until the mid-nineteenth century, collisions between trains were avoided solely by enforcing a minimum time interval between trains. Railway employees (known as "policemen") would stand at regular intervals ("blocks") along the line and signal trains with hand gestures to slow down if too little time had elapsed since the previous train had passed. In the case of a break-down of a train, the guard in the rear of the train would run back along the track to warn any oncoming trains of the danger ahead.

With ever-increasing traffic, growing lapses in this system eventually led to the installation of crude *block signalling* in particularly troublesome places, in which some *protocol* would be followed to ensure that only one train occupied a given stretch of track at any given moment. Such a protocol typically involved railway workers at each end of the section signalling each other via telegraph of the passage of trains: having let one.train proceed, the signalman would hold back any further trains until a message was received indicating that the first train had cleared the section ahead. The first commercial electric telegraph was constructed in Britain for use on the Great Western Railway, and the first section of track to be protected by block signalling using telegraph communication was the track through Clayton tunnel outside Brighton. However, on the morning of Sunday 25 August 1861, this protocol failed to prevent a catastrophic collision inside the tunnel which killed 23 people and injured a further 176.

In normal operation, when a train arrived at Clayton tunnel, it would meet a rail-side signal which would be set at "danger" unless the signalman at the entrance to the tunnel set it to "all right" authorising the train to enter the tunnel. This signalman would telegraph a "train in tunnel" message to the signalman at the other end of the tunnel, and the rail-side signal would be reset to "danger" to prevent any further trains from entering the tunnel until the signalman received a "tunnel clear" message by telegraph from the signalman at the other end of the tunnel, indicating that the train had emerged from the other side.

On the fateful morning in question, three trains left Brighton Station within a seven-minute period and steamed towards Clayton tunnel. These trains were scheduled to depart at 8:05, 8:15 and 8:30, respectively; however, the first train was running late, and the assistant stationmaster in charge that morning – one Charles Legg – opted to ignore the strict regulation of ensuring a minimum five-minute separation between trains by sending them off at 8:28, 8:31 and 8:35, respectively. The first train was given the "all right" signal to enter the tunnel, and the signalman – named Henry Killick – telegraphed the "train in tunnel" message to his counterpart – a man by the name of Brown – at the other end. He was then taken by surprise by the quick arrival of the second train, which passed the rail-side signal before he had had a chance to reset it to "danger." In desperation, he rushed out of his cabin furiously waving his red flag to stop the second train just as it was disappearing into the tunnel; there was no way for him to know, however, whether or not the driver had seen the flag.

Killick telegraphed to Brown a further "train in tunnel" message and waited tentatively for a response. Killick telegraphed a further message to Brown asking if the tunnel was clear; and to his relief he finally received the "tunnel clear" message from Brown. Unfortunately for Killick, Brown had not realised from Killick's repeated "train in tunnel" message that a second train had entered the tunnel; his "tunnel clear" message was in response to the passing of the first train. The driver of the second train – unbeknownst to Killick – had in fact seen the red flag; and having finally brought his heavy load to a stop, he was in the process of cautiously reversing back towards Killick. When, at that moment, the third train arrived at the entrance to the tunnel, Killick offered it the "all right" signal – with fatal consequences.

The Clayton tunnel accident is obviously not the result of a computer failure, but it is based on a poorly designed communication protocol between distributed agents, and therefore typical for computing systems. *Mutual exclusion algorithms,* which prevent more than one computing agent at a time accessing a resource such as a printer or a global variable, are instrumental parts of any operating system. The accident also shows a standard pitfall of systems design: the whole idea of the signalling protocol at Clayton tunnel was to ensure that two trains could not occupy the same block at the same time. But it couldn't handle the exceptional case it was supposed to prevent. (For details, see L.T.C. Rolt, *Red for Danger: The Classic History of Railway Disasters*, The History Press, 2009.)

0.1.2   USS Scorpion

In 1968, the nuclear submarine USS Scorpion was destroyed killing all of its 99 crew members. Though the cause of its destruction has long been steeped in mystery, evidence which was only declassified three decades later suggests that the submarine may in fact have been destroyed by one of its own torpedoes which had been accidentally activated and thus ejected. The torpedo had been cleverly designed to seek out its nearest target, which is precisely what it did on this occasion, with devastating consequences. (For details, see P G Neumann, *Computer Related Risks*, Addison Wesley, 1994.)

The negative implications of seemingly sensible and harmless design decisions often arise only in hindsight as unintended consequences after disaster has struck. Clearly, every eventuality needs to be accounted for, especially in safety-critical designs where failure of the system could lead to injury, illness or loss of life; serious environmental damage; or major financial loss.

0.1.3   Therac 25 Radiotherapy Machine

The Therac 25 was a radiation therapy machine that intermittently gave the wrong radiation doses over a period of three years (1985-87) due predominantly to errors in the software controlling its operation, as well as its poor interface design. The problems with the Therac 25 have been very thoroughly analysed, and six accidents – three of them fatal – have been attributed to its failures. (For details, see N Leveson and C S Turner, "An Investigation of the Therac-25 Accidents," *IEEE Computer* 26(7), pages 18–41, July 1993.)

The basic issue involved the replacement of hardware interlocks used in previous models by a software-only system. The machine had two modes of operation: electron mode and photon (or X-ray) mode, which were used for treating tumours at different depths in a patient's body. Electron mode involved a low-power electron beam, while photon (X-ray) mode involved a high power electron beam (three orders of magnitude more powerful), but with a metal plate between the device and the patient, to generate the X-rays. The electron beam had to be in low-power mode if the plate was not present, and in earlier designs (Therac 6 and Therac 20) there was a mechanical interlocking device which physically ensured this. This hardware interlock was removed from the Therac 25 which was left to rely on a (faulty) software interlock.

The software was poorly specified (there was no documentation on its software specification), designed and tested; and much of it was imported as-is from the previous models despite changes in requirements, without any form of integration testing. The problem was compounded by a complex user interface. In some cases, if the operator tried to enter certain control sequences (either in error or as shortcuts), the machine would operate incorrectly, using the high-power beam with no plate. It would then report an error, which it would normally do when no treatment had been delivered. Often in response to such an error report, operators would repeated the whole process, leading ultimately to fatal unintended consequences.

Leveson and Turner draw the following conclusion: "Virtually all complex software will behave in an unexpected or undesired fashion under some conditions – there will always be another bug. Accidents are seldom simple – they usually involve a complex web of interacting events with multiple contributing technical, human, and organisational factors." To improve the situation, they appeal to education: "Taking a couple of programming courses or programming a home computer does not qualify anyone to produce safety-critical software." The lesson is clear: the same rigorous standards should be applied to Software Engineering as to Engineering in general.

0.1.4   London Ambulance Service

In October 1992, the London Ambulance Service installed a computer aided dispatch (CAD) system, known as LASCAD, to control the dispatching of ambulances across London. It was to automatically match up each call to be responded to with the closest available ambulance. However, the system was unable to cope with real-time data, which was on the order of 5000 calls per day. As it became more and more swamped with information and requests, it generated more and more exception messages requiring human intervention. The volume of these messages caused the exception messages, together with information needing to be dispatched to ambulances, to scroll off the top of the controllers' screens. As many as thirty deaths have been attributed to failings of the system. For example, it was reported that one ambulance arrived to find the patient had died and long since been collected by the undertaker; and that another ambulance took 11 hours to reach its destination – five hours after the stroke victim had made their own way to the hospital.

The London Ambulance Service quickly reverted partially to its manual dispatching system. However, after eight days, the automated system crashed completely, leading the service to revert completely to the original manual system. Taking responsibility for the £1.5 million failure, the chief executive of the London Ambulance Service duly resigned from his post. (For details, see A Finkelstein and J Dowell, "A Comedy of Errors: The London Ambulance Service Case Study," in *The Eighth International Workshop on Software Specification and Design*, IEEE CS Press, pages 2-4, 1996.)

As in our previous examples, this disaster was caused by a complex web of managerial and economic pressure, incompetence and technical failure; but Finkelstein and Dowell conclude that "at the heart of the failure are breakdowns in specification and design common to many software development projects."

0.1.5 Intel Pentium

When the Intel Pentium PC was initially released in 1994, problems were found in its floating-point unit. With certain inputs, the unit gave inaccurate results when performing division, thus rendering it useless for mathematical or scientific work.

The error had been caused in the design stage of the chip when a new algorithm for floating-point division was implemented which was three to five times faster than previous methods. This algorithm is based on using look-up tables to calculate intermediate results. The hardware was implemented using a program to download values into the look-up tables; however, an error in this software caused five of the 1066 entries to be inadvertently omitted.

Because the calculations recursively use information from the look-up tables, the errors that can accrue magnify in scale. For example, performing the sum $x - (x/y) * y$ should return the answer 0 for any inputs x and y. Given that computers have to deal with approximations to real numbers, we typically have to settle for a value close to zero to be returned. But with input values $x = 4195835$ and $y = 3145727$ the first Pentium release gave the answer 256. (For details, see T R Halfhill, "The Truth Behind the Pentium Bug," *Byte* 20(3), pages 163-164, March 1995.)

0.1.6   Ariane 5

In June 1996, the maiden flight of the Ariane 5 satellite launch rocket, Flight 501, ended in disaster: the rocket veered off course and exploded 40 seconds after lift-off. Its self-destruct system was initiated when the rocket detected it was disintegrating. This damage was caused by friction with the atmosphere as the rocket was travelling at too shallow an angle.

The flight path of the rocket was controlled by two software components, one providing the flight data, and the other converting this data into signals which controlled nozzles that direct the rocket's boosters. The problem was found to be with the software providing the flight data, which was imported as-is from the earlier Ariane 4 (a similar problem underlying the Therac 25 failure).

The software executed an instruction to convert a 64-bit integer to a 16-bit representation on a number that was too big to be stored as a 16-bit integer. (Ariane 5 used a different flight path from Ariane 4 which involved a shorter period of vertical ascent before yawing over to accelerate, thus reaching shallower angles than Ariane 4 sooner in the flight; this problem thus never arose with Ariane 4.) As there was no code to deal with this exception, the program crashed, and the ensuing error messages generated by the system were interpreted by the guidance system as flight data. Ironically, the part of the software that failed was only needed by Ariane 4 before lift-off, and was only active during the first part of the flight due to the possibility of a short hold prior to lift-off. This piece of software was unnecessary for Ariane 5.

The Ariane 501 Inquiry Board reported that the failure was "due to specification and design errors in the software of the inertial reference system" because the Ariane 5 Development Programme "did not include adequate analysis and testing of the inertial reference system or of the complete flight control system." It recommended that the European Space Agency should in the future ascertain that "specification, verification and testing are of consistently high quality." (For details see "Ariane 501 Inquiry Board report," http://esamultimedia.esa.int/docs/esa-x-1819eng.pdf.)

0.1.7   Needham-Schroeder Protocol

When communicating over the Internets, where anyone can intercept and read the messages you send, it is important to securely encrypt any sensitive information that you may send out, such as your credit card details, so that only the intended recipient of your message can decrypt and read it. The

Needham-Schroeder protocol was devised to allow two parties – commonly referred to as "Alice" and "Bob" – to authenticate themselves over such an insecure channel: after executing such a protocol, Alice will believe that she is talking to Bob and vice versa, and hence they will have established mutual trust for further transactions.

The Needham-Schroeder protocol is based on *public-key cryptography*: Bob (and anyone else) can, for instance, use Alice's public key – which he can obtain from some trusted server – for encrypting messages to Alice which only Alice can decrypt and read using her private key which she keeps secret. The protocol then works as follows:

1. Alice sends a message to Bob – encrypted with his public key – consisting of a random number along with some statement about her identity.

2. Bob decrypts this message with his private key, and sends a message in response to Alice - encrypted with her public key – consisting of Alice's random number along with a random number of his own.

3. Alice decrypts Bob's message with her private key, and sends another message to Bob – again encrypted with his public key – consisting of Bob's random number.

When Alice receives Bob's response to her first message, she will believe that she is talking to Bob, as only Bob could have decrypted her message and discovered the random number that she had sent him. Equally, when Bob receives Alice's second message, he will believe that he is talking to Alice, as only she could have decrypted his message and discovered the random number that he had sent her. Hence, after executing this protocol, Alice and Bob will both have reason to trust each other's identities.

This protocol was devised in 1978, and for over 15 years it gave no cause for concern to the network community. Indeed, there were a variety of "proofs" attesting to the correctness and reliability of this protocol. Despite this evidence of the protocol's security, in 1995 it was discovered to be susceptible to a very basic *man-in-the-middle attack*: an intruder could participate in the protocol and convincingly impersonate another agent – even without breaking the encryption. Here is how it works:

1. The intruder masquerades as Bob so that Alice encrypts her initial message with the intruder's public key and sends her message to him.

2. The intruder decrypts Alice's message with his private key, then encrypts it with Bob's public key and sends this on to Bob.

3. Bob sends Alice's random number together with his own, encrypted with Alice's public key, to the intruder, who forwards it – unaltered – to Alice.

4. Alice decrypts Bob's message, encrypts Bob's random number with the intruder's public key and sends it to the intruder.

5. The intruder decrypts this message, encrypts it with Bob's public key and sends this on to Bob.

As far as Alice and Bob are concerned, the results of this interaction as interfered with by the intruder appear identical to those of the original interaction, so they will once again believe – this time incorrectly – that they are talking directly to each other. Their subsequent correspondence will all be via the intruder, who will be able to read all of Alice's messages, as they will continue to be encrypted using the intruder's public key and re-encrypted by the intruder with Bob's public key before being sent on to Bob. The intruder will still not be able to read Bob's messages, though, as these will all be encoded using Alice's public key, and the intruder will only be able to forward these unaltered to Alice. (For details, see G Lowe, "An attack on the Needham-Schroeder public key authentication protocol," *Information Processing Letters* 56(3), pages 131-136, November 1995.)

In contrast to the previous examples, this is a pure design error, which is again rather unexpected and surprising given the simplicity and stringency of the protocol. The difficulty with detecting this flaw is that intruders can behave in various unexpected ways that – being unpredictable – are very difficult to analyse. Even very simple protocols can lead to a wide variety of different system behaviours that need to be considered. It seems rather improbable that such diversity can be catered for simply by testing.

The various failures discussed above have complex and generally multiple causes, and most of them can be traced back to poor software development processes. What is lacking in the development process is a rigorous engineering discipline through which a thorough understanding of the system being developed is obtained before the system is constructed. In traditional engineering disciplines, the methods for obtaining such an understanding are well established and based on formally modelling an appropriately-abstract version of the system being developed. The challenge for Software Engineering is to mimic these methods; to do so requires an understanding of how to describe and analyse abstract models of software systems. Of course, this first requires an understanding of these terms.

0.2   System, Model, Abstraction and Notation

The notions of "system," "model," "abstraction" and "notation" are essential to this book. In this section we provide various dictionary-style definitions of these concepts, interspersed with some examples and thoughts.

System

> An assemblage of objects arranged in a regular subordination, or after some distinct method, usually logical or scientific; a complete whole of objects related by some common law, principle, or end; a complete exhibition of essential principles or facts, arranged in a rational dependence or connection; a regular union of principles or parts forming one entire thing.

How do we understand systems and put them together? In the object oriented approach to software design, one is guided by the above dictionary definition and methodically describes the whole by giving descriptions of the constituent parts along with how these parts are put together. If you have tried and trusted building blocks, then you can reliably use them again.

To understand and analyse the world in terms of systems is very important to science; and to build them is the fundamental task of Engineers. Systems can be described, for instance, in terms of their *structure* – how they can be decomposed into parts and how these parts are related to each other; or in terms of their *behaviour* – how they evolve and interact with their environment; or in terms of their *functionality* – what their goals and objectives are. Systems are often contrasted with the environments in which they are embedded and with which they interact. A prime example from the world of computers is the *operating system*, which manages our interactions with the computer hardware.

This book addresses *computing systems*. However, we understand the term "computing" in a rather loose sense. We do not identify computing systems with computers, but with all kinds of systems that access, store, process and communicate information. Many biological, physical, economical and social systems have recently been studied from this point of view, and many of the concepts and techniques introduced in this book can be used in these contexts.

Model

> (1) A miniature representation of a thing, with the several parts in due proportion; sometimes, a facsimile of the same size. (2) Something intended to serve, or that may serve, as a pattern of something to be made; a material representation of or embodiment of an ideal; sometimes a drawing or a plan; a description of observed behaviour, simplified by ignoring certain details.

Building models is at the core of any scientific and engineering discipline. Scientists need models to interpret their data and make predictions; and traditional engineering products such as bridges and aeroplanes are never built until models of them have been developed and studied to understand the characteristics of the product. These models may be small-scale versions of the product which are tested in wind tunnels; or they may be purely abstract models described on paper using some formal notation which are then analysed more formally, for instance through simulations on a computer.

Modelling techniques are also becoming more and more important in software engineering, as computing systems become ever more complex and ubiquitous, and their proper functioning is often extremely critical. It is no longer possible to rely on the cleverness of our programming skills when building computing systems. In this book we shall explore basic modelling techniques for software engineering; consider various simple illustrative yet sufficiently interesting computing systems; and describe models that capture those aspects of their behaviour that interest us.

Models come in all shapes and sizes, and are designed to capture specific aspects of the thing they represent. Consider, for example, the following two uses of simple railway models.

- If we are interested in teaching the history of the development of railway locomotives, then full scale working replicas would be fun, but probably inconvenient; small scale working replicas might do, or even non-working replicas. Meaning (1) is appropriate.

- If we are interested in developing strategies for safe shunting, then a child's train set might do. But we could also make do with a paper and pencil model with a sketch of the track and buttons representing the engines and rolling stock; or a computer model with a graphical interface and a simulator might even be more useful. Meaning (2) is appropriate.

Note that models allow complex systems to be understood, and their behaviour predicted, only within the scope of the model; they may give incorrect descriptions and predictions for situations outside the scope of their intended use. For example, a toy train set would not be much use if we were interested in the stresses and strains induced in real rolling stock when shunting. Building good models not only requires formal training, but also a lot of experience, and a critical mind.

Abstraction

The act or process of leaving out of consideration one or more properties of a complex object so as to attend to others.

Abstraction is an important part of model building: identifying those features that are essential for inclusion in the model and separating out those features that can be neglected since the essential elements do not rely on their presence.

As an example of an abstraction, OS (Ordnance Survey) maps are used by walkers in Britain who usually want to know where they are, where they are going (how far, which direction), and how long it will take. OS maps are to scale (typically 4 cm to 1 km), and include *(easting, northing)* grid reference pairs allowing the user to pinpoint locations very accurately. For example, Perriswood near Reynoldston has the grid reference (502, 888) on the Gower map (OS Explorer Map 164). By eye or by laying a piece of string along a proposed route, experienced walkers can estimate its length fairly accurately, and then use a simple formula such as Naismith's Rule of 5 kilometres per hour plus 10 minutes for each 100 metres of uphill to estimate their walking time. To be useful to walkers, OS maps are portable (they fold flat), are to scale, and use contours and shading to show heights and indicate steepness.

The London Underground train map and A to Z street atlas have different formats from OS maps, and from each other, as they serve different purposes. The first A to Z street atlas was designed in 1936 by Phyllis Pearsall, a portrait artist, due to her frustration at getting lost during her walks through London while trying to follow an OS map. The Underground train map, on the other hand, would be of very little use to a walker. It was originally designed in 1931 by Harry Beck, a draughtsman educated in electronics, and is reminiscent of an electronic circuit board diagram with only vertical, horizontal and 45-degree lines. Train stations are not depicted geographically accurately; the connections between stations are accurate, but the stations and routes of the trains are distorted to provide an æsthetically-pleasing image. As such, it provides an ideal model for using the Tube, when you don't need to know where you are geographically as you would if you were walking, but rather are only interested in where to get on, where to change lines, and where to get off. By distorting the geography, in particular by pulling in very remote stations located at the ends of lines, a balanced and concise diagram results which is easy to use and pleasant to look at.

Notation

Any particular system of characters, symbols, or abbreviated expressions used in art, or in science, to express briefly technical facts, quantities, etc. Especially the system of figures, letters, and signs used in arithmetic and algebra to express number, quantity, or operations.

Notation is one of the most undervalued idea in computer science. It is prevalent in the form of programming languages, but typically ignored at any higher level. A good notation provides the shortest distance between the idea in your head and a piece of paper.

Florian Cajori's two-volume masterpiece *A History of Mathematical Notations* (1928-1929) points out that scientific progress was sometimes held back for years, decades, or even centuries because there wasn't the right notation around in which to express the relevant ideas. Compare Roman and Arabic numerals for addition, subtraction, etc. Consider also zero, the decimal point, complex numbers, the calculus. Imagine expressing sameness in quantity before Robert Recorde's invention of the equality sign. The effect that notation has on facilitating problem-solving is aptly summarised by Alfred North Whitehead as follows: *"By relieving the mind of all unnecessary work, a good notation sets it free to concentrate upon more advanced problems"*.

(0.3) Specification, Implementation and Verification

The concepts and methods of computational modelling and thinking are relevant to many different fields, but their foremost domain is the development of high quality and dependable software. To set the scene, we briefly discuss three tasks that are central to software development and its formalisation through computational modelling.

1. *Specification* refers to the task of modelling a computing system together with its functionality and behaviour. This can be understood as a formal description of a problem to be solved.

2. *Implementation* refers to the task of programming the specification so that it can be executed on a computer. This can be understood as an effective solution to the problem posed by the specification.

3. *Verification* refers to the task of rigorously demonstrating that the implementation does indeed respect the specification. This can be understood as a proof that the implementation does indeed solve the problem specified.

The development of mathematical methods that formalise these three tasks is sometimes considered to be the *Holy Grail of Software Engineering*. In an ideal world, such methods could make software testing obsolete and software bugs history. But after four decades of research on such methods, this still remains an ideal, and there are mathematical results about decision problems, program termination and incompleteness of theories that suggest that this may be necessarily so.

However, while nobody would expect mathematical formalisation to solve all problems of science or engineering, mathematical methods and tools have significantly contributed to the success of these disciplines. The situation is similar in computing: many light-weight mathematical methods for modelling computing systems have already made their way into industrial applications from programming languages to design and analysis tools for the specification, implementation and verification of software systems. By developing and adopting such industrial-strength methods to avoid errors, the task of searching for and repairing software bugs will hopefully become more and more unnecessary – or at least simpler and more routine – thus making "Software's Chronic Crisis" of system failures something of mere historical interest.

Part I

Mathematics for
Computer Science

Chapter 1

Propositional Logic

Either this man is dead or my watch has stopped.

- Groucho Marx.

Like her three older brothers before her, little Amanda always wants to know "Why": "Why do I have to go to school?" "Why does it only snow in Winter?" As young as she is, she can understand that – logically – the responses she gets satisfy each and every one of her queries: "You go to school to learn things." "It only snows in Winter because that's the only time it gets cold." However, these answers rarely satisfy her – they merely open the way for yet more queries to explain the reasons she gets as answers to her previous questions: "Why do I have to learn things?" "Why does it only get cold in the Wintertime?" Her impatient father rarely wins this game; it inevitably ends either with a definitive "Just because!" or, more usually, with a simple "Gee, I don't know, that's a very good question! Go ask your mother."

This behaviour demonstrates more than mere curiosity; and in fact curiosity typically has little to do with it. It is the fun of the game of logical reasoning which motivates her: the pursuit of the absolute, unquestionable premises from which all the other points follow. Her father's goal in this game, of course, is to identify these premises as quickly as possible. (Her true goal, one can't help but feel, is to get her father to give up in exasperation.)

It is in our nature as human beings to reason about the world and our existence, to assimilate the knowledge which we accumulate and to make logical deductions based on this knowledge. Despite the fact that we are born with a built-in propensity to apply logical rules to make deductions from our knowledge – if we do something potentially dangerous such as step out into the street without looking for cars, then we may get hurt, and therefore we shouldn't do such things – it is nonetheless the case that we are very bad at doing this correctly consistently. The problem lies to a great extent with the ambiguities in our language.

In this chapter we shall see how logically correct reasoning manifests itself in a multitude of ways, and we shall learn how to tame our use of

F. Moller, G. Struth, *Modelling Computing Systems*,
Undergraduate Topics in Computer Science,
DOI 10.1007/978-1-84800-322-4_2, © Springer-Verlag London 2013

language in order to prevent the types of ambiguities and mismatches which lead to the sorts of invalid logical arguments which all too typically underly system failures. We will see that precise rules of logical reasoning can be written down and mechanically applied like the rules of chess. But, due to their universality as laws of thought, they are much more than a mere formal game. They can be applied to model and reason about a huge variety of systems and situations. In particular, they can be very useful in detecting unexpected misbehaviour or inconsistency of computing systems.

Logic in fact lies at the very core of computing. Historically, the concepts of computation and effective computability have been developed from a logical basis and they were motivated by questions about the mathematical foundations of logic. All computer programming languages rely on logical notions in their specifications, their implementations and their constructs. Logics are also among the most popular and effective methods for specifying and analysing computational systems in formally rigorous ways. And, last but not least, the design and implementation of digital systems is strongly based on logic.

1.1 Propositions and Deductions

Consider the following argument.

1. **Either this man is dead or my watch has stopped.**

2. **My watch is still ticking.**

   Therefore

3. **This man is dead.**

This is an example of the sort of reasoning which we (mostly unconsciously) perform constantly all day long. If we analyse the structure of the argument, we see the following elements.

A. The argument involves three *statements*, or *propositions*, by which we mean declarations which are either true or false (but not both). Each of the statements in the argument is declared to be true.

B. The first statement expresses an *option* between two simpler statements, namely

   1a. **This man is dead.**

   or

   1b. **My watch has stopped.**

C. A *deduction* or *inference* is made to infer the truth of the third statement from the truth of the first two statements. The third statement is referred to as the *conclusion* of the argument, while the first two statements from which we draw the conclusion are referred to as the *premises* of the argument.

Such arguments can be formalised in propositional logic. The *syntax* (structure) of propositional logic provides a language for modelling systems, situations and arguments. The *semantics* (meaning) of propositional logic gives an interpretation to the symbols of the language. The language of propositional logic starts with **atomic propositions**, such as "This man is dead", and builds up larger compound propositions using a variety of **propositional connectives**, such as "*or*". Each connective is given a precise prescribed meaning which aims to reflect its everyday use in natural language. The purpose of this formalization is to remove ambiguities which are prevalent in the use of English or any other natural language.

Example 1.1

The following rules, adapted from those specified by the World Chess Federation FIDE, describe the conditions for castling. *Castling* is a move of the king and either rook of the same colour, counting as a single move of the king, and executed as follows: the king is transferred from its original square two squares towards the rook in question, and then that rook is transferred to the square which the king has just crossed.

1. The right for castling with a particular rook has been lost:

   (a) *if* the king has already moved; *or*
   (b) *if* the rook in question has already moved.

2. Castling with a particular rook is prevented:

   (a) *if* the right for castling with that rook has been lost; *or*
   (b) *if* there is a piece between the king and the rook in question; *or*
   (c) *if* the square on which the king stands, *or* the square which it must cross, *or* the square which it is to occupy, is under attack by one or more of the opponent's pieces.

The conditions that permanently or temporarily prevent castling use the propositional connectives "*or*" and "*if*" to express constraints under which castling is prohibited.

Arguments are all about truth. Therefore, not all sentences can take part in arguments, simply because not all sentences express statements which can be true or false. This is the case with questions like *"Is that man dead?"* and requests like *"Bring me a watch that works."* To be true or false, a sentence must state a potential fact, hence be related to a potential bit of reality. This criterion distinguishes statements or propositions from all other kinds of sentences.

Exercise 1.1   (Solution on page 405)

Which of the following are statements (propositions)?

1. 2+3=5.

2. 2+3=6

3. Do your homework, Joel!

4. Joel didn't do his homework.

5. Is there life on Mars?

6. False

7. What Felix says is false.

8. What this sentence says is false.

Each atomic statement can, of course, be further analysed with respect to its grammatical structure – *Joel*, for instance, is a subject noun, *do* a verb, and *homework* an object noun – but this is of no relevance to propositional logic. It is concerned solely with the distinction between logical and non-logical components and, correspondingly, with the way in which the truth of simpler statements determines that of more complex ones.

Exercise 1.2   (Solution on page 405)

Which of the following are valid deductions?

1. If the fire alarm sounds, then everyone must leave the building. Everyone is leaving the building.
   Therefore the fire alarm has sounded.

2. If the fire alarm sounds, then everyone must leave the building. The fire alarm has sounded.
   Therefore everyone is leaving the building.

3. If the signal is green, then the train may proceed. The signal is red.
   Therefore the train must wait.

4. The right for castling with a particular rook has been lost if the king has already moved.
   Both rooks have already moved.
   Therefore the right for castling with a particular rook has been lost.

5. The right for castling with a particular rook has been lost if the king has already moved, or if the rook in question has already moved.
   One of the two rooks has already moved.
   Therefore the right for castling with a particular rook has been lost.

6. It is unlawful for any person to keep more than three dogs and three cats on their property within the city.
Charles keeps five dogs (but no cats) on his property in the city.
Therefore Charles is breaking the law.

Exercise 1.3   (Solution on page 406)

Which of the following are valid deductions?

1. Epimenides is a Cretan.
   All Cretans are liars.
   Therefore Epimenides is a liar.

2. Epimenides is a Cretan.
   Epimenides said that "All Cretans are liars."
   Therefore Epimenides is a liar.

3. Epimenides is a Cretan.
   Epimenides said that "All Cretans are liars."
   Therefore all Cretans are liars.

4. Epimenides is a Cretan.
   Epimenides said that "All Cretans are liars."
   Therefore not all Cretans are liars.

5. Epimenides is a Cretan.
   Aristotle said that "All Cretans are liars."
   Therefore Epimenides is a liar.

1.2   The Language of Propositional Logic

The *syntax* of propositional logic is the formal definition of the language, the *object language* of formal logic. This definition is given in a *meta-language* – natural language in this case – in which we speak *about* the language of propositional logic. The metalanguage itself will use logical notions and reasoning, albeit at an informal level; since the levels can be kept separate, there should be no conceptual confusion.

The definition of syntax has two steps. In the first step, the basic *symbols* of the language are defined. In the second step, the rules for writing *formulæ* with these symbols is defined; these represent statements or propositions. The precise definition of a formula will be given at the end of this section; we first introduce the components of this definition informally.

1.2.1   Propositional Variables

In propositional logic, the meaning of a particular atomic proposition is given solely by its truth or falsity. We therefore abstract from these propositions and introduce propositional variables instead.

In algebra, variables such as x, y and z are used to represent unknown numbers. The occurrences of the variable x in the quadratic equation $x^2 + 2x - 15 = 0$ are place holders for some value, in this case a number. The equation restricts the admissible values of x to being either 3 or -5. That is, if 3 is substituted for every occurrence of x in the equation, or if -5 is substituted for every occurrence of x in the equation, then the equation holds; and when any other number is substituted, it doesn't.

We use variables in a similar way in propositional logic. *Propositional variables* such as $P, Q, R, \ldots$ represent unknown propositions. In algebra we may assign a specific value to a variable; for example, we might write "let x=3" and then interpret every subsequent occurrence of the variable x by the value 3. Similarly we may let a propositional variable represent a specific proposition, for example writing "let Dead represent the statement: This man is dead." (Following good programming style, we will typically use meaningful words as propositional variables rather than mere letters to obtain more readable statements.)

In algebra, values (including unknown values represented by variables) can be combined using various operations, such as addition ($+$), subtraction ($-$), multiplication ($\times$) and division ($\div$). In propositional logic, we may combine propositions using various *propositional connectives*, specifically "not" ($\neg$), "or" ($\vee$), "and" ($\wedge$), "if ... then ..." ($\Rightarrow$), and "... if, and only if, ..." ($\Leftrightarrow$). An informal description of the connectives of propositional logic is given in the follow sections.

1.2.2   Negation

The *negation* $\neg p$ of a statement p, pronounced "not p", is a statement which is true if, and only if, p is false. This is typically expressed in English in one of the following ways:

- *not p; (more precisely, the statement p with "not" modifying the verb, typically by appearing immediately after it.)*
- *p does not hold / is not true / is false;*
- *it is not the case that p.*

Example 1.3

If Dead stands for the statement "This man is dead," then $\neg$Dead says "It is *not* the case that this man is dead," or, equivalently, "This man is *not* dead."

If a proposition is not true, then it must be false; and conversely, if it is not false, then it must be true. In particular then, if a proposition is *not* not true, then it is true: $\neg\neg p$ is the same as p. This is referred to as the *Law of Double Negation*.

Exercise 1.4   (Solution on page 406)

Rewrite the following statements without negations at the start.

1. $\neg$ "The Earth revolves around the sun."
2. $\neg$ "All of my children are boys."
3. $\neg(2 + 2 \leq 4)$.

1.2.3   Disjunction

The *disjunction* $p \vee q$ of two statements p and q, pronounced "p or q", is a statement which is true if, and only if, p is true or q is true (or indeed if both are true); that is, at least one of p and q is true. This is typically expressed in English in one of the following ways:

- *p or q;*
- *p or q or both;*
- *p and/or q;*
- *p unless q.*

In the context of the disjunction $p \vee q$, the propositions p and q are individually referred to as *disjuncts*.

Example 1.4

If Dead stands for the statement "This man is dead" and Watch stands for the statement "My watch has stopped," then Dead $\vee$ Watch says "Either this man is dead or my watch has stopped," or, equivalently, "If this man is alive, then my watch must have stopped." This does not preclude the possibility that the man is dead *and* my watch has stopped, in which case Dead $\vee$ Watch will still be true.

Example 1.5

In chess, the right for castling with a particular rook has been lost if the king has already moved, or if the rook in question has already moved. This condition can be formalised as KingMoved ∨ RookMoved, where KingMoved and RookMoved are propositional variables stand for the statements "The king has moved" and "The rook has moved," respectively. In particular, therefore, one may not castle with a particular rook if *both* the king *and* the rook in question have already been moved.

Recalling that $\neg p$ is true if p is not true, we can note that $p \lor \neg p$ must always be true regardless of what proposition p stands for: either p is true, or it is not true. This fact is referred to as the *Law of the Excluded Middle*: there is no middle ground when it comes to the truth of a propositional formula.

Exercise 1.5   (Solution on page 406)

Are the following disjunctions true or false?

1. $(3 < 2) \lor (3 < 5)$
2. $(5 < 4) \lor (7 < 5)$
3. $(5 < 6) \lor (6 < 8)$

Note that $p \lor q$ is true if (though not only if) *both* p and q are true. In propositional logic, there can be no ambiguity: the "or" is always taken in this *inclusive* sense. In some everyday circumstances, however, "or" is used in the *exclusive* sense: the statement *"Either you be quiet now or you won't get an ice cream!"* certainly is not supposed to be true in the case in which the child under consideration is quiet but still doesn't get the ice cream – that would be an unfair trick. Such an "exclusive or" is in fact provided by a different connective from the (inclusive) "or" used in propositional logic; it is written ⊕, and it has its own different truth conditions: $p \oplus q$ is true if, and only if, one of p and q is true and the other is false; that is, *precisely* one of p and q is true. Note that this connective is not formally a part of the definition of propositional logic; however, it can be expressed using the connectives of propositional logic (see Example 1.10 on page 29).

Exercise 1.6   (Solution on page 406)

For each of the following disjunctive statements, decide whether you think the speaker intends to use the inclusive or exclusive sense of the disjunction.

1. Joel came in last place in the round-robin competition; so that mean that either Felix beat him or Oskar beat him.
2. The light is either on or off.
3. You can have tea or coffee.

1.2.4   Conjunction

The *conjunction* $p \wedge q$ of two statements p and q, pronounced "p and q", is a statement which is true if, and only if, both p and q are true. This is typically expressed in English in one of the following ways:

- *p and q;*
- *p but q;*
- *not only p but also q.*

In the context of the conjunction $p \wedge q$, the propositions p and q are individually referred to as *conjuncts*.

Example 1.6

If Dead stands for the statement "This man is dead" and Watch stands for the statement "My watch has stopped," then Dead $\wedge$ Watch says "This man is dead *and* my watch has stopped," or, equivalently, "Not only is this man dead, but so is my watch!"

Recalling that $\neg p$ is false if p is true, we can note that $p \wedge \neg p$ must always be false regardless of what proposition p stands for: p and $\neg p$ cannot both be true at the same time.

Exercise 1.7   (Solution on page 407)

Are the following conjunctions true or false?

1. $(3 < 2) \wedge (3 < 5)$
2. $(5 < 4) \wedge (7 < 5)$
3. $(5 < 6) \wedge (6 < 8)$

1.2.5   Implication

Given two statements p and q, the *implication* $p \Rightarrow q$, pronounced "p implies q", is a statement which is true if, and only if, p is false, or q is true; that is, if p is true then q must also be true. In other words, $p \Rightarrow q$ is *false* if, and only if, p is true and q is false. This is typically expressed in English in one of the following ways:

- *p implies q;*
- *if p then q:*
- *q if p;*
- *p only if q;*

- *q whenever p;*
- *p is a sufficient condition for q;*
- *q is a necessary condition for p.*

In the context of the implication $p \Rightarrow q$, p is referred to as the *premise* and q is referred to as the *conclusion*.

Example 1.7

Let the variable SignalDanger stand for the statement "The signal shows danger," and let the variable TrainStop stand for the statement "The train stops." Then SignalDanger $\Rightarrow$ TrainStop stands for the statement "If the signal shows danger then the train stops."

The only event in which this statement can be false is when the signal shows danger and yet the train does not stop. Hence the rule allows the case that the signal does not show danger and yet the train nevertheless stops.

Exercise 1.8 (Solution on page 407)

Letting JoelHappy stand for "Joel is happy" and AmandaHappy stand for "Amanda is happy," each of the following statements translates as either JoelHappy $\Rightarrow$ AmandaHappy or as AmandaHappy $\Rightarrow$ JoelHappy. Determine which in each case.

1. "Joel is happy whenever Amanda is happy."
2. "Joel is happy only if Amanda is happy."
3. "Joel is happy unless Amanda is not happy."

Exercise 1.9 (Solution on page 407)

On the door of a particular house is the following warning to potential thieves:

Barking dogs don't bite.

My dog doesn't bark.

Should a potential thief *necessarily* be concerned?

1.2.6   Equivalence

The *equivalence* $p \Leftrightarrow q$ of two statements p and q, pronounced "p if, and only if, q", is a statement which is true if, and only if, both p and q are true, or both p and q are false; that is, if p and q have the same truth value. This is typically expressed in English in one of the following ways:

- *p if, and only if, q;*
- *p is equivalent to q;*
- *p is a necessary and sufficient condition for q.*

The symbol for equivalence $\Leftrightarrow$ looks like the symbol for implies $\Rightarrow$ pointing in both directions. This is very much by design since, with a bit of thought, it is evident that $p \Leftrightarrow q$ is true if, and only if, $p \Rightarrow q$ and $p \Leftarrow q$ (that is, $q \Rightarrow p$) are both true.

Example 1.9

Let the variable TrainEnter stand for the statement "The train enters the tunnel," and let the variable TunnelClear stand for the statement "The tunnel is clear." Then TrainEnter $\Leftrightarrow$ TunnelClear stands for the statement "The train enters the tunnel if, and only if, the tunnel is clear."

This statement is false if the train enters the tunnel while the tunnel is not clear, or if the tunnel is clear but the train does not enter.

1.2.7   The Syntax of Propositional Logic

We can now summarise the above discussion of propositional logic in the following formal definition. A statement written in propositional logic is called a *propositional formula*, and is either:

- an *atomic formula*, typically represented by a variable such as P, Q or R; or
- a *compound formula*, in which case it is built up using the above propositional connectives as summarised in Figure 1.1.

There are two special atomic propositional formulæ, true (representing the proposition which is always true) and false (representing the proposition which is always false).

The above defines the formal *syntax* of the language of propositional formulæ. To emphasise that a propositional formula must be written syntactically correctly according to Figure 1.1, it is also referred to as a *well-formed formula* (*wff*).

Note that in Figure 1.1 (as well as throughout this whole chapter) the letters p and q are *not* propositional variables, but rather *metavariables* which stand for arbitrary propositions.

If p and q are propositional formulæ, then so are the following:

true		*truth*
false		*falsity*
P		*atomic proposition*
$\neg p$	*not p*	*negation*
$p \vee q$	*p or q*	*disjunction*
$p \wedge q$	*p and q*	*conjunction*
$p \Rightarrow q$	*if p then q*	*implication*
$p \Leftrightarrow q$	*p if, and only if, q*	*equivalence*

Figure 1.1: The formulæ of propositional logic.

1.2.8 Parentheses and Precedences

It is common to use parentheses when writing mathematical expressions such as $(5 + 3) \times 2$, in order to disambiguate such expressions. Most mathematicians (as well as many hand-held calculators) will calculate $5 + 3 \times 2 = 11$, as it is standard to consider multiplication as binding more tightly than addition; that is, multiplications are applied before additions whenever possible. Multiplication is said to have a *higher precedence* than addition. However, with parentheses the meaning of this expression changes dramatically: $(5 + 3) \times 2 = 16$. Similarly, we would use parentheses to calculate $5 - (3 - 1) = 5 - 2 = 3$, as without them we would naturally apply the subtractions left-to-right and calculate $5 - 3 - 1 = 2 - 1 = 1$.

In a similar vein we can and will regularly make use of parentheses within propositional formulæ to ensure that the meaning of our formulæ is clear. For example, the formula $P \vee Q \Rightarrow R$ can be read either as $(P \vee Q) \Rightarrow R$ or as $P \vee (Q \Rightarrow R)$, so we shall write the formula with parentheses in one of the above ways in order to make sure it is read as intended. We shall thus extend our definition of a well-formed formula to include parentheses which enclose subformulæ.

However, to reduce the need for parentheses, we will consider $\neg$ as binding more tightly than $\wedge$, which will bind more tightly than $\vee$, which will bind more tightly than $\Rightarrow$, which will bind more tightly than $\Leftrightarrow$. Apart from this, the connectives will be applied right-to-left, so that for example an expression of the form

$$p \Rightarrow q \wedge r \Rightarrow s$$

would be interpreted as

$$p \Rightarrow (q \wedge r) \Rightarrow s$$

due to $\wedge$ binding more tightly than $\Rightarrow$, and thus as

$$p \Rightarrow \big((q \wedge r) \Rightarrow s\big)$$

due to the right-to-left application order of the $\Rightarrow$ connectives.

Omitting parentheses by adopting the above precedence and application orders on connectives will often make formulæ easier to read. However, parentheses can and should still be used despite these conventions in cases when confusions can easily arise. For example, we will typically write

$$p \Rightarrow \big((q \wedge r) \Rightarrow s\big)$$

despite the redundancy of the parentheses.

Example 1.10

We can express the "exclusive or" operation $p \oplus q$ – which says that one of p and q is true and the other is false – as a simple equivalence, by noting that $p \oplus q$ says that one of p and q is true if, and only if, the other is *not* true. It can thus be defined simply by:

$$p \oplus q \;=\; p \Leftrightarrow \neg q$$

or, equivalently, by

$$p \oplus q \;=\; \neg p \Leftrightarrow q.$$

Both of these options abide by the hint that $p \oplus q$ says that one of p and q is true if, and only if, the other is not true.

You may be tempted to define it as

$$p \oplus q \;=\; (p \Leftrightarrow \neg q) \;\wedge\; (q \Leftrightarrow \neg p)$$

which would be correct, but this would be overkill; with a little thought you should realise that $p \Leftrightarrow \neg q$ is the same as $q \Leftrightarrow \neg p$.

Exercise 1.10 (Solution on page 407)

Express the following connectives using the connectives of propositional logic.

1. The NAND connective $p \,|\, q$ which is true if, and only if, p and q are not both true.

2. The NOR connective $p \downarrow q$ which is true if, and only if, neither p nor q are true.

3. The *conditional connective* $q \triangleleft p \triangleright r$ which is true if, and only if, either p and q are both true, or $\neg p$ and r are both true. In other words: *"If p is true then q must be true; otherwise r must be true."*

1.2.9   Syntax Trees

It can be helpful to view a well-formed propositional formula as a tree-like diagram, called a *syntax tree*, in which the tree structure reflects the way in which the formula is constructed. For example, the formula $(P \vee Q) \Rightarrow \neg(P \wedge Q)$ corresponds to the following syntax tree:

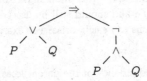

To recognise the expression $(P \vee Q) \Rightarrow \neg(P \wedge Q)$ as a well-formed propositional formula, we need only break it down to its constituent parts, and to reconstruct it from the inside out:

- P and Q, being propositional variables, are propositional formulæ.

- Since P and Q are propositional formulæ, so too are their disjunction $P \vee Q$ and conjunction $P \wedge Q$.

- Since $P \wedge Q$ is a propositional formula, so too is its negation $\neg(P \wedge Q)$.

- Since $P \vee Q$ and $\neg(P \wedge Q)$ are propositional formulæ, so too is their implication $(P \vee Q) \Rightarrow \neg(P \wedge Q)$.

This decomposition is directly reflected in the syntax tree, and also provides a method for determining whether or not the formula is true.

The syntax tree makes it clear how the expression should be parsed, without the need for parentheses or precedence rules to tell the reader how to interpret the formula. Without the rules of precedence, there are many different ways to read the expression $P \vee Q \Rightarrow \neg P \wedge Q$, all of which having completely different meanings and syntax trees.

Example 1.11

Consider the expression $P \Rightarrow \neg Q \vee R \Rightarrow Q$. According to the precedence rules, it is represented by the following syntax tree:

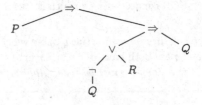

In order to evaluate this expression – that is, to determine its truth value – we first need to know the truth values of the propositional variables P, Q and R. We then compute $\neg Q$, as $\neg$ binds more tightly than the other connectives; then $(\neg Q) \vee R$ is computed, as $\vee$ binds more closely than $\Rightarrow$; then $((\neg Q) \vee R) \Rightarrow Q$ is computed followed by $P \Rightarrow \big(((\neg Q) \vee R) \Rightarrow Q\big)$, since the two $\Rightarrow$ connectives are computed in a left-to-right order.

Fully bracketed, the formula is thus interpreted as

$$P \Rightarrow \big(((\neg Q) \vee R) \Rightarrow Q\big).$$

Example 1.12

The string of symbols $\neg(P \wedge (Q \vee \neg))$ is *not* a well-formed propositional formula. This can be seen by applying the formation rules in Figure 1.1 backwards.

- $\neg(P \wedge (Q \vee \neg))$ is a formula only if $(P \wedge (Q \vee \neg))$ is a formula.

- $(P \wedge (Q \vee \neg))$ is a formula only if P and $(Q \vee \neg)$ are formulæ.

- P is a propositional variable and is therefore a formula.

- $(Q \vee \neg)$ is a formula only if Q and $\neg$ are formulæ.

- Q is a propositional variable and is therefore a formula.

- However, $\neg$ is a logical connective; it is neither a propositional variable nor a compound formula, so it is *not a* formula.

- Therefore, $\neg(P \wedge (Q \vee \neg))$ is not a well-formed formula.

Exercise 1.12   (Solution on page 407)

Which of the following are well-formed formulæ? Rewrite each well-formed formula using a minimal number of parentheses without changing its meaning, and draw its syntax tree.

1. $((P \Rightarrow Q) \Leftrightarrow (Q \Rightarrow P))$.

2. $P \vee Q(\wedge P)$.

3. $(P \vee Q) \wedge P$.

4. $(P \vee Q) \Leftrightarrow (R \neg S))$.

5. $\big(P \vee (Q \wedge R)\big) \Leftrightarrow \big(P \vee (Q \wedge (P \vee R))\big)$.

1.3   Modelling with Propositional Logic

Propositional logic is very important for modelling real-life scenarios, in which we define propositional variables to represent particular properties which may be true or false. Indeed we have described many such examples already above. We shall here consider a few further such examples.

Example 1.13

A particular computer program contains the following lines of code:

```
...
if CabinPressure < MinPressure then PrepareForLanding;
if FlightHeight < MinHeight then PrepareForLanding;
...
```

A software engineer assessing this code proposes that it could be optimised as follows:

```
...
if (CabinPressure < MinPressure and FlightHeight < MinHeight)
   then PrepareForLanding;
...
```

Is this correct?

Logically, we can use the variables Pressure and Height to express the two conditions that signal a need to land; and the variable Land to express the execution of PrepareForLanding. The program then gives rise to the following propositional formula:

$$(\text{Pressure} \Rightarrow \text{Land}) \land (\text{Height} \Rightarrow \text{Land})$$

while the suggested optimisation corresponds to

$$(\text{Pressure} \land \text{Height}) \Rightarrow \text{Land}.$$

The formula corresponding to the program is false if, and only if, either Pressure is true and Land is false, or Height is true and Land is false; this is the case if, and only if, either Pressure or Height is true while Land is false.

The formula for the suggested optimisation, on the other hand, would only be false if *both* Pressure and Height are true while Land is false; for example, having the cabin pressure drop below its minimum allowed value would wrongly *not* cause the aeroplane to prepare for landing if the aeroplane is cruising above its minimum allowed height.

The correct variant of the propositional formula – one which is equivalent to the formula corresponding to the program – would be

$$(\text{Pressure} \lor \text{Height}) \Rightarrow \text{Land}.$$

That is, the optimised code should should have a disjunction (or) in the condition, not a conjunction (and). Of course this logical analysis only confirms our intuition: The aeroplane should prepare for landing if *either* condition is satisfied, not if *both* of them hold.

Example 1.14

Consider the following four symbols: a white circle, a black circle, a white square, and a black square:

Let B represent the proposition that the symbol in question is black, and C represent the proposition that the symbol in question is a circle.

- B is true of the black circle and the black square, but false of the white circle and the white square.

- $\neg B$ is true of the white circle and the white square, but false of the black circle and the black square.

- $B \vee C$ is true of the white circle, the black circle and the black square, but false of the white square.

- $B \wedge C$ is true of the black circle, but false of the white circle, the white square and the black square.

- $B \Rightarrow C$ is true of the white circle, the black circle and the white square, but false of the black square.

- $B \Leftrightarrow C$ is true of the black circle and the white square, but false of the white circle and the black square.

These facts are summarised in the table in Figure 1.2. Almost all of them are self-evident, though you should spend time considering carefully when $B \Rightarrow C$ is true and when it is not true. Specifically, the only way that it can be false is if the symbol in question is black yet is not a circle.

The Oxford mathematician Charles Lutwidge Dodgson (1832-1898), better known as Lewis Carroll, the author of *Alice in Wonderland*, enjoyed inventing puzzles which required careful logical reasoning to solve. The following is a typical example.

Exercise 1.14   (Solution on page 408)

Lewis Carroll concludes that "Amos Judd loves cold mutton" from the following seven assumptions:

1. All the policemen on this beat sup with our cook.

	○	●	□	■	
B	×	√	×	√	*(it's black)*
$\neg B$	√	×	√	×	*(it's not black)*
$B \vee C$	√	√	×	√	*(it's black or it's a circle)*
$B \wedge C$	×	√	×	×	*(it's black and it's a circle)*
$B \Rightarrow C$	√	√	√	×	*(if it's black then it's a circle)*
$B \Leftrightarrow C$	×	√	√	×	*(it's black if and only if it's a circle)*

Figure 1.2: $B=$"the symbol is black", $C=$"the symbol is a circle".

2. No man with long hair can fail to be a poet.

3. Amos Judd has never been in prison.

4. Our cook's cousins all love cold mutton.

5. None but policemen on this beat are poets.

6. None but her cousins ever sup with our cook.

7. Men with short hair have all been in prison.

Explain how Lewis Carroll can draw his conclusion.

Exercise 1.15   (Solution on page 410)

Translate the rules for castling in chess presented in Example 1.1 into propositional logic using the following propositional variables:

- RightToCastleLeft / RightToCastleRight:
  You have the right to castle with the rook to the left / right.

- MayCastleLeft / MayCastleRight:
  You may perform a castling move with the rook to the left / right.

- KingMoved: The king has moved.

- LeftRookMoved / RightRookMoved:
  The left / right rook has moved.

- PieceBetweenLeft / PieceBetweenRight:
  There is a piece between the king and the rook to the left / right.

- KingAttack: The king is under attack.

- LeftSquareAttack / RightSquareAttack:
  The square to the left / right of the king is under attack.

- KingMoveLeftAttack / KingMoveRightAttack:
  The square two to the left / right of the king is under attack.

The following puzzle may appear hard at first sight, but it becomes surprisingly simple when approached logically.

Exercise 1.16   (Solution on page 410)

Joel, Felix and Oskar give Amanda the following puzzle. The three of them each write their name on a piece of paper, and then exchange the pieces of paper so that no one has the piece with their own name on it. They then hold these pieces of paper so that Amanda can't see what's on them, but tell her that each has the name of one of the others, and they challenge her to figure out who is holding each name. She is allowed to look at the name written on any one piece of paper.

1. Give a propositional formula which expresses the fact that each boy holds one of the pieces of paper but no one holds the piece of paper with their own name on it. Use the following propositional variables to do this.

   JonF: "Joel" is on Felix's paper.

   JonO: "Joel" is on Oskar's paper.

   FonJ: "Felix" is on Joel's paper.

   FonO: "Felix" is on Oskar's paper.

   OonJ: "Oskar" is on Joel's paper.

   OonF: "Oskar" is on Felix's paper.

2. Suppose Amanda looks at Joel's paper and sees "Oskar" written on it. Use the formula above to deduce what name is written on the other two pieces of paper.

1.4 Ambiguities of Natural Languages

Despite their intentionally obfuscated form, the statements in the Amos Judd puzzle in Exercise 1.14 are precise and unambiguous. There are, however, many common abuses of logical arguments arising from the ambiguities of a natural language such as English. In the following examples we consider particular difficulties which beginning logicians often find problematic.

Example 1.16

Children can get very unruly in the back seat of the family car during long drives. In such instances, an increasingly exasperated father in the driving seat might find himself making promises such as the following:

> "Everyone who sits quietly for the next hour
> will get an ice cream when we stop for petrol."

What exactly does this statement say? And more importantly, does it express what the father means to say? You might well imagine that he wants to suggest that:

> "Anyone who misbehaves will not get ice cream."

However, this does not follow from his statement: the children who get ice cream will include those who sit quietly, but may well include the noisy ones as well. In fact, he knows that even greater problems of retribution will arise later on during the drive if only some of the children get the promised ice cream, so it is always his unspoken intention that all of the children will get ice cream, regardless of their behaviour (within reason).

His aim in making the statement was to manipulate language to his benefit, as well as to provide a lesson for his children in its logical use. He was being intentionally vague, relying on his children to misinterpret his statement as saying something more than it actually does, namely that any misbehaving children will not get ice cream. When in the end even the misbehaving children get ice cream, those that sat quietly in anticipation of their reward would be mildly upset at the unfairness of it all, but they could not argue with their father's explanation that he did not actually say that the unruly children would lose out. Without a doubt he spoke the truth.

Needless to say, this strategy would not work for very long, as the children will quickly become keen interpreters of any statements that their father makes.

Example 1.17

Suppose a menu at a restaurant states the following:

> "You may have coffee or tea with your meal."

This clearly expresses a disjunction of two atomic propositions:

> "You may have coffee with your meal
> or you may have tea with your meal."

However, does it really do this? Clearly the intention is that if you ask for coffee, then you will be served coffee. But consider the following scenarios.

1. Suppose the coffee maker is broken on the day you visit, and only tea is available that day; is the menu wrong in this case? Certainly not logically, assuming that you may still have tea.

2. Suppose the restaurant doesn't have a coffee maker, and never actually serves coffee at all; is the menu wrong in this case? Still as certainly not logically, assuming that it serves tea.

The real intention of the proposition on the menu is something more akin to conjunction rather than disjunction, as follows.

"You may have coffee with your meal
and you may have tea with your meal."

However, this is still not true either, as it is unlikely that the restaurant intends to allow you to order both beverages with your meal. The following proposition might be a more accurate interpretation of the intended option on the menu.

"You may have coffee with your meal
and you may have tea with your meal,
but not both."

Are you satisfied with this? There is in fact still something seriously wrong with this proposition. To see this clearly, let us introduce the following two atomic propositions.

A = You may have coffee with your meal.
B = You may have tea with your meal.

Then the above proposition is

$(A \wedge B) \wedge \neg(A \wedge B).$

However, this proposition is of the form $p \wedge \neg p$; and recalling the fact noted after Example 1.6 that no proposition p (such as $A \wedge B$) can be true at the same time as its negation $\neg p$, this means that the menu is giving no option whatsoever!

The problem here is one of *modality*. That we *may* have a coffee, and that we indeed *do* have a coffee, are different propositions, and we need to be careful how we treat such modalities.

To correctly formulate the option, we might introduce the following two atomic propositions.

C = You have coffee with your meal.
T = You have tea with your meal.

Then the option stated on the menu would stipulate that one, and only one, of these atomic propositions are true. This can be rendered in many (equivalent) ways, such as

$$(C \lor T) \land \neg(C \land T)$$

> "You have coffee with your meal
> *or* you have tea with your meal,
> *but* not both."

or

$$(C \land \neg T) \lor (\neg C \land T)$$

> "You have coffee but not tea with your meal
> *or* you have tea but not coffee with your meal."

But this is still not the end of the story. Perhaps a particular diner drinks neither coffee nor tea. The menu surely doesn't force the diner to accept one of these beverages; the diner surely has the option of having neither. The option on the menu thus is merely stipulating the following

$$\neg(C \land T)$$

> "You do not have both coffee and tea with your meal."

or equivalently

$$\neg C \lor \neg T$$

> "You do not have coffee with your meal
> *or* you do not have tea with your meal."

From this simple English proposition has sprouted a plethora of complications. This is the greatest problem in formulating the design of systems, and hence of getting such designs correct.

(Example 1.18)

If p is false then by definition $p \Rightarrow q$ is true *regardless of the truth of q*. This observation gives rise not so much to a problem of ambiguity, but to one of misunderstanding and confusion. For example, assuming that Carlos is an ordinary man who is not the King of Spain, the following proposition is false:

> "If Carlos is a man, then Carlos is the King of Spain."

However, the following statement is true:

> "If Carlos is a woman, then Carlos is the King of Spain."

Do not be distracted by the falsity of the conclusion; the only way that the above statement can be false is if the premise is true whilst the conclusion is false. It is unfortunately a common misconception that the above implication is false, as the implication should be as follows:

"If Carlos is a woman, then Carlos is the *Queen* of Spain."

This statement is true as well, for precisely the same reason that the previous one is true: the premise of the implication is false.

Though this is a common confusion, it is well understood and properly applied in several instances of natural language. For example, the statement

"If I told you once, I told you a hundred times!"

is meant to convey that you have been told something a hundred times (assuming that you've been told once). This statement, of course, is typically false due to an intended use of hyperbole – it is highly unlikely that you have been told something so many times.

As another example, the statement

"If he ever pays me back, then I'll be a monkey's uncle!"

expresses the doubt (i.e., falsity) that money lent will ever be returned, by concluding an obviously-false conclusion from the premise which is being denied. As I can never be a monkey's uncle, the only way that this statement can be true is if he never pays me back.

Example 1.19

Suppose your teacher says the following to you:

"If you understand implication, then you will pass the exam."

There are four scenarios to consider:

1. Suppose you understand implication, and you pass the exam. Clearly you would consider the above statement to be true.

2. Suppose you *don't* understand implication, and you *fail* the exam. Again you would consider the above statement to be true, and you might even think your teacher to be a wise sage. However, this thought would just go to show that you indeed don't understand implication. The reason you failed the exam is not (necessarily) because you don't understand implication. To understand this point, consider the next scenario.

3. Suppose you *don't* understand implication, but nonetheless you *pass* the exam, because you understand enough of the rest of the material. This does *not* contradict your teacher's claim; it is still true.

4. Suppose, finally, that you understand implication, but you *fail* the exam nonetheless. In this case you may feel angry towards your teacher, since he was obviously lying to you. (Of course, your teacher would maintain that it is *you* who are lying, in claiming that you understand implication.)

In summary, the *only* way for the teacher's statement to be false is if the premise is true (i.e., you understand implication) while the conclusion is false (you fail the exam).

Exercise 1.19   (Solution on page 411)

Consider the following four symbols: a white circle, a black circle, a white square, and a black square:

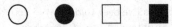

I have in mind one of these four symbols. I will *accept* any symbol which either has the same colour or the same shape (or both) as the one I have in mind, and otherwise I will *reject* it. If I accept the black square, what does this suggest to you about whether I accept or reject the other three symbols?

Exercise 1.20   (Solution on page 411)

If two's a company and three's a crowd, what's four and five?

1.5   Truth Tables

By thinking carefully about the logical connectives, we can informally understand their intended meanings. However, we still need to express these meanings precisely; that is, we need to *define* the meaning of the connectives. In doing this, the *semantics* of propositional logic is formally, rigorously and unambiguously defined.

One way in which we can do this concisely is by explicitly listing out the truth values which a compound formula takes for each of the possible combinations of truth values of its constituent propositions. A table which contains this listing is called a *truth table*.

For example, *negation* $\neg p$ can be defined by specifying its truth value for each of the two possible truth values of p: if the truth value of p is true, then the truth value of $\neg p$ will be false; and if the truth value of p is false, then the truth value of $\neg p$ will be true. For ease of presentation, we shall reserve the symbols **T** for true and **F** for false. The truth table for negation is thus as follows.

p	$\neg p$
F	T
T	F

The remaining four connectives are similarly defined by the following truth tables, which all have four rows corresponding to the four distinct combinations of truth values for the two propositions p and q being combined using the connectives.

p	q	$p \vee q$
F	F	F
F	T	T
T	F	T
T	T	T

p	q	$p \wedge q$
F	F	F
F	T	F
T	F	F
T	T	T

p	q	$p \Rightarrow q$
F	F	T
F	T	T
T	F	F
T	T	T

p	q	$p \Leftrightarrow q$
F	F	T
F	T	F
T	F	F
T	T	T

Truth tables can also be used to understand far more complicated formulæ, such as in the following example.

Example 1.20

Consider the statement from Example 1.16 made by a certain father:

"Everyone who sits quietly for the next hour
will get an ice cream when we stop for petrol."

Let us define the following atomic propositions.

Quiet = You sit quietly.
Ice = You get an ice cream.

For you, as a perfectly logical child, the above statement translates to Quiet $\Rightarrow$ Ice – if you remain quiet then you will get an ice cream – which has the following truth table:

Quiet	Ice	Quiet $\Rightarrow$ Ice
F	F	T
F	T	T
T	F	F
T	T	T

The *only* scenario in which the above statement can be considered false is if Quiet is true and Ice is false – that is, if you do not get an ice cream despite being quiet; in this instance you would be justified in being angry with your father for lying to you. However, your father, being trustworthy, would never allow this scenario.

It is tempting to be angry that your noisy siblings *also* get ice cream. However, there is no justification in this based on the above statement. As is clear from the second row of the truth table, the statement is true even in the instance that a noisy child gets an ice cream. It is a common pitfall to interpret $p \Rightarrow q$ as $p \Leftrightarrow q$ (that is, to understand from the above statement that you will get an ice cream if, and only if, you are quiet), and to believe that $p \Rightarrow q$ implies that $q \Rightarrow p$ (that is, to understand from the above statement that you will *not* get an ice cream if you are not quiet).

The above statement is giving you a guarantee that you will get an ice cream if you are quiet – and therefore you best be quiet. If you are not quiet, then there is no guarantee that you will get an ice cream, but there is no guarantee that you won't!

Exercise 1.21   (Solution on page 411)

Recall the statement from Example 1.19 made by a certain teacher:

> **"If you understand implication, then you will pass the exam."**

Translate this statement into a propositional formula, and use its truth table to justify when it is true or false.

Example 1.21

Catherine wishes to go to a party tonight, and would be happy to go with either Jim or Jules. However, as she is currently dating both Jim and Jules, she doesn't want to go to the party if they will both be there.

Let us define the following atomic propositions.

Cat = Catherine goes to the party.
Jim = Jim goes to the party.
Jules = Jules goes to the party.

Catherine's predicament then can be formalised as follows.

$$\text{Cat} \Rightarrow \neg(\text{Jim} \wedge \text{Jules}).$$

This proposition states that Catherine goes to the party *only if* Jim and Jules don't both go to the party. We can determine when this proposition is true or false by building up a truth table based on all possible values of the atomic propositions Cat, Jim and Jules, and the values of the constituent propositions which make up the complete proposition. The resulting truth table is as follows.

| | | | | ¬(Jim ∧ Jules) | |
Cat	Jim	Jules	Jim ∧ Jules		Cat ⇒ ¬(Jim ∧ Jules)
F	F	F	F	T	T
F	F	T	F	T	T
F	T	F	F	T	T
F	T	T	T	F	T
T	F	F	F	T	T
T	F	T	F	T	T
T	T	F	F	T	T
T	T	T	T	F	F

The first three columns systematically list out the eight distinct combinations of truth values for the three propositions Cat, Jim and Jules; the next column applies the rules from the truth table for ∧ to the columns for Jim and Jules; the next column applies the rules for ¬ to the column just constructed; and the final column applies the rules for ⇒ to the columns for Cat and ¬(Jim ∧ Jules). From this we can discover that the proposition is true in all cases *except* when all three atomic propositions are true; that is, it is *false* if, and only if, all three participants in this love triangle go to the party.

As a point of interest, we can build truth tables in a more concise way which entails writing the proposition of interest along the top row of the truth table, and filling in columns defined by the propositional variables and connectives, working from the "inside out." The truth table for the above example would then be rendered as follows:

Cat	Jim	Jules	Cat	⇒	¬	(Jim	∧	Jules)
F	F	F	F	T	T	F	F	F
F	F	T	F	T	T	F	F	T
F	T	F	F	T	T	T	F	F
F	T	T	F	T	F	T	T	T
T	F	F	T	T	T	F	F	F
T	F	T	T	T	T	F	F	T
T	T	F	T	T	T	T	F	F
T	T	T	T	F	F	T	T	T
(0)	(0)	(0)	(1)	(4)	(3)	(1)	(2)	(1)

The bottom row of numbers is included in this example to indicate at what stage each column was filled in:

(0) The three initial columns are filled in, representing all 8 possible combinations of truth values for the atomic propositions Cat, Jim and Jules.

(1) The columns for the propositional variables are then filled in during the first stage.

(2) After this the column for Jim ∧ Jules is filled in (under the ∧ symbol) during the second stage.

(3) Then the column for ¬(Jim ∧ Jules) is filled in (under the ¬ symbol) during the third stage.

(4) Finally the column for Cat ⇒ ¬(Jim ∧ Jules) is filled in (under the ⇒ symbol) during the fourth stage.

Each column is computed by referring to columns which have been computed in earlier stages.

Exercise 1.22   (Solution on page 412)

How many rows will there be in a truth table involving four propositional variables P, Q, R and S? What if there are five propositional variables? What if there are n propositional variables?

Exercise 1.23   (Solution on page 412)

Construct truth tables for the following propositions.

1. $\neg(P \Leftrightarrow \neg Q)$.

2. $(P \wedge Q) \vee (\neg P \wedge \neg Q)$.

3. $(P \wedge Q) \Rightarrow (\neg R \vee S)$.

Exercise 1.24   (Solution on page 413)

The "exclusive or" operation $p \oplus q$ has the following truth table:

p	q	$p \oplus q$
F	F	F
F	T	T
T	F	T
T	T	F

That is, $p \oplus q$ is true if, and only if, one of p and q is true and the other is false.

Confirm that the formula you gave in Example 1.10 (page 29) for expressing $p \oplus q$ in propositional logic gives the same truth table.

1.6 Equivalences and Valid Arguments

We have seen that a given proposition can be expressed as a formula in propositional logic in different yet equivalent ways. As a further example, the formula

$$\text{Cat} \Rightarrow \neg(\text{Jim} \wedge \text{Jules}) \qquad \text{"If Cat then not both of Jim and Jules."}$$

from Example 1.21 is equivalent to the formula

$$\neg(\text{Cat} \wedge \text{Jim} \wedge \text{Jules}) \qquad \text{"Cat, Jim and Jules cannot all be true."}$$

as well as

$$\neg\text{Cat} \vee \neg\text{Jim} \vee \neg\text{Jules}. \qquad \text{"One of Cat, Jim or Jules is false."}$$

To verify that two compound formulæ p and q are equivalent, we could construct truth tables for p and q and observe that they have the same truth values under all interpretations of their respective atomic propositions. Alternatively we could build the truth table for the formula $p \Leftrightarrow q$ and observe that it is true under all interpretations. If so, the two propositions p and q are said to be *logically equivalent*.

A proposition which is true regardless of the truth values of its atomic propositions is called a *tautology*, and the proposition is said to be *valid*. A *contradiction* on the other hand is a proposition which is false regardless of the truth values of its atomic propositions, and is said to be *unsatisfiable*. A proposition which is true under *some* interpretation of its atomic propositions – that is, one that is *not* a contradiction – is said to be *satisfiable*.

Example 1.24

Any formula of the form $p \vee \neg p$ is a tautology, while any of the form $p \wedge \neg p$ is a contradiction. These facts were noted already in Section 1.2, and can be verified formally by constructing the truth tables for these formulæ.

p	$\neg p$	$p \vee \neg p$
F	T	T
T	F	T

p	$\neg p$	$p \wedge \neg p$
F	T	F
T	F	F

Each entry in the column for $p \vee \neg p$ is true, confirming that $p \vee \neg p$ is a tautology, while each entry in the column for $p \wedge \neg p$ is false, confirming that $p \wedge \neg p$ is a contradiction.

Note that if we take $p = A \wedge B$, then the contradiction

$$p \wedge \neg p = (A \wedge B) \wedge \neg(A \wedge B)$$

is precisely the formula which appeared in Example 1.17 (page 37).

Exercise 1.25   (Solution on page 413)

Construct truth tables for each of the following formulæ to determine which are tautologies and which are contradictions.

1. $p \vee (\neg p \wedge q)$.

2. $(p \wedge q) \wedge \neg (p \vee q)$.

3. $(p \Rightarrow \neg p) \Leftrightarrow \neg p$.

4. $(p \Rightarrow q) \Rightarrow p$.

5. $p \Rightarrow (q \Rightarrow p)$.

Tautologies are important in ascertaining the validity of arguments. Consider, for example, our first argument from Section 1.1 (page 18):

1. **Either this man is dead or my watch has stopped.**

2. **My watch is still ticking.**

   Therefore

3. **This man is dead.**

This argument is valid if the conjunction of the two premises implies the conclusion, that is, if the following implication is valid:

(Dead ∨ Watch) ∧ ¬Watch ⇒ Dead

Again, this means that the proposition is a tautology, that it is true regardless of the truth values of its atomic propositions. We can easily confirm this by constructing a truth table for this proposition:

Dead	Watch	(Dead	∨	Watch)	∧	¬	Watch	⇒	Dead
F	F	F	F	F	F	T	F	**T**	F
F	T	F	T	T	F	F	T	**T**	F
T	F	T	T	F	T	T	F	**T**	T
T	T	T	T	T	F	F	T	**T**	T

In contrast, consider the argument suggested by Exercise 1.9 (page 26):

1. **If my dog barks, then my dog doesn't bite.**

2. **My dog doesn't bark.**

   Therefore

3. **My dog bites.**

Its formalisation yields the following truth table:

Barks	Bites	(Barks	$\Rightarrow$	$\neg$ Bites)	$\wedge$	$\neg$ Barks	$\Rightarrow$	Bites
F	F	F	T T	F	T T F	**F**		F
F	T	F	T F	T	T T F	**T**		T
T	F	T	T T	F	F F T	**T**		F
T	T	T	F F	T	F F T	**T**		T

The first row of this truth table shows that the proposition – and hence the argument it represents – is not valid. It presents a scenario in which the proposition may be false: a dog that neither barks nor bites satisfies both premises, but not the conclusion. Such a dog provides a *counterexample* to the validity of the argument.

Exercise 1.26   (Solution on page 414)

In Example 1.13 we represented a piece of computer program in propositional logic as:

$$p = (\text{Pressure} \Rightarrow \text{Land}) \wedge (\text{Height} \Rightarrow \text{Land}).$$

We also considered two optimisations of this program represented as

$$q = \text{Pressure} \wedge \text{Height} \Rightarrow \text{Land};$$
$$r = \text{Pressure} \vee \text{Height} \Rightarrow \text{Land}.$$

Of course, an optimisation is only correct if the representation of the optimised program code is equivalent to the original one. Explain which of the two optimisations is correct and which is not.

★ **1.7**   ## Algebraic Laws for Logical Equivalences

Using truth tables to prove properties about propositions, specifically that two propositions are equivalent, can quickly become tedious. However, we can avoid relying on truth tables by reasoning equationally much as we would do in algebra and arithmetic.

For example, we might conclude that $3 \times (4+5) = 27$ in the following way:

$$3 \times (4+5) = (3 \times 4) + (3 \times 5)$$
$$= \quad 12 \quad + \quad 15 \quad = 27.$$

In the first line of this calculation we used the algebraic law that says that multiplication distributes over addition: $a(b+c) = ab+ac$; and in the second line we used the principle that we can replace equals by equals: if $a = b$ and $c = d$ then $a + c = b + d$.

A similar kind of reasoning is possible with propositional logic, with equivalence $\Leftrightarrow$ playing the role of equality $=$. Once we have determined that two propositions p and q are equivalent, that $p \Leftrightarrow q$, we can then replace one with the other. First, though, we need to know what equivalences we can use as our "algebraic laws". A large number of these are given as follows.

Commutativity Laws

$p \vee q \;\Leftrightarrow\; q \vee p$ $\qquad\qquad\qquad$ $p \wedge q \;\Leftrightarrow\; q \wedge p$

Associativity Laws

$p \vee (q \vee r) \;\Leftrightarrow\; (p \vee q) \vee r$ $\qquad$ $p \wedge (q \wedge r) \;\Leftrightarrow\; (p \wedge q) \wedge r$

Idempotence Laws

$p \vee p \;\Leftrightarrow\; p$ $\qquad\qquad\qquad$ $p \wedge p \;\Leftrightarrow\; p$

Distributivity Laws

$p \vee (q \wedge r) \;\Leftrightarrow\; (p \vee q) \wedge (p \vee r)$ $\qquad$ $p \wedge (q \vee r) \;\Leftrightarrow\; (p \wedge q) \vee (p \wedge r)$

De Morgan's Laws

$\neg(p \vee q) \;\Leftrightarrow\; \neg p \wedge \neg q$ $\qquad\qquad$ $\neg(p \wedge q) \;\Leftrightarrow\; \neg p \vee \neg q$

Double Negation Law

$\neg \neg p \;\Leftrightarrow\; p$

Tautology Laws

$p \vee \text{true} \;\Leftrightarrow\; \text{true}$ $\qquad\qquad$ $p \wedge \text{true} \;\Leftrightarrow\; p$

Contradiction Laws

$p \vee \text{false} \;\Leftrightarrow\; p$ $\qquad\qquad$ $p \wedge \text{false} \;\Leftrightarrow\; \text{false}$

Excluded Middle Laws

$p \vee \neg p \;\Leftrightarrow\; \text{true}$ $\qquad\qquad$ $p \wedge \neg p \;\Leftrightarrow\; \text{false}$

Absorption Laws

$p \vee (p \wedge q) \;\Leftrightarrow\; p$ $\qquad\qquad$ $p \wedge (p \vee q) \;\Leftrightarrow\; p$

Implication Law

$p \Rightarrow q \;\Leftrightarrow\; \neg p \vee q$

Contrapositive Law

$$p \Rightarrow q \iff \neg q \Rightarrow \neg p$$

Equivalence Law

$$p \Leftrightarrow q \iff (p \Rightarrow q) \land (q \Rightarrow p)$$

You can (and should) show that all of the above laws are valid tautologies by constructing appropriate truth tables. However, some laws can be shown to be valid by using laws that have already been previously confirmed. For example, we can verify the validity of the Contrapositive Law as follows:

$$
\begin{aligned}
p \Rightarrow q \; &\iff \; \neg p \lor q && \textit{(Implication)} \\
&\iff \; q \lor \neg p && \textit{(Commutativity)} \\
&\iff \; \neg\neg q \lor \neg p && \textit{(Double Negation)} \\
&\iff \; \neg q \Rightarrow \neg p && \textit{(Implication)}
\end{aligned}
$$

Of course, this derivation relies on the Implication, Commutativity and Double Negation Laws being verified first.

More importantly, we can use the above equivalences to derive ever more equivalences, bypassing the need to construct truth tables to justify them.

Example 1.26

We can derive the equivalence $p \lor (\neg p \land q) \iff p \lor q$ using the following sequence of steps:

$$
\begin{aligned}
p \lor (\neg p \land q) \; &\iff \; (p \lor \neg p) \land (p \lor q) && \textit{(Distributivity)} \\
&\iff \; \text{true} \land (p \lor q) && \textit{(Excluded Middle)} \\
&\iff \; (p \lor q) \land \text{true} && \textit{(Commutativity)} \\
&\iff \; p \lor q && \textit{(Tautology)}
\end{aligned}
$$

We can equally use this technique to verify that a proposition p is a tautology by demonstrating that $p \iff \text{true}$.

Example 1.27

We can demonstrate that $(p \Rightarrow q) \lor (q \Rightarrow r)$ is a tautology as follows:

$$(p \Rightarrow q) \lor (q \Rightarrow r)$$

$$\Leftrightarrow \quad (\neg p \lor q) \lor (\neg q \lor r) \qquad \textit{(Implication, twice)}$$

$$\Leftrightarrow \quad \neg p \lor ((q \lor \neg q) \lor r) \qquad \textit{(Associativity, twice)}$$

$$\Leftrightarrow \quad \neg p \lor (\text{true} \lor r) \qquad \textit{(Excluded Middle)}$$

$$\Leftrightarrow \quad (\neg p \lor r) \lor \text{true} \qquad \textit{(Commutativity, Associativity)}$$

$$\Leftrightarrow \quad \text{true} \qquad \textit{(Tautology)}$$

As in algebra, we will usually not mention applications of associativity and commutativity, and write formulæ like $p \lor q \lor r$ instead of $p \lor (q \lor r)$ or $(p \lor q) \lor r$. This allows us to represent the above calculation in a more compact way as follows:

$$(p \Rightarrow q) \lor (q \Rightarrow r)$$

$$\Leftrightarrow \quad \neg p \lor q \lor \neg q \lor r \qquad \textit{(Implication, twice)}$$

$$\Leftrightarrow \quad \neg p \lor \text{true} \lor r \qquad \textit{(Excluded Middle)}$$

$$\Leftrightarrow \quad \text{true} \qquad \textit{(Tautology)}$$

Exercise 1.27   (Solution on page 415)

Give derivations of the following equivalences.

1. $p \land (\neg p \lor q) \iff p \land q$.
2. $\neg(p \Rightarrow q) \iff p \land \neg q$.
3. $p \Rightarrow (q \lor r) \iff (p \Rightarrow q) \lor (p \Rightarrow r)$.
4. $p \Rightarrow (q \land r) \iff (p \Rightarrow q) \land (p \Rightarrow r)$.
5. $(p \land q) \Rightarrow r \iff (p \Rightarrow r) \lor (q \Rightarrow r)$.
6. $(p \lor q) \Rightarrow r \iff (p \Rightarrow r) \land (q \Rightarrow r)$.

1.8   Additional Exercises

1. Which of the following are statements?

   (a) "17 is an odd integer."
   (b) "Manchester is the capital of Great Britain."
   (c) "Unload the dishwasher if it has completed its washing cycle."
   (d) "Are all roses red?"

(e) "All roses are red."

2. Negate each of the items from above that you determine to be statements.

3. Which of the following are valid deductions?

(a) Mammals are warm-blooded animals.
   Whales are mammals.
   Therefore whales are warm-blooded animals.

(b) Mammals are warm-blooded animals.
   Fish are not mammals.
   Therefore fish are not warm-blooded animals.

(c) Some doctors are surgeons.
   Some women are doctors.
   Therefore some women are surgeons.

(d) All horses are animals.
   Therefore all horses' heads are animal heads.

(e) Some girls are better than others.
   Therefore some girls' mothers are better than other girls' mothers.

4. Formalise the following statement of Sherlock Holmes in propositional logic:

   "If I'm not mistaken Watson, that was the Dore and Totley tunnel through which we have just come, and if so we shall be in Sheffield in a few minutes."

5. Let E and T and W represent the following propositions.

   E: Your laptop's warranty has expired.
   T: You have tampered with the electronics in your laptop.
   W: Your laptop is covered by its warranty.

(a) Translate the following statements into propositional logic.

   W_1: Your laptop is covered by its warranty as long as the warranty has not expired and you have not tampered with the laptop's electronic components.

   W_2: Your laptop is not covered by its warranty if the warranty has expired or if you have tampered with the laptop's electronic components.

(b) How do these two statements differ? Which one would you prefer to see on the warranty of your new laptop?

6. Given that P and R are true while Q is false, determine the truth values of the following formulæ. Verify these by building truth tables for the given formulæ.

(a) $P \wedge (Q \vee R)$

(b) $(P \wedge Q) \vee R$

(c) $\neg(P \wedge Q) \wedge R$

(d) $\neg P \vee \neg(\neg Q \wedge R)$

7. Write each of the following statements symbolically in the form $P \Rightarrow Q$ (using the suggested propositional variables), and then express them in English in the form "If ... then ...."

(a) I will play golf tomorrow (G) unless it rains (R).

(b) I'll do it (D) if you ask me nicely (N).

(c) Ann cries (C) every time she watches *The Titanic* (W).

(d) I never leave the house (L) without locking the door (D).

(e) A rectangle is a square (S) only if all four of its sides are the same length (L).

(f) A rectangle is a square (S) if all four of its sides are the same length (L).

8. Letting CatAway stand for "The cat's away" and MicePlay stand for "The mice will play," translate each of the following into propositional logic.

(a) "The mice will play whenever the cat's away."

(b) "The mice will play only if the cat's away."

(c) "The mice will play unless the cat's not away."

9. Suppose I lay the following four cards on the table, each of which has a shape on one side (either a circle or a square) and a pattern on the other side (either stripes or dots).

I claim that:

> "Every card with a circle on one side
> always has stripes on the other side."

Which card(s) do you need to turn over in order to be certain that I am telling the truth?

This exercise is known as a **Wason Selection Test** after the psychologist Peter Wason who first described it in 1966. Be careful with your answer: studies rarely result in a reported success rate of over 20%!

10. Explain the difference between the following three offers:

    (a) You can watch TV if you tidy your room.

    (b) You can watch TV only if you tidy your room.

    (c) You can watch TV if, and only if, you tidy your room.

    Which offer should a logical parent make to their children?

11. Give the truth tables defining the NAND, NOR and conditional connectives $p \mid q$, $p \downarrow q$ and $q \lhd p \rhd r$ defined in Exercise 1.10, and show that these are the same as the truth tables for the formulæ you gave in Exercise 1.10 for these connectives.

12. Propositional Logic is based on the three connectives $\neg$, $\vee$ and $\wedge$; the *Implication Law* and the *Equivalence Law* show that the two connectives $\Rightarrow$ and $\Leftrightarrow$ can be defined in term of the other three.

    (a) Show how to express $\neg p$, $p \vee q$ and $p \wedge q$ using only the NAND connective $\mid$.

    (b) Show how to express $\neg p$, $p \vee q$ and $p \wedge q$ using only the NOR connective $\downarrow$.

13. A friend proposes the following game to you. You keep tossing a coin over and over until one of the following two things happens:

    • if two heads occur in a row, then the game ends; you win, and your friend will give you £2;

    • if a tail occurs followed immediately by a head, then the game ends; your friend wins, and you must give your friend £1.

    Is it worth playing this game?

14. In a certain country, every inhabitant is either a truth teller who always tells the truth, or a liar who always lies. While travelling in this country, you meet two people, Abe and Ben. Abe says, "Ben and I are both liars." Is Abe a truth teller or a liar? What about Ben?

15. Argue that Superman doesn't exist. To do this, start by making the following four assumptions:

    X_1: If Superman were able and willing to prevent evil, he would do so.

    X_2: Superman does not prevent evil.

    X_3: If Superman were unable to prevent evil, he would be impotent; and if he were unwilling to prevent evil, he would be malevolent.

    X_4: If Superman exists, he is neither impotent nor malevolent.

    Argue as follows. First introduce the following variables:

A: "Superman is able to prevent evil."

W: "Superman is willing to prevent evil."

I: "Superman is impotent."

M: "Superman is malevolent."

P: "Superman prevents evil."

E: "Superman exists."

(a) The first assumption translates into the following formal logical statement:

$$X_1 : (A \wedge W) \Rightarrow P.$$

Translate the remaining assumptions X_2, X_3 and X_4 into formal logical statements.

(b) Use assumptions X_1 and X_2 to argue that $\neg A \vee \neg W$.

(c) Use assumption X_3, and the fact from (b), to argue that $I \vee M$.

(d) Use assumption X_4, and the fact from (c), to draw your conclusion.

16. Which of the following statements is true?

(a) All of the below.

(b) None of the below.

(c) All of the above.

(d) One of the above.

(e) None of the above.

(f) None of the above.

17. The following famous puzzle is referred to as the *Einstein Riddle* as Albert Einstein is sometimes credited with inventing it as a boy. He is also credited with claiming that only two percent of the world's population can solve it.

You are given the following information about five houses sitting in a row on some street which are each painted a different colour, and whose inhabitants are of different nationalities, own different pets, drink different beverages, and smoke different brands of American cigarettes. In statement (e), right refers to the reader's right.

(a) The Englishman lives in the red house.

(b) The Spaniard owns the dog.

(c) Coffee is drunk in the green house.

(d) The Ukrainian drinks tea.

(e) The green house is immediately to the right of the ivory house.

(f) The Old Gold smoker owns snails.

(g) Kools are smoked in the yellow house.

(h) Milk is drunk in the middle house.

(i) The Norwegian lives in the first house.

(j) Chesterfields are smoked next door to the man with the fox.

(k) Kools are smoked next door to the house where the horse is kept.

(l) The Lucky Strike smoker drinks orange juice.

(m) The Japanese smokes Parliaments.

(n) The Norwegian lives next to the blue house.

The question is: Who drinks water? Who owns the zebra?

18. Verify the Laws of Equivalence from Section 1.7, either directly by using truth tables, or by deriving them from previous laws which have already been verified.

19. Verify the following laws for implication and equivalence.

(a) $p \Rightarrow p$

(b) $(p \Rightarrow q) \wedge (q \Rightarrow r) \Rightarrow (p \Rightarrow r)$.

(c) $(p \Rightarrow q) \Rightarrow (p \vee r \Rightarrow q \vee r)$.

(d) $(p \Rightarrow q) \Rightarrow (p \wedge r \Rightarrow q \wedge r)$.

(e) $(p \Rightarrow q) \Leftrightarrow (\neg q \Rightarrow \neg p)$.

(f) $p \Leftrightarrow p$.

(g) $(p \Leftrightarrow q) \Rightarrow (q \Leftrightarrow p)$.

(h) $(p \Leftrightarrow q) \wedge (q \Leftrightarrow r) \Rightarrow (p \Leftrightarrow r)$.

(i) $(p \Leftrightarrow q) \Rightarrow (p \vee r \Leftrightarrow q \vee r)$.

(j) $(p \Leftrightarrow q) \Rightarrow (p \wedge r \Leftrightarrow q \wedge r)$.

(k) $(p \Leftrightarrow q) \Leftrightarrow (\neg p \Leftrightarrow \neg q)$.

Chapter 2

Sets

I refuse to join any club that would have me as a member.

- Groucho Marx.

Propositional logic allows us to reason about the world by inferring new facts from facts that we already know. However, we also need to structure our knowledge by grouping things together and by relating such collections of things with each other. In the parlance of Computer Science, we don't only need algorithms that process information, but also data structures that collect and store it.

There are many words in English for describing a collection of things (especially animals) such as: a *pack* (of wolves), a *school* (of fish), a *gaggle* (of geese), a *host* (of angels), a *den* (of thieves), a *crowd* (of onlookers), or a *fleet* (of cars). The idea of regarding a collection of things as a single entity is fundamental in mathematics as well as in everyday parlance. However, mathematics usually restricts itself to using a single collective noun: *set*.

2.1 Set Notation

A *set* is a collection of objects which typically share a property. The objects belonging to the collection are individually referred to as its *elements*, or *members*. The number of objects in a set A is referred to as its *cardinality*, and is written $|A|$. If there are not too many elements in the set, then it is most typically described by writing its elements in a comma-separated list between curly braces, as in the following four examples of sets:

- { false, true };
- { 3, 7, 14 };
- { red, blue, yellow };
- { Joel, Felix, Oskar, Amanda }.
- { Aberystwyth, Bangor, Cardiff, Lampeter, Newport, Swansea };

F. Moller, G. Struth, *Modelling Computing Systems*,
Undergraduate Topics in Computer Science,
DOI 10.1007/978-1-84800-322-4_3, © Springer-Verlag London 2013

The above sets all contain a small number of elements – their cardinalities are 2, 3, 3, 4 and 6, respectively – and as such are easily written out. Larger sets which aren't so easily written out explicitly are often informally described using an *ellipsis* "...", as in the following three examples:

- $\{1, 3, 5, \ldots, 99\}$ *(the set of 50 odd positive integers below 100)*;

- $\{a, b, c, \ldots, z\}$ *(the set of 26 letters of the alphabet)*;

- $\{2, 3, 5, 7, 11, 13, 17, \ldots\}$ *(the infinite set of prime numbers)*.

Though we shall freely use this notation, it is generally inadequate. For example, how confident are you that the final set above denotes the set of prime numbers? Having an infinite number of elements, it would be impossible to list them all inside curly braces, so we would have to stop somewhere. But perhaps the next element we have in mind in the sequence after 17 is 21. Perhaps it isn't even a number; perhaps the next element in the sequence is Groucho Marx!

To avoid any ambiguity, sets are typically describe not by explicitly listing the elements between curly braces, but rather by describing the property that the elements share. In general, we shall describe sets using the following *set-builder notation*:

$$\{x : x \text{ has property } P\}.$$

That is, this set consists of exactly those objects x which satisfy the property P. We may, of course, use a more appropriate variable than x.

Example 2.1

The following are all examples of sets:

1. The collection of all beaches on the Gower Peninsula:

   $\{b : b \text{ is a beach on the Gower Peninsula}\}$.

2. The collection of all people who have climbed Mount Kailash:

   $\{p : p \text{ has climbed Mount Kailash}\}$.

3. The collection of all prime numbers:

   $\{n : n \text{ is a prime number}\}$.

4. The collection of all sets of people who have a common grandmother:

   $\{A : A \text{ is a set of people who share a common grandmother}\}$.

The first set is finite, and its members can be explicitly listed by referring to a map of the Gower Peninsula. The second set – as far as we know – has no members. The third set has infinitely many members, and so could not be explicitly listed. The members of the fourth set are themselves sets.

You will likely be familiar with many standard mathematical sets such as the following.

$\emptyset = \{\,\}$ *(the empty set)*

$\mathbb{B} = \{0, 1\}$ *(the binary digits, or bits)*

$\mathbb{N} = \{0, 1, 2, 3, \dots\}$ *(the natural numbers)*

$\mathbb{Z} = \{\dots, -3, -2, -1, 0, 1, 2, 3, \dots\}$ *(the integers)*

$\mathbb{Q} = \{\frac{m}{n} : m, n \in \mathbb{Z}, n \neq 0\}$ *(the rational numbers)*

$\mathbb{R} = \{x : x \text{ is a real number}\}$ *(the real numbers)*

Note that $\emptyset$ and $\{\emptyset\}$ are *different* sets; the set $\emptyset$ contains *no* elements, while the set $\{\emptyset\}$ contains *one* element, namely the set $\emptyset$ itself, and hence is not the same as the empty set $\emptyset$.

Also note that each set in the above list is bigger than the one above it, in the sense that it includes all of the elements of the set above it plus other elements not in the set above.

Exercise 2.1   (Solution on page 416)

Write out the following sets explicitly, by listing their elements within curly braces.

1. $\{x : x \text{ is an odd integer with } 0 < x < 8\}$.

2. $\{x : x \text{ is a day of the week not containing the letter n}\}$.

3. $\{x : x \text{ was a wife of Henry VIII}\}$.

4. $\{x : x \text{ starred as James Bond in the official series of films}\}$.

2.2   Membership, Equality and Inclusion

A set is defined solely by its members, so clearly the most basic question we can pose is to ask if an object x is a member of a set A. *Membership* is denoted by $\in$, pronounced *"is an element (or a member) of"*, as for example in

$$7 \in \{3, 7, 14\} \quad (\textit{"7 is an element of the set } \{3, 7, 14\}\textit{"}),$$

or

Felix $\in$ { Joel, Felix, Oskar },

whilst non-membership is denoted by $\notin$, as for example in

$8 \notin \{ 3, 7, 14 \}$ ("*8 is* not *an element of the set* $\{ 3, 7, 14 \}$"),

or

Amanda $\notin$ { Joel, Felix, Oskar }.

That is, $x \notin A$ is the same as $\neg(x \in A)$.

Exercise 2.2 (Solution on page 416)

Write out the following sets explicitly, by listing their elements within curly braces.

1. $\{ x : x$ is an integer with $x = 2y$ where $y \in \{ 1, 2, 3, 4, 5 \} \}$.
2. $\{ x : x$ is an integer with $2x = y$ where $y \in \{ 1, 2, 3, 4, 5 \} \}$.

Exercise 2.3 (Solution on page 416)

Which of the following propositions are true?

1. $2 \in \{ 1, 2, 3 \}$.
2. $\{ 2 \} \in \{ 1, 2, 3 \}$.
3. $\{ 2 \} \in \{ \{ 1 \}, \{ 2 \}, \{ 3 \} \}$.
4. $\emptyset \in \{ \}$.
5. $\emptyset \in \{ \emptyset \}$.

Since a set is defined solely by its members, two sets are *equal* if, and only if, they have the same elements. So when you list the elements of a set, the order in which you list them, and the number of times you list each element, doesn't matter. Thus, for example,

$\{ 3, 7, 14 \} = \{ 7, 14, 3, 7, 3 \}$

while

{ Joel, Felix, Oskar } $\neq$ { Joel, Felix, Oskar, Amanda }.

If you want to show that two sets are different, it suffices to find a witness to this fact; that is, an element of one set which is not in the other.

Exercise 2.4 (Solution on page 416)

Which of the following sets are equal?

- $A = \{1, \{1, 2\}\}$
- $B = \{1, \{2\}\}$
- $C = \{1, \{1\}\}$
- $D = \{\{1, 1\}, 1\}$
- $E = \{\{2, 1\}, 1\}$

One set A is a *subset* of another set B if, and only if, each element of A is an element of B; in such a case we write $A \subseteq B$. We also say that A is *included*, or *contained*, in B; or that B is a *superset* of A, written $B \supseteq A$; or that B *includes*, or *contains*, A. Reflecting on the description of equality of sets above, two sets A and B are thus equal, $A = B$, if, and only if, each is included in the other:

$$A = B \iff A \subseteq B \wedge B \subseteq A;$$

that is, if any element of one is an element of the other.

As further notation, we write $A \nsubseteq B$ to denote that A is *not* a subset of B, that is, if there is an element of A which is *not* an element of B. In other words, $A \nsubseteq B$ is the same as $\neg(A \subseteq B)$. Finally, we write $A \subset B$ if A is a *proper* subset of B, that is, if $A \subseteq B$ but $A \neq B$.

Example 2.4

As already noted above, the binary digits form a proper subset of the natural numbers; the natural numbers form a proper subset of the integers; the integers form a proper subset of the rational numbers; and the rational numbers form a proper subset of the real numbers:

$$\emptyset \subset \mathbb{B} \subset \mathbb{N} \subset \mathbb{Z} \subset \mathbb{Q} \subset \mathbb{R}.$$

A useful graphical way of depicting sets, and in particular the relationship between them, is by so-called *Venn diagrams*. Such a diagram is obtained by laying out the elements of a set on a piece of paper and then encircling them. For example, we can depict the sets

$$X = \{1, 2, 3, 4, 5\}$$
$$Y = \{2, 3, 4\}$$
$$Z = \{3, 4, 5, 6\}$$

by the Venn diagram in Figure 2.1. The rectangle represents some understood universal set $\mathcal{U}$, referred to as the *universe of discourse* consisting of all elements under consideration, which in this example we take to be the integers from 1 to 10:

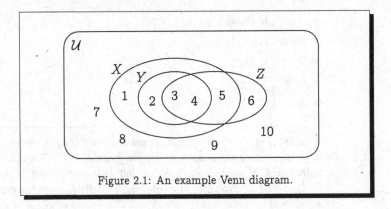

Figure 2.1: An example Venn diagram.

$$\mathcal{U} = \{1, 2, 3, 4, 5, 6, 7, 8, 9, 10\},$$

and the sets X, Y and Z are represented by encircling the relevant elements, depicted in the Venn diagram in Figure 2.1, The diagram clearly shows that $Y \subseteq X$, and indeed $Y \subset X$, since $1 \in X$ but $1 \notin Y$; whereas Z is incomparable to both X and Y: $X \not\subseteq Z$ and $Z \not\subseteq X$; and $Y \not\subseteq Z$ and $Z \not\subseteq Y$.

Furthermore, it is clear that for any set A: $\emptyset \subseteq A$ and $A \subseteq \mathcal{U}$; and that for any sets A, B and C: if $A \subseteq B$ and $B \subseteq C$ then $A \subseteq C$.

Exercise 2.5 (Solution on page 416)

Which of the following propositions are true?

1. $\{2\} \subseteq \{1, 2, 3\}$.

2. $\{1, 2, 3\} \subseteq \{\{1\}, \{2\}, \{3\}\}$.

3. $\{\{1, 2\}\} \subseteq \{\{1, 2, 3\}\}$.

As a final observation, we can note the following special properties of the subset relation, all of which are obvious using Venn diagrams.

1. It is *reflexive*, meaning that $A \subseteq A$ holds for every set A.

2. It is *antisymmetric*, meaning that if $A \subseteq B$ and $B \subseteq A$ then $A = B$.

3. It is *transitive*, meaning that if $A \subseteq B$ and $B \subseteq C$ then $A \subseteq C$.

Moreover, the empty set is the least set with respect to inclusion; that is, it is contained in any other set: $\emptyset \subseteq A$ holds for each set A.

2.3   Sets and Properties

We have already seen that listing elements is not appropriate for defining sets with infinitely many elements. Instead of writing

$$\text{Primes} \; = \; \{\, 2, 3, 5, 7, 11, 13, 17, 19, \ldots \,\}$$

for the set of all prime numbers, we use the set-builder notation

$$\text{Primes} \; = \; \{\, x \; : \; x \text{ is a prime number} \,\}.$$

to define Primes as the set of all objects x such that x is a prime number. More generally, we use the notation

$$\{\, x \; : \; x \text{ has the property } P \,\}$$

to indicate that we are building (defining) the set of all objects x which satisfy the property P.

This set-builder notation is typically used to define a subset B of an existing set A, in which case we write:

$$B \; = \; \{\, x \in A \; : \; x \text{ has the property } P \,\}$$

instead of

$$B \; = \; \{\, x \; : \; x \in A \text{ and } x \text{ has the property } P \,\}$$

The set-builder notation used in this way *separates* the objects in set A which satisfy a given property from those that do not.

Example 2.5

Philosophers have classified humans as rational animals (albeit a reasonable rationality criterion might be to disagree with this classification). Accordingly, the property of being rational separates humans from all other animals; it holds of all humans, and of no other animals. Letting Animals denote the set of all animals and Humans the set of all humans, we can write

$$\text{Humans} \; = \; \{\, x \in \text{Animals} \; : \; x \text{ is rational} \,\}.$$

Thus, $x \in \text{Humans}$ if, and only if, $x \in \text{Animals}$ and x is rational.

Example 2.6

Given two real numbers $a, b \in \mathbb{R}$ the following four intervals frequently occur in mathematics:

$$[a, b] = \{ x \in \mathbb{R} \; : \; a \leq x \leq b \};$$
$$(a, b] = \{ x \in \mathbb{R} \; : \; a < x \leq b \};$$
$$[a, b) = \{ x \in \mathbb{R} \; : \; a \leq x < b \};$$
$$(a, b) = \{ x \in \mathbb{R} \; : \; a < x < b \}.$$

Given two integers $m, n \in \mathbb{Z}$ the interval between them is defined as

$$[m..n] = \{ k \in \mathbb{Z} \; : \; m \leq k \leq n \}.$$

In all of the above intervals, if the first (left-hand) value is greater than the second (right-hand) value, then the interval defined is the empty set $\emptyset$.

Example 2.7

Obviously, x is in the set

{ Joel, Felix, Oskar, Amanda }

if, and only if,

$x =$ Joel or $x =$ Felix or $x =$ Oskar or $x =$ Amanda,

which is a property of x. Therefore, the above set can be rewritten, somewhat tediously, using set-builder notation as:

$\{ x \; : \; x =$ Joel or $x =$ Felix or $x =$ Oskar or $x =$ Amanda $\}$.

2.3.1 Russell's Paradox

The set-builder notation is very powerful; however, it must be used with some care.

We have seen that sets can contain any type of object, including sets themselves. Normally a set will not be a member of itself, but there is nothing to preclude us considering abnormal sets that *are* elements of themselves. Consider, then, the set of normal sets: those sets that are not elements of themselves; this set, which we call R, can be defined using the set-builder notation as follows:

$$R = \{ A \; : \; A \notin A \}.$$

We can then ask: is R itself a normal set? That is, do we have $R \in R$? Or do we have $R \notin R$? Certainly one of these two must be true: either R is a normal set, or it isn't.

- Suppose that $R \in R$. Then R must satisfy the property required of being an element of R, namely we must have that $R \notin R$.

- Suppose that $R \notin R$. Then R must *fail* to satisfy the property required of being an element of R, namely we must *not* have that $R \notin R$; that is, we must have that $R \in R$.

By the Law of the Excluded Middle, one of the above two cases must hold. This means that we must have *both* $R \in R$ and $R \notin R$; that is, R is *both* a normal set and an abnormal set. This is a contradiction, and as such cannot be true.

This anomaly is known as **Russell's Paradox**, after the philosopher Bertrand Russell who devised it to demonstrate the need to be vigilant in how you define sets. In particular, it should not be possible to speak of the set of all sets, as such circularity leads directly to contradictions. Fortunately, this anomaly cannot arise as long as we restrict the use of the set-builder notation to the restricted form

$$\{ x \in A \; : \; x \text{ has the property } P \}$$

in which we define the set as a subset of another given set which has been previously defined. We also need not worry about using the general set-builder notation if we have an implicit underlying universe of discourse.

Exercise 2.7   (Solution on page 416)

Let A be any set, and define the set R by

$$R = \{ X \in A \; : \; X \notin X \}.$$

Do we now have $R \in R$? Or do we have $R \notin R$? Why is Russell's Paradox not a problem here?

2.4   Operations on Sets

In the previous sections we have seen that sets can be constructed directly by putting curly braces around a listing of its elements, or indirectly using the set-builder notation. In this section we will consider a variety of operations which can be used to construct new sets from old.

2.4.1   Union

The *union* $A \cup B$ of two sets A and B consists of exactly those elements of the universe of discourse which are in either A or B (or both):

$$A \cup B = \{ x \; : \; x \in A \text{ or } x \in B \}.$$

Thus,

$$x \in A \cup B \quad \Leftrightarrow \quad x \in A \vee x \in B.$$

This is depicted by the following Venn diagram, where the gray area represents $A \cup B$.

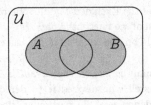

Example 2.8

$$\{1, 2, 3, 4, 5\} \cup \{2, 4, 6, 8, 10\} = \{1, 2, 3, 4, 5, 6, 8, 10\}$$

Example 2.9

The union of the set of people who can speak English and the set of people living in France is the set of people who can either speak English or who live in France (or both).

2.4.2   Intersection

The *intersection* $A \cap B$ of two sets A and B consists of exactly those elements of the universe of discourse which are in both A and B:

$$A \cap B = \{x : x \in A \text{ and } x \in B\}.$$

Thus,

$$x \in A \cap B \quad \Leftrightarrow \quad x \in A \wedge x \in B.$$

This is depicted by the following Venn diagram, where the gray area represents $A \cap B$.

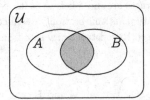

Example 2.10

$$\{1, 2, 3, 4, 5\} \cap \{2, 4, 6, 8, 10\} = \{2, 4\}$$

Example 2.11

The intersection of the set of people who can speak English and the set of people living in France is the set of people living in France who can speak English.

Two sets A and B are said to be *disjoint* if they have no elements in common; that is to say, if their intersection is empty: $A \cap B = \emptyset$. In terms of Venn diagrams, this means that the regions depicting A and B do not overlap.

There will typically be fewer elements in the union of two finite sets A and B, $|A \cup B|$, than $|A| + |B|$; the whole will generally be less than the sum of the parts. This is due to the fact that $|A| + |B|$ counts the members of the intersection $A \cap B$ twice. To balance this, we have the the following principle.

Theorem 2.11 Inclusion-Exclusion Principle

For finite sets A, B and C: $|A \cup B| = |A| + |B| - |A \cap B|$.

2.4.3   Difference

The *difference* $A \setminus B$ of two sets A and B consists of exactly those elements of the universe of discourse which are in A but not in B:

$$A \setminus B = \{x \in A : x \notin B\}.$$

Thus,

$$x \in A \setminus B \iff x \in A \wedge x \notin B.$$

This is depicted by the following Venn diagram, where the gray area represents $A \setminus B$.

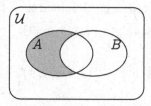

Example 2.12

$$\{1, 2, 3, 4, 5\} \setminus \{2, 4, 6, 8, 10\} = \{1, 3, 5\},$$

and

$$\{2, 4, 6, 8, 10\} \setminus \{1, 2, 3, 4, 5\} = \{6, 8, 10\}.$$

Example 2.13

The difference of the set of people who can speak English and the set of people living in France is the set of English-speaking people who do not live in France.

Conversely, the difference of the set of people living in France and the set of people who can speak English is the set of non-English-speaking people living in France.

2.4.4 Complement

The *complement* $\overline{A}$ of a set A is the set consisting of exactly those elements of the universe of discourse which are *not* elements of A:

$$\overline{A} = \{x : x \notin A\}.$$

Thus,

$$x \in \overline{A} \iff x \notin A.$$

The set $\overline{A}$ is thus the same as $\mathcal{U} \setminus A$, and is depicted by the following Venn diagram, where the gray area represents $\overline{A}$.

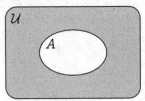

Example 2.14

Assuming the universe of discourse is $\mathcal{U} = \{1, 2, 3, 4, 5, 6, 7, 8, 9, 10\}$,

$$\overline{\{1, 2, 3, 4, 5\}} = \{6, 7, 8, 9, 10\},$$

and

$$\overline{\{2, 4, 6, 8, 10\}} = \{1, 3, 5, 7, 9\}.$$

Example 2.15

Assuming the universe of discourse is the set of people in the world, the complement of the set of people who can speak English is the set of non-English-speaking people; and the complement of the set of people living in France is the set of people who do not live in France.

Exercise 2.15   (Solution on page 416)

Consider the following sets:

$$\mathcal{U} = \{\,1,\, 2,\, 3,\, 4,\, 5,\, 6,\, 7,\, 8,\, 9,\, 10\,\}, \quad \textit{(the universe of discourse)}$$
$$A = \{\,1,\, 3,\, 5,\, 7,\, 9\,\},$$
$$B = \{\,3,\, 4,\, 5\,\},$$
$$C = \{\,5,\, 6,\, 7,\, 8,\, 9\,\}.$$

Draw a Venn diagram depicting these sets, and compute the following sets:

1. $A \cap C$.
2. $(A \cap B) \cup C$
3. $A \cap (B \cup C)$
4. $(A \cup B) \setminus C$.
5. $\overline{(A \cup B)} \cap C$.

Exercise 2.16   (Solution on page 417)

Let A, B and C be sets.

1. If $A \subseteq B$, what can you say about $A \cup B$ and $A \cap B$?
2. If $A \subseteq B$, what can you say about $\overline{A}$ and $\overline{B}$?
3. What is $\overline{\overline{A}}$, the complement of the complement of A?
4. If $C \subseteq A$ and $C \subseteq B$, how is C related to $A \cap B$?
5. If $A \subseteq C$ and $B \subseteq C$, how is C related to $A \cup B$?

2.4.5   Powerset

The *powerset* $\mathcal{P}(A)$ of a set A is the set consisting of all subsets of A:

$$\mathcal{P}(A) = \{\,X \,:\, X \subseteq A\,\}.$$

Thus,

$$x \in \mathcal{P}(A) \iff x \subseteq A.$$

In particular, $\emptyset \in \mathcal{P}(A)$ and $A \in \mathcal{P}(A)$.

We might only be interested in *finite* subsets. In this case we shall denote by $\mathcal{P}_{\text{fin}}(A)$ the set consisting of all *finite* subsets of A:

$$\mathcal{P}_{\text{fin}}(A) = \{\,X \,:\, X \subseteq A \text{ and } X \text{ is finite}\,\}.$$

Example 2.16

1. The set $\{0, 1\}$ has four subsets:

$$\mathcal{P}(\{0, 1\}) = \{\emptyset, \{0\}, \{1\}, \{0, 1\}\}.$$

More specifically, there are the following subsets:

- one subset with no elements (the empty set);
- two singleton subsets (one for each element in the set); and
- one subset with two elements (the whole set itself).

2. The set $\{$cola, fanta, sprite$\}$ has eight subsets:

$$
\begin{aligned}
\mathcal{P}(\{\text{cola, fanta, sprite}\}) \\
= \{\emptyset, \\
\{\text{cola}\}, \{\text{fanta}\}, \{\text{sprite}\}, \\
\{\text{cola, fanta}\}, \{\text{cola, sprite}\}, \{\text{fanta, sprite}\}, \\
\{\text{cola, fanta, sprite}\}\}.
\end{aligned}
$$

More specifically, there are the following subsets:

- one subset with no elements (the empty set);
- three singleton subsets (one for each element in the set);
- three subsets with two elements (one for each element left out); and
- one set with three elements (the whole set itself).

3. The set $\{$Joel, Felix, Oskar, Amanda$\}$ has 16 subsets:

$$
\begin{aligned}
\mathcal{P}(\{\text{Joel, Felix, Oskar, Amanda}\}) \\
= \{\emptyset, \\
\{\text{Joel}\}, \{\text{Felix}\}, \{\text{Oskar}\}, \{\text{Amanda}\}, \\
\{\text{Joel, Felix}\}, \{\text{Joel, Oskar}\}, \{\text{Joel, Amanda}\}, \\
\{\text{Felix, Oskar}\}, \{\text{Felix, Amanda}\}, \{\text{Oskar, Amanda}\}, \\
\{\text{Joel, Felix, Oskar}\}, \{\text{Joel, Felix, Amanda}\}, \\
\{\text{Joel, Oskar, Amanda}\}, \{\text{Felix, Oskar, Amanda}\}, \\
\{\text{Joel, Felix, Oskar, Amanda}\}\}.
\end{aligned}
$$

More specifically, there are the following subsets:

- one subset with no elements (the empty set);
- four singleton subsets (one for each element in the set);
- six subsets with two elements (one for each pair);

- four subsets with three elements (one for each element left out); and
- one set with four elements (the whole set itself).

4. In general, if $|A| = n$ then $|\mathcal{P}(A)| = 2^n$: a set with n elements has 2^n different subsets.

Example 2.17

Amanda has invited the following six friends to her birthday party: Daniel, Ella, Mia, Rhodri and Zoe. However, some of them might not show up. If we let

$$\text{Friends} = \{\text{Daniel, Ella, Mia, Rhodri, Zoe}\}$$

then the collection of combinations of friends that might come to Amanda's party is given by $\mathcal{P}(\text{Friends})$. For example, perhaps Ella and Rhodri are busy that day, but the others all come; then the set of friends that come to Amanda's party is:

$$\{\text{Daniel, Mia, Zoe}\} \in \mathcal{P}(\text{Friends}).$$

Exercise 2.17   (Solution on page 417)

List the elements of $\mathcal{P}(\text{Friends})$, where Friends is the set defined in Example 2.17 above. How many sets of each size are there?

Exercise 2.18   (Solution on page 418)

Form the following sets from the empty set $\emptyset$:

1. the set $A = \mathcal{P}(\emptyset)$;
2. the set $B = \mathcal{P}(A)$;
3. the set $C = \mathcal{P}(B)$.

How many elements are in each of these sets?

Exercise 2.19   (Solution on page 418)

Given an arbitrary set A, what are $\mathcal{P}(A) \cap \emptyset$ and $\mathcal{P}(A) \cap \{\emptyset\}$?

★ 2.4.6   Generalised Union and Intersection

It makes perfect sense to take the union or intersection of any number of sets, not just two. For example, we can consider the union

$$A \cup B \cup C$$

of three sets A, B and C, meaning the set whose elements are those objects which are members of *any* of the sets A, B or C; or the intersection

$$A \cap B \cap C \cap D \cap E$$

of five sets A, B, C, D and E, meaning the set whose elements are those objects which are members of *all* of the sets A, B, C, D and E. We don't have to worry about which order we take the sets; for example, the set $A \cup (B \cup C)$ is clearly the same as $(C \cup A) \cup B$. This is because the union and intersection operations are associative:

$$A \cup (B \cup C) = (A \cup B) \cup C \text{ and } A \cap (B \cap C) = (A \cap B) \cap C;$$

and commutative:

$$A \cup B = B \cup A \text{ and } A \cup B = B \cup A.$$

In fact, we can extend union and intersection to apply to arbitrary families (sets) of sets: if $\mathcal{F}$ is a set of sets, then

$$\bigcup \mathcal{F} = \{x : x \in A \text{ for } some \ A \in \mathcal{F}\}$$

$$\bigcap \mathcal{F} = \{x : x \in A \text{ for } all \ A \in \mathcal{F}\}$$

In particular, $A \cup B = \bigcup\{A, B\}$ and $A \cap B = \bigcap\{A, B\}$. With a little thought, the following identities become apparent:

1. $A = \bigcup\{A\}$ and $A = \bigcap\{A\}$.
2. $A = \bigcup \mathcal{P}(A)$ and $\emptyset = \bigcap \mathcal{P}(A)$.
3. $\emptyset = \bigcup \emptyset$ and $\mathcal{U} = \bigcap \emptyset$, where $\mathcal{U}$ is the universe of discourse.

The final two identities are worth further explanation. By definition, $x \in \bigcup \emptyset$ if, and only if, $x \in A$ for some $A \in \emptyset$; but since there can be no such $A \in \emptyset$, there can be no such $x \in A$.

Similarly, by definition, $x \in \bigcap \emptyset$ if, and only if, $x \in A$ for all $A \in \emptyset$; but since there can be no such $A \in \emptyset$, it is vacuously true that $x \in A$ for each of these $A \in \emptyset$.

Example 2.19

Suppose, e.g., that CS101 is the set of all students enrolled on the course *Computer Science 101*, and that ClassLists is the set of all class lists, so that, for example, CS101 ∈ ClassLists. Then the set Students of all students, that is, all people who are enrolled on some course, would be

Students $=$ $\bigcup$ ClassLists.

The set $\bigcap$ ClassLists would *likely* be empty, as it would contain those students who are enrolled on *all* courses.

Exercise 2.20   (Solution on page 418)

Given an arbitrary set A, what are $\bigcap \mathcal{P}_{\text{fin}}(A)$ and $\bigcup \mathcal{P}_{\text{fin}}(A)$?

2.5 Ordered Pairs and Cartesian Products

An *ordered pair* is simply a pair of objects (a, b) with *first coordinate a* and *second coordinate b*. For example, points in the xy-plane are denoted by ordered pairs; the ordered pair $(4, 9)$, for example, denotes the point with x-coordinate 4 and y-coordinate 9. The ordered pair (a, b) is different from the set $\{a, b\}$ in that it is ordered; $(a, b) \neq (b, a)$ (unless, of course, $a=b$), whereas $\{a, b\} = \{b, a\}$. More precisely,

$$(a, b) = (c, d) \text{ if, and only if, } a = c \text{ and } b = d.$$

The *Cartesian product* $A \times B$ of two sets A and B is the set of all ordered pairs in which the first coordinate a is an element of A and the second coordinate b is an element of B.

$$A \times B = \{ (a, b) : a \in A \text{ and } b \in B \}.$$

Thus,

$$(a, b) \in A \times B \iff a \in A \land b \in B.$$

For example, $\mathbb{R} \times \mathbb{R}$, typically written as $\mathbb{R}^2$, denotes the set of points in the xy-plane.

Example 2.20

The Cartesian product $[1..m] \times [1..n]$ of the intervals $[1..m]$ and $[1..n]$ can model a finite grid, such as the points of an LCD screen or the squares on a chess board.

$$
\begin{array}{cccc}
(1,1) & (1,2) & \cdots & (1,n) \\
(2,1) & (2,2) & \cdots & (1,n) \\
\vdots & \vdots & \ddots & \vdots \\
(m,1) & (m,2) & \cdots & (m,n)
\end{array}
$$

Example 2.21

Many programming languages offer abstract data types that allow you to store and retrieve data using *key-value pairs*. These data types have different names in different programming languages, such as associative array, dictionary, map, or table. But key-value pairs are always ordered pairs from a Cartesian product Keys × Values, where Keys is a set of keys and Values is a set of values. The values are the pieces of information that are stored in the data type, and the keys – which are unique for each value – allow you to retrieve the value.

As an example, we may have a national database in which each person is assigned a unique identification number. In this case, names serve as keys and the values are the identification numbers associated with each name:

$$
\begin{aligned}
\text{IDNumbers} \ = \ \{ \ &(\text{Joel}, 7613), \\
&(\text{Felix}, 8217), \\
&(\text{Oskar}, 6457), \\
&(\text{Amanda}, 9601), \\
&\ \cdots \qquad\qquad\qquad \}.
\end{aligned}
$$

As another example, a correspondence can be made between countries and their capital cities:

$$
\begin{aligned}
\text{CapitalCities} \ = \ \{ \ &(\text{France}, \text{Paris}), \\
&(\text{Peru}, \text{Lima}), \\
&(\text{Japan}, \text{Tokyo}), \\
&(\text{Mali}, \text{Bamako}), \\
&\ \cdots \qquad\qquad\qquad \}.
\end{aligned}
$$

We can form the Cartesian product of *any* number $n \in \mathbb{N}$ of sets, whose elements are n-tuples. For example

$$
A \times B \times C \ = \ \{ (a,b,c) \ : \ a \in A, \ b \in B \text{ and } c \in C \}
$$

represents the set of triples (a, b, c) in which the first coordinate a is an element of A, the second coordinate b is an element of B, and the third coordinate c is an element of C. In general, we write A^n to denote $A \times A \times \cdots \times A$, that is, the Cartesian product of n copies of the set A. Three-dimensional space, thus, is defined by $\mathbb{R}^3 = \mathbb{R} \times \mathbb{R} \times \mathbb{R}$.

Note that the number of elements in a product is the product of the number of elements in the individual sets. In particular, for any set A,

$$
A \times \emptyset \ = \ \emptyset \ = \ \emptyset \times A.
$$

Example 2.22

Let S represents all students, C represents all courses, and G represents possible grades. Then $S \times C \times G$ represents all triples (s, c, g) where $s \in S$ is a student, $c \in C$ is a course and $g \in G$ is a grade. A University student database would be represented as a subset of this set, recording the grades for all students registered in each course.

Example 2.23

A *pixel* is a point on a computer screen, and these are laid out in a rectangular grid $[1..h] \times [1..v]$ as in Example 2.20, with the number of pixels dependent on the size and resolution of the screen.

Each pixel is displayed as a dot of a certain colour. In the **RGB model**, a colour is specified by a triple

$$(r, g, b) \in [0, 1]^3 \text{ where } [0, 1] = \{x \in \mathbb{R} : 0 \le x \le 1\}$$

representing an intensity of *red*, *green* and *blue*, respectively, with 0 being no intensity at all and 1 being maximum intensity. For example, black is represented by $(0, 0, 0)$ (no colours) while white is represented by $(1, 1, 1)$ (maximum intensity of all colours); and *red*, *green* and *blue* are obviously represented by $(1, 0, 0)$, $(0, 1, 0)$ and $(0, 0, 1)$, respectively. We can thus define the following two sets:

Pixel $= [1..h] \times [1..v]$ and

Colour $= [0, 1]^3$,

and use them to define a point on the screen as a member of the set

Point $=$ Pixel $\times$ Colour

which assigns a colour to a pixel. Each point is therefore represented by an ordered pair $((x, y), (r, g, b))$ whose first coordinate is the ordered pair (x, y), and whose second coordinate is the ordered triple (r, g, b).

Exercise 2.23   (Solution on page 418)

Every rational number can be represented as an ordered pair of integers. The number $3/4$, for example, corresponds to the ordered pair $(3, 4)$. Define the operations of addition and multiplication on ordered pairs of integers such that they correspond to the standard operations on fractions.

2.6   Modelling with Sets

As the fundamental data structures of mathematics, sets inevitably occur in the specifications of systems. In many cases, sets capture system properties more concisely than propositional logic. In this section, we explore a number of examples, starting with revisiting Amos Judd.

Example 2.24

Consider the following three assumptions:

1. All candy has sugar.

2. John eats only healthy foods.

3. No healthy food contains sugar.

We can reason about these assumptions by introducing the sets H, S, J and C to represent, respectively, the set of healthy foods, the set of sugary foods, the set of foods that John eats, and the set of candy. The above assumptions can be expressed, equationally and with a Venn diagram, as follows:

1. $C \subseteq S$

2. $J \subseteq H$

3. $S \cap H = \emptyset$

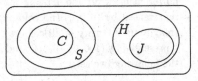

From this picture it is clear that no candy is healthy, and as such that John doesn't eat candy.

Exercise 2.24   (Solution on page 418)

Recall the situation regarding Amos Judd from Exercise 1.14 (page 33), in which the fact that Amos Judd loves cold mutton could be inferred from the following assumptions:

1. All the policemen on this beat sup with our cook.

2. No man with long hair can fail to be a poet.

3. Amos Judd has never been in prison.

4. Our cook's cousins all love cold mutton.

5. None but policemen on this beat are poets.

6. None but her cousins ever sup with our cook.

7. Men with short hair have all been in prison.

Demonstrate how to solve this problem by reasoning about appropriately-defined sets.

Exercise 2.25   (Solution on page 419)

Another one of Lewis Carroll's famous puzzles has the following premises:

> *All babies are illogical.*
>
> *Nobody is despised who can manage a crocodile.*
>
> *Illogical persons are despised.*

Use an appropriate Venn diagram to deduce from these premises that no baby can manage a crocodile.

Exercise 2.26   (Solution on page 420)

Use an appropriate Venn diagram to determine whether or not the following argument is valid.

> *All oceans are full of water.*
>
> *No ponds are oceans.*
>
> *Therefore no ponds are full of water.*

Example 2.26

At a certain hospital, 40 patients each have at least one of the following symptoms: a fever, a sore throat, or an earache. 18 of them have an earache and 25 of them have a sore throat, while eight of them have both an earache and a sore throat. Of the fever sufferers, 11 of them have sore throats, nine have earaches, and two have both a sore throat and an earache. How many fever sufferers are there?

We can use a Venn diagram to solve this problem, by drawing the three sets of patients as follows:

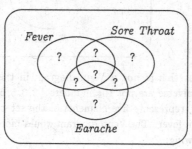

The question marks represent the numbers of patients in the relevant subsets, and these numbers must add up to 40. We merely need to replace these question marks with the relevant numbers based on the information given in the problem, which we can do by working from the inside out.

- We first put a 2 in the intersection of all three sets, depicting the two patients who are suffering from all three symptoms.
- Since eight patient are suffering from both a sore throat and an earache, six of these must not have fever, so we can put a 6 in the relevant place in the diagram.
- Next, there are 11 fever sufferers who have a sore throat; we know that two of these also have an earache, so nine of these must not have an earache, so we can thus put a 9 in the relevant place in the diagram.
- Also, there are nine fever sufferers who have an earache; we know that two of these also have a sore throat, so seven of these must not have a sore throat, so we can thus put a 7 in the relevant place in the diagram.
- There are 18 patients with earaches, 15 of which have other symptoms; thus three have no other symptoms, so we can put a 3 in the relevant place in the diagram.
- There are 25 patients with sore throats, 17 of which have other symptoms; thus eight have no other symptoms, so we can put an 8 in the relevant place in the diagram.
- As there are 40 patients in total, and 35 are accounted for as having either a sore throat or an earache, there are five patients who only suffer from fever, so we can put a 5 in the relevant place in the diagram.

The Venn diagram thus looks as follows:

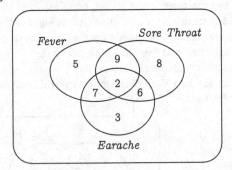

Note that this is *not* a Venn diagram in the usual sense: the elements of the universe are not the numbers { 2, 3, 5, 6, 7, 8, 9 }. Rather, the 5, for example, represents five elements of the set of patients who are suffering only from fever. The Venn diagram would more rightly look something like the following:

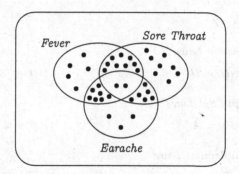

Here, each dot represents a distinct patient. However, the original Venn diagram, with just the numbers, is far easier to read.

With this, a final simple count tells us that 23 patients suffer from fever.

★ **(2.7) Algebraic Laws for Set Identities**

We can often represent the same set in a variety of ways; for example, we've already noted that it doesn't matter whether we write $A \cup B$ or $B \cup A$ as these give the same set. In this section we list a variety of identities, which will allow us to reason algebraically about sets. All of the laws presented can be verified informally by considering the appropriate Venn diagrams.

Commutativity Laws

$$A \cup B = B \cup A \qquad\qquad A \cap B = B \cap A$$

Associativity Laws

$$A \cup (B \cup C) = (A \cup B) \cup C \qquad A \cap (B \cap C) = (A \cap B) \cap C$$

Idempotence Laws

$$A \cup A = A \qquad\qquad A \cap A = A$$

Distributivity Laws

$$A \cup (B \cap C) = (A \cup B) \cap (A \cup C) \qquad A \cap (B \cup C) = (A \cap B) \cup (A \cap C)$$

De Morgan's Laws

$$\overline{(A \cup B)} = \overline{A} \cap \overline{B} \qquad\qquad \overline{(A \cap B)} = \overline{A} \cup \overline{B}$$

Double Complement Law

$$\overline{\overline{A}} = A$$

Universe Laws

$$A \cup \mathcal{U} = \mathcal{U} \qquad\qquad A \cap \mathcal{U} = A$$

Empty Set Laws

$$A \cup \emptyset = A \qquad\qquad A \cap \emptyset = \emptyset$$

Complement Laws

$$A \cup \overline{A} = \mathcal{U} \qquad\qquad A \cap \overline{A} = \emptyset$$

Absorption Laws

$$A \cup (A \cap B) = A \qquad\qquad A \cap (A \cup B) = A$$

You can (and should!) convince yourself of all of the above identities by constructing appropriate Venn diagrams.

Exercise 2.27   (Solution on page 420)

Draw the Venn diagrams which justify the two Distributive Laws.

We can use the above identities to derive even more identities, bypassing the need to construct Venn diagrams to justify them.

Example 2.27

We can derive the identity $A \cup (\overline{A} \cap B) = A \cup B$ using the following sequence of steps:

$$
\begin{aligned}
A \cup (\overline{A} \cap B) &= (A \cup \overline{A}) \cap (A \cup B) && \textit{(Distributivity)} \\
&= \mathcal{U} \cap (A \cup B) && \textit{(Complement)} \\
&= (A \cup B) \cap \mathcal{U} && \textit{(Commutativity)} \\
&= A \cup B && \textit{(Universe)}
\end{aligned}
$$

Exercise 2.28   (Solution on page 420)

Give a derivation of the identity $A \cap (\overline{A} \cup B) = A \cap B$.

The above laws allow us to reason about set inclusions as well as identities, by observing first that the set inclusion $X \subseteq Y$ can be expressed as a set identity, in any of the following ways:

$$X \cup Y = Y, \ X \cap Y = X, \ X \setminus Y = \emptyset, \ \overline{X} \cup Y = \mathcal{U}.$$

That each of the above are equivalent to the proposition that $X \subseteq Y$ can be readily be checked by considering the appropriate Venn diagram:

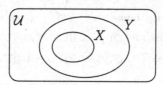

Example 2.28

We can derive the new law $A \subseteq A \cup B$ as follows:

- By Associativity and Idempotence, $A \cup (A \cup B) = A \cup B$.
- Letting $X = A$ and $Y = A \cup B$, this says that $X \cup Y = Y$.
- By the above, this means that $X \subseteq Y$; that is, that $A \subseteq A \cup B$.

Exercise 2.29   (Solution on page 421)

Derive the law $A \cap B \subseteq A$.

★ **2.8**   **Logical Equivalences versus Set Identities**

The astute reader will have noticed that there is a direct correspondence between the Equivalence Laws for Propositional Logic from Section 1.7 and the Set Identities from the previous Section 2.7. For convenience, these laws are listed once again here, side-by-side.

Commutativity Laws

$P \vee Q \Leftrightarrow Q \vee P$

$P \wedge Q \Leftrightarrow Q \wedge P$

Commutativity Laws

$A \cup B = B \cup A$

$A \cap B = B \cap A$

Associativity Laws

$P \vee (Q \vee R) \Leftrightarrow (P \vee Q) \vee R$

$P \wedge (Q \wedge R) \Leftrightarrow (P \wedge Q) \wedge R$

Associativity Laws

$A \cup (B \cup C) = (A \cup B) \cup C$

$A \cap (B \cap C) = (A \cap B) \cap C$

Idempotence Laws

$P \vee P \Leftrightarrow P$

Idempotence Laws

$A \cup A = A$

$$P \wedge P \Leftrightarrow P \qquad\qquad A \cap A = A$$

Distributivity Laws	**Distributivity Laws**
$P \vee (Q \wedge R) \Leftrightarrow (P \vee Q) \wedge (P \vee R)$	$A \cup (B \cap C) = (A \cup B) \cap (A \cup C)$
$P \wedge (Q \vee R) \Leftrightarrow (P \wedge Q) \vee (P \wedge R)$	$A \cap (B \cup C) = (A \cap B) \cup (A \cap C)$

De Morgan's Laws	**De Morgan's Laws**
$\neg (P \vee Q) \Leftrightarrow \neg P \wedge \neg Q$	$\overline{(A \cup B)} = \overline{A} \cap \overline{B}$
$\neg (P \wedge Q) \Leftrightarrow \neg P \vee \neg Q$	$\overline{(A \cap B)} = \overline{A} \cup \overline{B}$

Double Negation Law	**Double Complement Law**
$\neg \neg P \Leftrightarrow P$	$\overline{\overline{A}} = A$

Tautology Laws	**Universe Laws**
$P \vee \text{true} \Leftrightarrow \text{true}$	$A \cup \mathcal{U} = \mathcal{U}$
$P \wedge \text{true} \Leftrightarrow P$	$A \cap \mathcal{U} = A$

Contradiction Laws	**Empty Set Laws**
$P \vee \text{false} \Leftrightarrow P$	$A \cup \emptyset = A$
$P \wedge \text{false} \Leftrightarrow \text{false}$	$A \cap \emptyset = \emptyset$

Excluded Middle Laws	**Complement Laws**
$P \vee \neg P \Leftrightarrow \text{true}$	$A \cup \overline{A} = \mathcal{U}$
$P \wedge \neg P \Leftrightarrow \text{false}$	$A \cap \overline{A} = \emptyset$

Absorption Laws	**Absorption Laws**
$P \vee (P \wedge Q) \Leftrightarrow P$	$A \cup (A \cap B) = A$
$P \wedge (P \vee Q) \Leftrightarrow P$	$A \cap (A \cup B) = A$

Each law of equivalence for propositions gives rise to a set identity by replacing $\vee$ by $\cup$, $\wedge$ by $\cap$, and $\neg$ by $^-$ (as well as false by $\emptyset$ and true by $\mathcal{U}$). This exploits a tight analogy between logical equivalence $P \Leftrightarrow Q$ and equality of sets $A = B$, which can be extended to logical implication $P \Rightarrow Q$ and subset inclusion $A \subseteq B$ as in the following example.

Example 2.29

The *Implication Law* from Section 1.7:

$$P \Rightarrow Q \quad \Leftrightarrow \quad \neg P \vee Q.$$

gives rise to the following property of sets:

$$A \subseteq B \text{ if, and only if, } \overline{A} \cup B = \mathcal{U}.$$

This property is arrived at by translating $P \Rightarrow Q$ into $A \subseteq B$, and expressing $\neg P \vee Q$ as $\neg P \vee Q \Leftrightarrow$ true before translating it in the above fashion. (The equivalence symbol itself is translated merely into English.)

Exercise 2.30 (Solution on page 421)

Find properties of sets corresponding to the following laws for propositions taken from Section 1.7.

1. *Contrapositive Law*: $P \Rightarrow Q \Leftrightarrow \neg Q \Rightarrow \neg P$.

2. *Equivalence Law*: $P \Leftrightarrow Q \Leftrightarrow (P \Rightarrow Q) \wedge (Q \Rightarrow P)$.

Although the analogy between propositions and sets is tight, care must be taken when trying to use it. You should always check the validity of a property of sets which is so derived, for example by considering the relevant Venn diagrams.

Exercise 2.31 (Solution on page 421)

What property of sets is suggested by the following law for propositions:

$$\neg(P \Rightarrow Q) \quad \Leftrightarrow \quad P \wedge \neg Q$$

If you do this exercise carefully, you may well arrive at a property which is generally *not* true of sets. This exercise thus serves to point out that it is dangerous to rely on informal, intuitively-correct arguments.

2.9 Additional Exercises

1. Let $A = \{1, 2, 3, 4, 5\}$, $B = \{4, 5, 6, 7, 8, 9\}$, and $C = \{2, 4, 6, 8\}$.

   What are the following sets?

   (a) $A \cup B \cup C$.

   (b) $A \cap B \cap C$.

(c) $(A \cap B) \cup C$.

(d) $A \cap (B \cup C)$.

2. What sets are defined by $\{x : x \neq x\}$ and $\{x \in A : x = x\}$?

3. Draw Venn diagrams to justify the two De Morgan Laws $\overline{A \cup B} = \overline{A} \cap \overline{B}$ and $\overline{A \cap B} = \overline{A} \cup \overline{B}$.

4. Draw the Venn diagrams which justify the following laws.

   (a) $(A \cup B) \setminus C = (A \setminus C) \cup (B \setminus C)$.

   (b) $A \cap (B \setminus C) = (A \cap B) \setminus (A \cap C)$.

   (c) $A \setminus (B \cap C) = (A \setminus B) \cup (A \setminus C)$.

   (d) $(A \setminus B) \setminus C = A \setminus (B \cup C)$.

   (e) $A \cup (B \setminus C) = A \cup ((A \cup B) \setminus (A \cup C))$.

5. What can you say about the sets A and B if we know the following to be true?

   (a) $A \cup B = A$.

   (b) $A \cap B = A$.

   (c) $A \setminus B = A$.

   (d) $A \setminus B = B \setminus A$.

6. Form the following sets from the set $A = \{a\}$:

   (a) the set $B = \mathcal{P}(A)$;

   (b) the set $C = \mathcal{P}(B)$;

   (c) the set $D = \mathcal{P}(C)$.

7. Let $A = \{1, \{2, 3\}, \{4, 5, \{6\}\}\}$.

   (a) What is $\mathcal{P}(A)$?

   (b) State whether the following are true or false.

       i. $\emptyset \in A$.

       ii. $1 \in A$.

       iii. $\{2, 3\} \subseteq A$.

       iv. $\{\{2, 3\}\} \subseteq A$.

       v. $\{4, 5, \{6\}\} \subseteq A$.

8. The *symmetric difference* of two sets A and B, denoted $A \oplus B$ is the set which contains those elements which are in A or B but not in both A and B.

   (a) Draw a Venn diagram depicting $A \oplus B$.

   (b) Draw Venn diagrams to justify the following laws.

       i. $A \oplus B = (A \setminus B) \cup (B \setminus A)$.

    ii. $A \oplus B = (A \cup B) \setminus (A \cap B)$.

(c) What propositional connective does $\oplus$ correspond to?

9. Use the Inclusion-Exclusion Principle of Fact 2.11 to show the following three-set version: for finite sets A, B and C,

$$|A \cup B \cup C| = |A| + |B| + |C|$$
$$- |A \cap B| - |A \cap C| - |B \cap C|$$
$$+ |A \cap B \cap C|.$$

Explain informally why this principle holds.

10. Use the three-set version of the Inclusion-Exclusion Principle from the previous exercise to solve the hospital problem of Example 2.26.

11. Felix, Oskar and Amanda play a game to see who can list the most countries in five minutes. They each make a list, and after five minutes they compare these lists, crossing off any countries that are on more than one list.

Felix had listed the most countries, 29, but they were mostly common countries that the other two got: in fact, 23 of them were on Oskar's list, and 12 of these 23 were also on Amanda's list.

Amanda had listed the fewest countries, 22, but – with more than a little help from Joel – she had come up with many obscure countries: she had listed seven countries that were not on Felix's list, and nine countries that were not on Oskar's list.

After crossing out all of the duplicated countries, they were left with a total of 13 countries on their lists.

Who won the game?

12. The ordered pair (x, y) can be defined as the set $\{\{x\}, \{x, y\}\}$.

(a) With this definition, show that $(x, y) = (u, v)$ if, and only if, $x=u$ and $y=v$.

(b) Why can we not define the ordered pair as $(x, y) = \{x, \{y\}\}$?

13. In a certain town lives a barber, who is a man, who shaves every man in the town who does not shave himself.

The question is: Who shaves the barber? Explain your answer.

14. An adjective is *autological* if it describes itself. For example, "short" is autological since it is short; and "pentasyllabic" is autological since it is pentasyllabic; that is, it has five syllables. Any adjective that is *not* autological is said to be *heterological*. For example, "long" and "monosyllabic" are heterological.

The question is: Is "heterological" autological, or is it heterological? Explain your answer. What about "autological"?

Chapter 3

★ # Boolean Algebras and Circuits

There are 10 types of people in this world: those who understand binary numbers and those who don't.

- Anonymous.

At the end of the last chapter we noted a close analogy between Equivalence Laws for Propositional Logic on the one hand, and Set Identities on the other. In this chapter we explore this connection by looking at *Boolean algebras*, the mathematical structures underlying both propositional logic and sets.

This analogy extends to the world of digital computers and other electronic devices, which are built from circuits which have binary inputs and outputs; that is, they manipulate values from the set $\mathbb{B} = \{0, 1\}$. At the implementation level these binary inputs and outputs are delivered by voltages on wires, with a low voltage being interpreted as 0 and a high voltage being interpreted as 1. The simplest components of digital circuits, *logic gates*, are based on the connectives of propositional logic, with 0 (low voltage) and 1 (high voltage) being interpreted as **F** (false) and **T** (true), respectively. Composing logic gates together to create ever more complicated electronic components can thus be done in a way which is amenable to analysis via propositional logic. In this chapter we shall examine the fundamental role of Boolean algebra in underlying the building blocks of digital computers.

3.1 Boolean Algebras

A *Boolean algebra* is a set B which contains (at least) two distinct special elements 0 and 1, referred to as *zero* and *unit*, respectively, along with two binary operators $+$ and $\cdot$, referred to as *sum* and *product*, as well as a unary operator $'$, referred to as *complementation*. That is, for every pair (x, y) of elements of B there are three further (but not necessarily different) elements of B denoted $x+y$, $x \cdot y$, and x'. These operators must all satisfy the ten *Laws of Boolean Algebra* given in Figure 3.1.

F. Moller, G. Struth, *Modelling Computing Systems*,
Undergraduate Topics in Computer Science,
DOI 10.1007/978-1-84800-322-4_4, © Springer-Verlag London 2013

Commutativity:	$x + y = y + x$	(Comm1)
	$x \cdot y = y \cdot x$	(Comm2)
Associativity:	$(x + y) + z = x + (y + z)$	(Assoc1)
	$(x \cdot y) \cdot z = x \cdot (y \cdot z)$	(Assoc2)
Distributivity:	$x + (y \cdot z) = (x + y) \cdot (x + z)$	(Distr1)
	$x \cdot (y + z) = (x \cdot y) + (x \cdot z)$	(Distr2)
Identity:	$x + 0 = x$	(Ident1)
	$x \cdot 1 = x$	(Ident2)
Complement:	$x + x' = 1$	(Compl1)
	$x \cdot x' = 0$	(Compl2)

Figure 3.1: The Laws of Boolean Algebra.

Boolean algebras provide an abstract representation of familiar ideas in various areas of study. Indeed we have already met concrete examples of Boolean algebras in the form of sets and propositions.

Example 3.1   The Boolean Algebra of Sets

The power set $\mathcal{P}(U)$ of a set U gives rise to a Boolean algebra, with the roles of 0, 1, +, $\cdot$ and $'$ taken by $\emptyset$, U, $\cup$, $\cap$ and $^-$, respectively.

In this case, the laws give rise to the following set identities, which we confirmed in Section 2.7:

Commutativity:	$A \cup B = B \cup A$	(Comm1)
	$A \cap B = B \cap A$	(Comm2)
Associativity:	$(A \cup B) \cup C = A \cup (B \cup C)$	(Assoc1)
	$(A \cap B) \cap C = A \cap (B \cap C)$	(Assoc2)
Distributivity:	$A \cup (B \cap C) = (A \cup B) \cap (A \cup C)$	(Distr1)
	$A \cap (B \cup C) = (A \cap B) \cup (A \cap C)$	(Distr2)
Identity:	$A \cup \emptyset = A$	(Ident1)
	$A \cap U = A$	(Ident2)
Complement:	$A \cup \overline{A} = U$	(Compl1)
	$A \cap \overline{A} = \emptyset$	(Compl2)

Example 3.2   The Boolean Algebra of Propositions

The set of propositions gives rise to a Boolean algebra, with the roles of 0, 1, $+$, $\cdot$ and $'$ taken by false, true, $\vee$, $\wedge$ and $\neg$, respectively. (Equality $p = q$ is interpreted by equivalence $p \Leftrightarrow q$.)

In this case, the laws give rise to the following equivalences, which we confirmed in Section 1.7:

Commutativity:	$p \vee q \;\Leftrightarrow\; q \vee p$	*(Comm1)*
	$p \wedge q \;\Leftrightarrow\; q \wedge p$	*(Comm2)*
Associativity:	$(p \vee q) \vee r \;\Leftrightarrow\; p \vee (q \vee r)$	*(Assoc1)*
	$(p \wedge q) \wedge r \;\Leftrightarrow\; p \wedge (q \wedge r)$	*(Assoc2)*
Distributivity:	$p \vee (q \wedge r) \;\Leftrightarrow\; (p \vee q) \wedge (p \vee r)$	*(Distr1)*
	$p \wedge (q \vee r) \;\Leftrightarrow\; (p \wedge q) \vee (p \wedge r)$	*(Distr2)*
Identity:	$p \vee \text{false} \;\Leftrightarrow\; p$	*(Ident1)*
	$p \wedge \text{true} \;\Leftrightarrow\; p$	*(Ident2)*
Complement:	$p \vee \neg p \;\Leftrightarrow\; \text{true}$	*(Compl1)*
	$p \wedge \neg p \;\Leftrightarrow\; \text{false}$	*(Compl2)*

Example 3.3   The two-valued Boolean Algebra

The two-element set $\mathbb{B} = \{0, 1\}$ itself gives rise to an important Boolean algebra, with the operations defined as follows:

x	y	$x+y$
0	0	0
0	1	1
1	0	1
1	1	1

x	y	$x \cdot y$
0	0	0
0	1	0
1	0	0
1	1	1

x	x'
0	1
1	0

As we shall see, this particular algebra is of fundamental importance in the design of digital circuits.

Exercise 3.3   (Solution on page 421)

Verify that the laws of Boolean algebra hold for the two-valued Boolean algebra $\mathbb{B}$.

From now on we shall typically omit $\cdot$ and write xy rather than $x \cdot y$, and freely omit parentheses by allowing $\cdot$ to bind tighter than $+$ and $'$ to bind tighter than $\cdot$; thus for example, we shall write $x + (y \cdot (z'))$ simply as $x + yz'$.

(3.2)   Deriving Identities in Boolean Algebras

From the Laws of Boolean Algebra, we can derive very many identities which must be true in any Boolean algebra (in particular, as set identities and logical equivalences). In this section we derive some important identities, and leave it as an exercise to consider what these identities say as set identities and logical equivalences, many of which were derived already in previous chapters. We shall state our new identities as *Theorems* (true statements), and justify their truth using *proofs* (step-by-step derivations of their truth); the appearance of the box symbol "□" indicates the end of a proof.

Theorem 3.3   Further Distributive Laws

$$(x + y)z = xz + yz \qquad (Distr3)$$
$$xy + z = (x + z)(y + z) \qquad (Distr4)$$

Proof: We prove only the first identity, and leave the second as an exercise.

$$
\begin{aligned}
(x + y)z &= z(x + y) &&(Comm2)\\
&= zx + zy &&(Distr2)\\
&= xz + yz &&(Comm2,\ twice)
\end{aligned}
$$
□

Theorem 3.4   Idempotence Laws

$$x + x = x \qquad (Idemp1)$$
$$xx = x \qquad (Idemp2)$$

Proof: We prove only the first identity, and leave the second as an exercise.

$$
\begin{aligned}
x + x &= (x + x)1 &&(Ident2)\\
&= (x + x)(x + x') &&(Compl1)\\
&= x + xx' &&(Distr1)\\
&= x + 0 &&(Compl2)\\
&= x &&(Ident1)
\end{aligned}
$$
□

Theorem 3.5   Domination Laws

$$x + 1 = 1 \qquad (Dom1)$$
$$x0 = 0 \qquad (Dom2)$$

Proof: We prove only the first identity, and leave the second as an exercise.

$$
\begin{aligned}
x + 1 &= x + (x + x') & (Compl1)\\
&= (x + x) + x' & (Assoc1)\\
&= x + x' & (Idemp1)\\
&= 1 & (Compl1)
\end{aligned}
$$

$\square$

Theorem 3.6   Absorption Laws

$$
\begin{aligned}
x + xy &= x & (Absorp1)\\
x(x + y) &= x & (Absorp2)
\end{aligned}
$$

Proof: We prove the first identity here, and present the proof of the second in Example 3.11.

$$
\begin{aligned}
x + xy &= x1 + xy & (Ident2)\\
&= x(1 + y) & (Distr2)\\
&= x(y + 1) & (Comm1)\\
&= x1 & (Dom1)\\
&= x & (Ident2)
\end{aligned}
$$

$\square$

Next, we prove a law that we shall find useful in further calculations.

Theorem 3.7

If $x + y = x + z$ and $xy = xz$ then $y = z$.

Proof:

$$
\begin{aligned}
y &= y(x + y) & (Comm1,\ Absorp2)\\
&= y(x + z) & (Assumption\ 1)\\
&= yx + yz & (Distr2)\\
&= zx + zy & (Comm2,\ Assumption\ 2)\\
&= z(x + y) & (Distr2)\\
&= z(x + z) & (Assumption\ 1)\\
&= z & (Comm1,\ Absorp2)
\end{aligned}
$$

$\square$

Next, we consider a few results about complementation. The first of these is the observation that the two Complementation Laws $x + x' = 1$ and $xx' = 0$ uniquely determine the complement: there is no value y different from x' which satisfies these two equations.

Theorem 3.8   Uniqueness of Complement

If $x + y = 1$ and $xy = 0$ then $y = x'$. That is to say, x' is the only element which satisfies $x + x' = 1$ and $xx' = 0$.

Proof: Suppose that $x + y = 1$ and $xy = 0$. Then

$$x + y = 1 \qquad \text{(Assumption 1)}$$
$$= x + x' \qquad \text{(Compl1)}$$

and

$$xy = 0 \qquad \text{(Assumption 2)}$$
$$= xx' \qquad \text{(Compl2)}$$

Thus, by Theorem 3.7, $y = x'$. $\qquad\qquad\square$

Theorem 3.9   Involution Law

$(x')' = x$.

Proof:

$$x' + (x')' = 1 \qquad \text{(Compl1)}$$
$$= x' + x \qquad \text{(Comm1, Compl1)}$$

and

$$x'(x')' = 0 \qquad \text{(Compl2)}$$
$$= x'x \qquad \text{(Comm2, Compl2)}$$

Thus, by Theorem 3.7, $(x')' = x$. $\qquad\qquad\square$

Exercise 3.9   (Solution on page 422)

Prove that $0' = 1$ and $1' = 0$.

Theorem 3.10   De Morgan Laws

$$(x + y)' = x'y' \qquad \text{(DeMorgan1)}$$
$$(xy)' = x' + y' \qquad \text{(DeMorgan2)}$$

Proof: We prove only the first identity, and leave the second as an exercise. We first note that it suffices to show that

$$(x+y)(x'y') = 0 \quad \text{and} \quad (x+y)+(x'y') = 1$$

as then, by the Uniqueness of Complement Theorem 3.8, we would get that $(x+y)' = x'y'$.

$$
\begin{aligned}
(x+y)(x'y') &= x(x'y')+y(x'y') && (Distr3)\\
&= 0y' + 0x' && (Assoc2,\ Comm2,\ Compl2)\\
&= 0+0 && (Dom2)\\
&= 0 && (Idemp1)
\end{aligned}
$$

$$
\begin{aligned}
(x+y)+(x'y') &= ((x+y)+x')((x+y)+y') && (Distr1)\\
&= (y+1)(x+1) && (Assoc1,\ Comm1,\ Compl1)\\
&= 1 \cdot 1 && (Dom1)\\
&= 1 && (Idemp2) \qquad \square
\end{aligned}
$$

Exercise 3.10   (Solution on page 422)

Prove the following theorems.

1. $(xy + x'y')' = xy' + x'y$.
2. If $x+y = x+z$ and $x'+y = x'+z$ then $y = z$.
3. If $x+y = 0$ then $x = y = 0$.
4. $x = 0$ if, and only if, $y = xy' + x'y$ for all y.

3.3   The Duality Principle

Given any formula in a Boolean algebra, its *dual* is formed by interchanging 0 and 1, and $+$ and $\cdot$, throughout. More generally, the dual of a statement involving Boolean algebra is that statement with every formula replaced with its dual. Thus for example, the dual of $x + y'z = 1$ is $x(y'+z) = 0$.

The following is a fundamental principle of Boolean algebras.

Theorem 3.11   The Principle of Duality

The dual of every theorem of Boolean algebra is also a theorem.

Proof: To see that this is a valid principle, we merely need realise that a proof of a theorem becomes a proof of the dual of the theorem simply by replacing each formula used in the proof by its dual. This is so since the Laws of Boolean Algebra consist of five statements and their duals. □

Example 3.11

Consider the following derivation of the second Absorption Law $x(x+y) = x$:

$$
\begin{aligned}
x(x + y) &= (x + 0)(x + y) & (Ident1)\\
&= x + 0y & (Distr1)\\
&= x + y0 & (Comm2)\\
&= x + 0 & (Dom2)\\
&= x & (Ident1)
\end{aligned}
$$

If we compare this derivation with that given in the proof of the first Absorption Law $x + (xy) = x$ in Theorem 3.6, the duality is immediately apparent: the two derivations are identical, but for the fact that each expression is replaced by its dual, and the identity justifying each step in the above derivation is the dual of the identity justifying the same step in the first derivation.

This principle allows us to infer the validity of the dual of any theorem that we prove, since a proof of the dual theorem can be constructed automatically from the proof of the theorem, simply by replacing every formula and identity with its dual, as in the above example. Throughout the previous section we provided theorems presenting pairs of identities; and in each case we only proved the first of each identity, leaving the proof of the second as an exercise. In fact, the second identity in each case is the dual of the first; so by using the Duality Principle, proofs of these are unnecessary. The Duality Principle guarantees that they are valid.

Exercise 3.11 (Solution on page 423)

Write out the dual of each of the following theorems from Exercise 3.10.

1. $(xy + x'y')' = xy' + x'y$.

2. If $x+y = x+z$ and $x'+y = x'+z$ then $y = z$.

3. If $x+y = 0$ then $x = y = 0$.

4. $x = 0$ if, and only if, $y = xy' + x'y$ for all y.

3.4   Logic Gates and Digital Circuits

Computers manipulate all forms of information: numbers, names, sounds, pictures, videos; but an electronic computer can only reliably represent data in essentially one way: either a wire has a high voltage, or it has a low voltage. By interpreting a high voltage as the number 1 and a low voltage as the number 0, every piece of data represented and manipulated by an electronic computer is reduced within the electronics of the machine to combinations of the binary digits (the *bits*) 0 and 1.

At its lowest level, a computer manipulates this binary data using digital circuits which transform voltages on wires feeding into the circuit into voltages on wires leading out from it. How the electronics works (using transistors) to cause the output voltages to reflect the correct values according to the input voltages is not a question that will concern us here; such concerns are left to physicists and electronics engineers.

Example 3.12

Consider a circuit HA with two input wires, labelled x and y, and two output wires, labelled s and c. Such a circuit might be represented as follows:

$$x \quad \boxed{\text{HA}} \quad s$$
$$y \qquad\qquad c$$

Note that when we draw a circuit, we will assume that its input lines enter from the left and its output lines exit from the right.

Such a picture may represent the circuit simply as a black box as above, with no indication as to how the output values relate to the input values. However, we can describe the behaviour of the circuit by indicating what output values are produced from each of the possible input values. To do this, we can list all of the possibilities in the form of a truth table. For example, the circuit HA which we have in mind above behaves as follows:

x	y	s	c
0	0	0	0
0	1	1	0
1	0	1	0
1	1	0	1

Thus, for example, if both input wires x and y hold high voltages, thus both representing the value 1, then the output wire s will be given a low voltage, representing the value 0, and the output wire c will be given a high voltage, representing the value 1.

Computer circuits can be extremely complicated – far more complicated than the above example. However, all circuits, including the one above, can be built up from three very basic building blocks (which can all be easily implemented using transistors): *OR gates*, *AND gates* and *NOT gates*.

An *OR gate* is a simple component circuit which takes two inputs x and y and produces the single output $x+y$ defined by

$$x+y = \begin{cases} 1 & \text{if } x=1 \text{ or } y=1; \\ 0 & \text{otherwise.} \end{cases}$$

Graphically it is drawn as follows:

$$x \,—\!\!\!\!\!\!\!\!\!\!\begin{array}{c} \\ y \end{array}\!\!\!\!\!\!\!\!\!\!\!\!\!\!\!\!\!\!\!\Big\rangle\!\!\!—\, x+y$$

An *AND gate* is a simple component circuit which takes two inputs x and y and produces the single output $x \cdot y$ defined by

$$x \cdot y = \begin{cases} 1 & \text{if } x=1 \text{ and } y=1; \\ 0 & \text{otherwise.} \end{cases}$$

Graphically it is drawn as follows:

$$x \,—\!\!\!\!\begin{array}{c} \\ y \end{array}\!\!\!\!\!\!\!\!\!\Big)\!—\, x \cdot y$$

A *NOT gate* is a simple component circuit which takes one input x and produces the single output x' defined by

$$x' = \begin{cases} 1 & \text{if } x=0; \\ 0 & \text{if } x=1. \end{cases}$$

Graphically it is drawn as follows:

$$x \,—\!\!\!\!\triangleright\!\circ\!—\, x'$$

Truth tables defining these three gates are as follows:

x	y	$x+y$
0	0	0
0	1	1
1	0	1
1	1	1

x	y	$x \cdot y$
0	0	0
0	1	0
1	0	0
1	1	1

x	x'
0	1
1	0

We can observe from the above definitions that the three basic gates compute exactly the functions of the two-valued Boolean algebra $\mathbb{B}$ defined in Example 3.3. (Note that, as before, we shall typically write xy instead of $x \cdot y$.) This section makes clear, then, the fundamental importance of this particular Boolean algebra. It is absolutely essential in the design of digital computers.

We can build large complicated circuits from these three basic gates by stringing them together – always in a left-to-right fashion. (Allowing feedback wires provides its own uses – and complications – which we shall not explore.)

Example 3.13

Consider the following circuit:

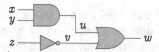

There are three inputs x, y and z to this circuit. The inputs x and y feed into an AND gate which outputs an intermediate value $u = xy$. Meanwhile, the input z feeds into a NOT gate which outputs a second intermediate value $v = z'$. The two intermediate values u and v output by the first two gates then feed as inputs into an OR gate which outputs the final value $w = u + v$. The effect of the whole circuit, therefore, is to output the value $w = xy + z'$. The value that is output is thus given according to the following table:

x	y	z	u	v	w
0	0	0	0	1	1
0	0	1	0	0	0
0	1	0	0	1	1
0	1	1	0	0	0
1	0	0	0	1	1
1	0	1	0	0	0
1	1	0	1	1	1
1	1	1	1	0	1

For example, if the inputs have values $x=1$, $y=0$ and $z=1$, then the output of the AND gate will be 0, as will the output of the NOT gate; and since both of the inputs to the OR gate will be 0, the value of the output w will also be 0:

$$x = 1$$
$$y = 0$$
$$0$$
$$z = 1$$
$$0$$
$$w = 0$$

The relationship between the table defining the function $w = xy + z'$ and the truth table for the proposition $(P \wedge Q) \vee \neg R$ is, hopefully, obvious.

Example 3.14

Consider the following circuit with three input lines a, b and c, and one output line m:

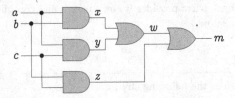

Note that we have used a dot to split a line, directing the same value (voltage) to two different inputs; and we have allowed lines to cross without interference (as if they were insulated from each other).

We can analyse the behaviour of this circuit as follows:

- the inputs a and b feed into the first AND gate to produce an intermediate value $x = ab$;

- the inputs a and c feed into the second AND gate to produce an intermediate value $y = ac$;

- the inputs b and c feed into the third AND gate to produce an intermediate value $x = bc$;

- the values x and y then feed into the first OR gate to produce a further intermediate value $w = x + y$;

- finally, the values w and z feed into the second OR gate to produce the final value $m = w + z$.

We can tabulate the value that is output by this circuit on any set of inputs as follows:

a	b	c	x	y	z	w	m
0	0	0	0	0	0	0	0
0	0	1	0	0	0	0	0
0	1	0	0	0	0	0	0
0	1	1	0	0	1	0	1
1	0	0	0	0	0	0	0
1	0	1	0	1	0	1	1
1	1	0	1	0	0	1	1
1	1	1	1	1	1	1	1

Algebraically, the effect of the circuit is to output the value

$$m = w + z = x + y + z = ab + ac + bc.$$

In other words, this circuit computes the majority function: the output m will be 1 exactly when at least two of the input values are 1.

Exercise 3.14   (Solution on page 423)

The *exclusive-OR gate*, or **XOR gate**, has the following definition (and gate symbol):

x	y	$x \oplus y$
0	0	0
0	1	1
1	0	1
1	1	0

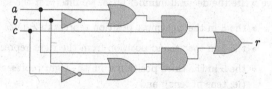

That is, the output $z = x \oplus y$ has the value 1 when exactly one of the inputs x or y has the value 1 (and the other has the value 0).

Build a circuit which realises this gate.

Exercise 3.15   (Solution on page 423)

Describe the behaviour of the following circuit by providing a Boolean expression and a truth table defining the output value r.

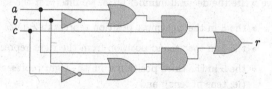

Exercise 3.16   (Solution on page 424)

Consider a car safety system in which a warning bell rings whenever the motor is running while a door is open or a seat belt is unbuckled. This is to be implemented as a Boolean circuit which takes three inputs M, D and B, respectively representing the states of the motor, doors and seat belts:

- M will be 1 if the motor is running and 0 otherwise;
- D will be 1 if the doors are closed and 0 otherwise;
- B will be 1 if the seat belts are fastened and 0 otherwise.

The circuit is to produce a single output R which should be 1 if the warning bell should ring and 0 otherwise. Build a circuit for this system.

3.5   Making Computers Add

In this section we consider the problem of constructing a circuit which will add two integers. To do this, we must first understand how integers are represented and manipulated by a computer using just the binary digits.

3.5.1   Binary Numbers

People have ten fingers, and children learn early on to count using the ten digits on their hands. When counting beyond ten on your fingers, the natural thing to do is to keep track of how many times you run through your fingers, which you can ask someone else to do using their ten fingers. Then a third person in turn can use their ten fingers to keep track of how many times the second person runs through all of their fingers, which happens every time you reach 100 (i.e., each time you run through your own ten fingers ten times). When the third person runs out of fingers, you will have counted ten lots of 100, i.e. up to 1000. If you are still counting, a fourth person can use their ten fingers to keep track of how many lots of 1000 you have counted. A fifth person can then keep track of how many lots of 10000 are counted; a sixth how many lots of 100000; etc.

This mechanism for counting is reflected in our use of decimal numbers, which is a positional notation for expressing quantities. For example, when we write the decimal number 6538, we interpret the four digits as follows:

- the 8 in the rightmost position represents 8 ones;

- the 3 in the second position from the right represents 3 lots of tens;

- the 5 in the third position from the right represents 5 lots of hundreds (ie, tens of tens); and

- the 6 in the fourth position from the right represents 6 lots of thousands (ie, tens of hundreds, or tens of tens of tens).

That is,

$$
\begin{aligned}
6538 = \quad & 6 \times 10^3 & = & \; 6 \times 1000 & = & \; 6000 \\
+ & 5 \times 10^2 & = & \; 5 \times 100 & = & 500 \\
+ & 3 \times 10^1 & = & \; 3 \times 10 & = & 30 \\
+ & 8 \times 10^0 & = & \; 8 \times 1 & = & \underline{8} \\
& & & & & 6538
\end{aligned}
$$

Digital computers have access to only the two binary digits, 0 and 1, not the ten decimal digits. Therefore, they naturally represent quantities as binary numbers rather than decimal numbers, which are sequences of binary digits (bits) rather than decimal digits. For example, the binary number 11101 is interpreted as follows:

- the 1 in the rightmost position represents 1 one;
- the 0 in the second position from the right represents 0 lots of twos;
- the 1 in the third position from the right represents 1 lot of fours (ie, twos of twos);
- the 1 in the fourth position from the right represents 1 lot of eights (ie, twos of fours, or twos of twos of twos); and
- the 1 in the fifth position from the right represents 1 lot of sixteens (ie, twos of eights, or twos of twos of twos of twos).

That is,

$$
\begin{aligned}
11101 = & \quad 1 \times 2^4 &=& \quad 1 \times 16 &=& \quad 16 \\
& + 1 \times 2^3 &=& \quad 1 \times 8 &=& \quad 8 \\
& + 1 \times 2^2 &=& \quad 1 \times 4 &=& \quad 4 \\
& + 0 \times 2^1 &=& \quad 0 \times 2 &=& \quad 0 \\
& + 1 \times 2^0 &=& \quad 1 \times 1 &=& \quad \underline{1} \\
& & & & & \quad 29
\end{aligned}
$$

Any natural number can be represented as a binary number, just as easily as it can be represented as a decimal number. The method for translating from binary to decimal can be extracted from the above description; and the method for translating from decimal to binary is almost as easy: we merely have to keep subtracting from the number in question the largest power of two that we can.

(**Example 3.16**)

To translate the decimal value 51 into binary:

- subtract $2^5 = 32$ from 51 to give a remainder of 19;
- subtract $2^4 = 16$ from 19 to give a remainder of 3;
- subtract $2^1 = 2$ from 3 to give a remainder of 1;
- subtract $2^0 = 1$ from 1 to give a remainder of 0.

We can thus express the decimal number $51 = 32 + 16 + 2 + 1$ in binary as:

$$
\begin{aligned}
110011 = & \quad 1 \times 2^5 &=& \quad 1 \times 32 &=& \quad 32 \\
& + 1 \times 2^4 &=& \quad 1 \times 16 &=& \quad 16 \\
& + 0 \times 2^3 &=& \quad 0 \times 8 &=& \quad 0 \\
& + 0 \times 2^2 &=& \quad 0 \times 4 &=& \quad 0 \\
& + 1 \times 2^1 &=& \quad 1 \times 2 &=& \quad 2 \\
& + 1 \times 2^0 &=& \quad 1 \times 1 &=& \quad \underline{1} \\
& & & & & \quad 51
\end{aligned}
$$

3.5.2   Adding Binary Numbers

Consider how we would naturally add two (decimal) numbers by hand. We would first line the numbers up one on top of the other. Then we would add the units, writing down the unit sum digit and moving the carry digit (if there is one) to the top of the tens column; then we would add the three numbers in the tens column, writing down the tens sum digit and moving the carry digit to the top of the hundreds column; then we would add the three numbers in the hundreds column, writing down the hundreds sum digit and moving the carry digit to the top of the thousands column; and continue doing this same calculation with each column from right to left.

This same method works equally well for binary numbers, and is the basis for how digital computers add numbers represented in binary.

Example 3.17

To add the two binary numbers 11101 and 10110, write them one over the other and add the bits column-wise from right to left, including carries where necessary, as indicated:

$$
\begin{array}{r}
1\ 1\ 1 \\
1\ 1\ 1\ 0\ 1 \\
1\ 0\ 1\ 1\ 0 \\
\hline
1\ 1\ 0\ 0\ 1\ 1
\end{array}
$$

The first two columns on the right each gives a sum of 1 with no carry; but the third column from the right give a 0 sum with a carry, as then does the fourth column. The fifth column gives a sum of 1 with a carry, which gives a sum of 1 for the new sixth column.

Exercise 3.17   (Solution on page 424)

What decimal sum is being calculate in the above example?

We are now in a position to design a digital circuit which adds two integers represented as binary numbers. More specifically, we shall build a circuit which will have 8 input lines representing two 4-bit binary numbers $a_3a_2a_1a_0$ and $b_3b_2b_1b_0$, and 5 output lines representing the 5-bit binary number $s_4s_3s_2s_1s_0$ resulting from adding $a_3a_2a_1a_0$ and $b_3b_2b_1b_0$:

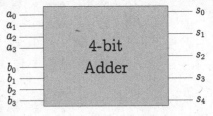

The construction we give can be easily scaled up to add arbitrarily-long bit strings.

3.5.3   Building Half Adders

The basic component from which we shall build our 4-bit adder is the circuit HA from Example 3.12 (page 95), which takes the two inputs x and y and produces the two outputs s and c representing the sum of x and y, with s being the sum bit and c being the carry bit. Such a circuit is called a *half adder*.

Our first task is to express the outputs in terms of the functions of the basic gates. For a start, computing the carry bit c is obvious: being 1 exactly when both x and y are 1, it is their product $c = xy$. The sum bit s is only slightly more cumbersome. It is 1 when one of the inputs is 1 and the other is 0: $s = x'y + xy'$.

Towards building these functions in a circuit using the three basic gates, we can first note the following two circuits that compute $x'y$ and xy', respectively:

These can then be combined to give a circuit for $x'y + xy'$ as follows:

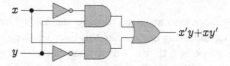

The above circuit computes the sum bit of the half adder; it only remains to add a further AND gate which computes the carry bit to complete the circuit:

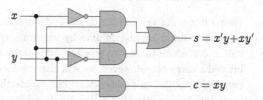

This circuit consists of six gates: three AND gates, two NOT gates and one OR gate. The question then arises: is it possible to build a simpler circuit which performs the computation of a half adder. Such questions are important when contemplating fitting ever-more computing power on a computer chip; you would certainly want to find the smallest possible circuits to compute the functions that are implemented on the chip.

Using the laws of Boolean algebra, we can make the following calculation:

$$
\begin{aligned}
x'y + xy' &= xy' + x'y && \textit{(commutativity)} \\
&= (xy + x'y')' && \textit{(Exercise 3.10(1))} \\
&= (xy)'(x'y')' && \textit{(DeMorgan1)} \\
&= (xy)'(x'' + y'') && \textit{(DeMorgan2)} \\
&= (xy)'(x + y) && \textit{(Involution, twice)} \\
&= (x+y)(xy)' && \textit{(commutativity)}
\end{aligned}
$$

This complicated calculation, in fact, tells us something natural: that having one input line holding the value 1 and the other holding the value 0: $x'y + xy'$ is the same as having one of the input lines holding the value 1 and not having both input lines holding the value 1: $(x+y)(xy)'$.

Indeed, such intuitive observations are where ideas for optimisations typically arise. The above derivation was a necessary step in the design process, in justifying the intuition which suggested the optimisation.

Importantly, the final expression $(x+y)(xy)'$ is simpler to evaluate than $x'y + xy'$, requiring only four basic operations rather than five; moreover, the product xy, is calculated in the process, so we need no further operations to complete the half adder circuit. The corresponding circuit is as follows:

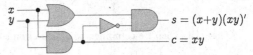

We have thus managed to build the half adder using four gates instead of six. Improving designs like this in order to reduce the number of gates – in this case by a third – is of obvious importance when it comes to fitting more power within the limited space on a circuit board. Reasoning with Boolean algebra is a crucial activity in the design of computer processors.

3.5.4   Building Full Adders

We are designing a circuit which will add two binary numbers using the usual method of summing bits column-by-column. So far we have constructed a half adder which takes two bits and adds these together, producing a sum bit and a carry bit. However, we will also need a circuit which adds not just two digits, as the half adder does, but rather three digits, to cater for the carry bit. Such a circuit is called a *full adder* and has the following form:

$$
\begin{array}{c}
x \\
y \\
z
\end{array}
\boxed{\text{FA}}
\begin{array}{c}
s \\
c
\end{array}
$$

The input wires x, y and z each have the value 0 or 1, and sum up to either 0, 1, 2 or 3, which is reflected in the output wires s and c.

The sum bit s will be 1 if exactly one or all three of the input bits x, y and z are 1:

$$s = xy'z' + x'yz' + x'y'z + xyz;$$

and the carry bit will be 1 if at least two of the input bits are 1:

$$c = xyz' + xy'z + x'yz + xyz.$$

Letting $t = yz'+y'z$ be the value of the sum bit from a half adder with inputs y and z, and noting from Exercise 3.10(1) that $t' = yz+y'z'$, we can note that

$$\begin{aligned} s &= xy'z' + x'yz' + x'y'z + xyz \\ &= x'(yz'+y'z) + x(yz+y'z') \quad (commutativity/distributivity) \\ &= x't + xt' \end{aligned}$$

and

$$\begin{aligned} c &= xyz' + xy'z + x'yz + xyz \\ &= x(yz'+y'z) + yz(x+x') \quad (distributivity) \\ &= xt + yz \quad (identity) \end{aligned}$$

These outputs are generated by combining two half adders and an OR gate as follows:

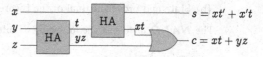

3.5.5 Putting It All Together

Having defined a full adder, adding two n-bit numbers is then achieved by stringing n such full adders together. In particular, to build our 4-bit adder, which adds together the two 4-bit binary numbers $a_3a_2a_1a_0$ and $b_3b_2b_1b_0$ to produce the 5-bit binary number $s_4s_3s_2s_1s_0$, we would use the following circuit:

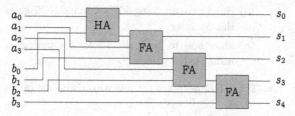

We start with a half adder, as we don't have to worry about a carry bit for the first two bits a_0 and b_0. Of course, stringing more full adders together would allow larger values to be added, meaning that this circuit can be easily scaled up.

3.6 Additional Exercises

1. Prove the second Further Distributive Law, the second Idempotence Law, the second Absorption Law, and the second Domination Law, all using the Laws of Boolean Algebra. (That is, do not rely on the Duality Principle.)

2. Prove that the set $S = \{1, 2, 5, 10\}$ of divisors of 10 is a Boolean algebra with zero 1 and unit 10, with $x + y$ interpreted as the least common multiple of x and y, lcm(x, y); xy interpreted as the greatest common factor of x and y, gcd(x, y); and $x' = 10/x$.

3. Prove that if we take $S = \{1, 2, 3, 6, 12\}$ to be the set of divisors of 12 in Exercise 2 above, then we would *not* get a Boolean algebra.

4. Does the finite powerset $\mathcal{P}_{\text{fin}}(U)$ of a set U give rise to a Boolean algebra (with, as usual, the roles of 0, 1, $+$, $\times$ and $\cdot'$ taken by $\emptyset$, U, $\cup$, $\cap$ and $\bar{\ }$, respectively)? Justify your answer.

5. (a) Prove that $xy' = 0$ if, and only if, $x' + y = 1$.

   (b) State and prove the dual of the theorem in part (a).

6. (a) Prove that $x = y$ if, and only if, $xy' + x'y = 0$.

   (b) State and prove the dual of the theorem in part (a).

7. The **NAND** *gate* has the following definition (and symbol):

x y	$x \mid y$
0 0	1
0 1	1
1 0	1
1 1	0

$$z = x \mid y$$

That is, the output $z = x \mid y$ has the value 0 if both of the inputs x or y have the value 1; otherwise it has the value 1.

   (a) Build a circuit using AND, OR and NOT gates which implements this operator.

   (b) Show how to build circuits for computing x', $x+y$ and xy only using NAND gates.

8. The **NOR** *gate* has the following definition (and symbol):

x y	$x \downarrow y$
0 0	1
0 1	0
1 0	0
1 1	0

$$z = x \downarrow y$$

That is, the output $z = x \downarrow y$ has the value 1 if neither of the inputs x nor y has the value 1; otherwise it has the value 0.

(a) Build a circuit using AND, OR and NOT gates which implements this operator.

(b) Show how to build circuits for computing x', $x+y$ and xy only using NOR gates.

9. Build circuits which implement the following Boolean expressions.

(a) $(a + b)(b + c)$

(b) $a'b + (b + c)'$

(c) $(ab)' + (bc)'$

10. Describe the behaviour of the following circuits by providing a Boolean expression and a truth table defining the output value X.

(a)

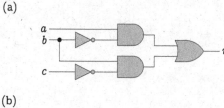

(b)

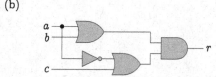

11. Joel, Felix and Oskar are using a simple voting machine to cast secret ballots to decide which DVD to watch tonight, the choice being between the latest Final Destination film and the new Fockers film. Each of them will vote either "0" for Final Destination or "1" for the Fockers; and they will then watch whichever film receives the majority of the three votes.

Build a circuit which accepts three inputs J, F and O representing their respective votes, and produces one output X representing the outcome of the election.

12. A *multiplexer* is a circuit with three input lines x_0, x_1 and s, and one output line r, defined as follows:

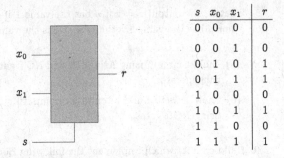

s	x_0	x_1	r
0	0	0	0
0	0	1	0
0	1	0	1
0	1	1	1
1	0	0	0
1	0	1	1
1	1	0	0
1	1	1	1

The s line acts as a *selector*; the value of the output r will be either that of x_0 or that of x_1, depending on the value of s.

Build a circuit which implements this multiplexer.

(Hint: First argue that $r = s'x_0 + sx_1$.)

Chapter 4

Predicate Logic

Death is more universal than life; everyone dies but not everyone lives.

- Andrew Sachs.

Propositional logic allows us to express and reason about simple propositions. However, we quickly run into its limitations. For example, Augustus De Morgan put made the following deduction:

All horses are animals.

Therefore, all horse-heads are animal-heads.

This deduction is certainly valid. However, this cannot be demonstrated using propositional logic, as there is no way to discuss the properties of individual horses or animals, let alone their heads.

In this section we extend propositional logic to include *predicates* – properties which may be true or false of particular elements in a given universe – and *quantifiers* – the means by which we refer to elements which satisfy such properties.

4.1 Predicates and Free Variables

Recall how we defined the set of prime numbers:

$\{\, x \,:\, x \text{ is a prime number}\,\}$.

We used this example to introduce the general scheme for defining sets as the collection of all objects which satisfy some property:

$\{\, x \,:\, x \text{ has property } P \,\}$

denotes the set of all objects x which satisfy the property P. Such a property is referred to as a *predicate*, and we write $P(x)$ to say that "the object x has property P." A predicate is an indeterminate proposition which is true or false of any particular element x of a given universe. Thus, for example,

F. Moller, G. Struth, *Modelling Computing Systems*,
Undergraduate Topics in Computer Science,
DOI 10.1007/978-1-84800-322-4_5, © Springer-Verlag London 2013

$$Prime(x) = \text{“}x \text{ is a prime number ”}$$

denotes the predicate which stipulates that the element x is a prime number; the universe of discourse in this instance, that is, the set of values which x may range over, would most naturally be the set of natural numbers $\mathbb{N}$ (though it could be anything; in the case that x were not a natural number, the predicate $Prime(x)$ would be false).

Predicates differ from propositions in that they do not have a fixed truth value, since we do not know the value of the object to which it refers: $Prime(x)$ may be true or false, depending on what value x refers to. The variable x is referred to as a *free variable*. If we instantiate the free variable in such a predicate, we would get a proposition. For example, $Prime(7)$ is a true proposition (7 *is* a prime number), while $Prime(9)$ is a false proposition ($9 = 3 \cdot 3$ is *not* a prime number). The set of objects which satisfy a predicate, that is, which make the predicate true, is called the *truth set* of the predicate. Thus, for example, the truth set of the predicate $Prime(x)$ is the set of prime numbers. When we define a set by $\{\, x \; : \; P(x) \,\}$, we are defining it to be the truth set of the predicate $P(x)$.

Example 4.1

Let the universe of discourse be the Duck family:

$$\text{Ducks} = \Big\{ \text{Quackmore, Hortense, Scrooge,}$$
$$\text{Donald, Della, Huey, Louis, Dewey} \Big\},$$

and define the following predicate:

$$Female(x) = \text{“}x \text{ is a female.”}$$

Then

- $Female$(Hortense) and $Female$(Della) are both true;
- $Female$(Quackmore), $Female$(Scrooge), $Female$(Huey), $Female$(Louis) and $Female$(Dewey) are all false;
- the truth set of the predicate $Female(x)$ is $\{\,$Hortense, Della$\,\}$.

Predicates may range over more than one element. As familiar examples, *equality* and *set inclusion* are predicates that range over two elements. In these cases, *infix notation* "$x = y$" and "$x \subseteq y$" is more natural to use than prefix notation "$=(x, y)$" and "$\subseteq(x, y)$." The statement $5 = 5$, for example, is true, whereas the statement $\{\emptyset\} = \emptyset$ is false.

The truth set of a predicate which ranges over more than one element consists of tuples of values; the number of coordinates of the tuples is equal to the number of free variables in the predicate. The tuples in the truth set represent those values that we can instantiate the free variables with in order to turn the predicate into a true proposition.

Example 4.2

We may use $Divides(x, y)$ to denote the two-place predicate over integers which stipulates that x divides evenly into y. In this case,

$$Divides(3, 15)$$

is true, since 3 divides evenly into 15 (5 times), while

$$Divides(4, 15)$$

is false, since 4 does not divide evenly into 15. The truth set of the predicate *Divides* is the set of pairs (x, y) such that x divides evenly into y:

$$\{ (x, y) : x \ divides \ evenly \ into \ y \}.$$

The standard mathematical symbol for this predicate is | and is written in infix notation, as in $3 \mid 15$ and $4 \nmid 15$.

Exercise 4.2   (Solution on page 424)

What are the truth sets of the following predicates?

1. $Even(x) = $ "x is an even integer."
2. $EvenPrime(x) = $ "x is an even prime number."
3. $DeadlySin(x) = $ "x is a deadly sin."
4. $Sum(x, y, z) = $ "x, y and z are integers, and $x + y = z$."
5. $Sum(u, 5, v)$, where $Sum(x, y, z)$ is the predicate defined above.

4.2   Quantifiers and Bound Variables

Before Joel, Felix, Oskar and Amanda go to school in the morning, they have to remember to brush their teeth; that is, the predicate

$$Teeth(x),$$

which denotes that child x has brushed their teeth, must be true of *each* of them. To this end, each child is asked in turn if they have brushed their teeth, in order to ensure that the compound proposition

$$Teeth(\text{Joel}) \wedge Teeth(\text{Felix}) \wedge Teeth(\text{Oskar}) \wedge Teeth(\text{Amanda})$$

is true. The universe of discourse is the set consisting of the four children:

Children = { Joel, Felix, Oskar, Amanda }.

After this final check, they get into the car and head off to school. One of the children has to sit in the front passenger's seat, as there is only room for three passengers in the back seat. Thus, the predicate

$Front(x)$,

which denotes that child x sits in the front seat, must be true of some *one* of them. They regularly argue over who this will be – either for, if they want to get away from their siblings, or against, to continue a joint activity – but the compound proposition

$Front(\text{Joel}) \lor Front(\text{Felix}) \lor Front(\text{Oskar}) \lor Front(\text{Amanda})$

must somehow be true.

In fact, as there is only room for one child in the front seat, the predicate $Front(x)$ must be true of *exactly* one child; that is, it must be true of one and false of all of the others. This means that the following proposition must be true:

$(\quad Front(\text{Joel}) \land \neg Front(\text{Felix}) \land \neg Front(\text{Oskar}) \land \neg Front(\text{Amanda}))$

$\lor$

$(\neg Front(\text{Joel}) \land \quad Front(\text{Felix}) \land \neg Front(\text{Oskar}) \land \neg Front(\text{Amanda}))$

$\lor$

$(\neg Front(\text{Joel}) \land \neg Front(\text{Felix}) \land \quad Front(\text{Oskar}) \land \neg Front(\text{Amanda}))$

$\lor$

$(\neg Front(\text{Joel}) \land \neg Front(\text{Felix}) \land \neg Front(\text{Oskar}) \land \quad Front(\text{Amanda}))$

That is: either Joel sits in the front seat and none of the others do; or Felix sits in the front seat and none of the others do; or Oskar sits in the front seat and none of the others do; or Amanda sits in the front seat and none of the others do.

These propositions are lengthy already when there are only four elements in the universe of discourse. Furthermore, we would not be able to write out formulæ to check if some or all elements satisfy a property if the universe of discourse is infinite. For example, to express the fact that every prime number greater than 2 is odd, using the predicate $Odd(x)$ to mean that x is an odd number, would require an infinitely-long conjunction:

$Odd(3) \land Odd(5) \land Odd(7) \land Odd(11) \land Odd(13) \land \cdots$

Similarly, to express the statement that some primes are square, using the predicate $Square(x)$ to mean that x is a prefect square, would require an infinitely-long disjunction:

$$Square(2) \lor Square(3) \lor Square(5) \lor Square(7) \lor \cdots$$

However, we cannot express infinite conjunctions and disjunctions in propositional logic.

Predicate logic provides two forms of *quantification* which allow you to express when properties are true of all elements in the universe of discourse, or true of at least some elements in the universe. These are outlined as follows.

4.2.1 Universal Quantification

When we want to express that a predicate $P(x)$ is true of *all* elements x of the universe of discourse, we can write:

$$\forall x \, P(x)$$

which is pronounced as

"*for all x, $P(x)$*";

it is true if, and only if, the predicate $P(x)$ is true of *all* possible values of x. This is called *universal quantification*.

For example, instead of writing

$$Teeth(\text{Joel}) \land Teeth(\text{Felix}) \land Teeth(\text{Oskar}) \land Teeth(\text{Amanda})$$

to express that $Teeth(x)$ is true of all four children, we can simply write

$$\forall x \, Teeth(x)$$

which says the same thing, that *everyone* has brushed their teeth (assuming the universe of discourse is the set of the four children).

Notice that $\forall x \, Teeth(x)$ is a proposition: it has a definite truth value. The variable x is not a free variable in this case; it is a *bound variable*; it is bound by the quantifier "$\forall x$".

Example 4.3

The statement

"*Nobody did the homework*"

is expressed as:

$$\forall x \, \neg H(x)$$

where $H(x) = $ "*x did the homework*".

The universe of discourse is (assumed to be) the set of students who were assigned the homework to do.

Notice that saying something is true of *nobody* is a universal quantification: it is the same as saying that this something is *not* true of *everybody*. In this case, we are saying that everybody did *not* do their homework.

Example 4.4

The statement

> *"Every dog that has stayed in the kennel will have to go into quarantine"*

is expressed as:

$$\forall x \, (\, K(x) \Rightarrow Q(x) \,)$$

where $K(x) = $ *"x has stayed in the kennel"*

$Q(x) = $ *"x will have to go into quarantine"*.

The universe of discourse is (assumed to be) the set of dogs, only some of which have stayed in the kennel in question.

This example demonstrates how to quantify universally over a subset of the universe of discourse: we simply stipulate that a property holds of something whenever it is a member of the subset of interest (that is, if it satisfies the predicate defining this subset). In this case, by using the implication

$$K(x) \Rightarrow Q(x)$$

we are not stating that every dog will have to go into quarantine, but only those dogs that have stayed in the kennel. If a particular dog x has not stayed in the kennel – that is, if $K(x)$ is not true – then that dog x need not go into quarantine – that is, $Q(x)$ need not be true. (Of course, this dog x might have to go into quarantine for some other reason; it is not necessarily the case that $Q(x)$ is false.)

Note that universal quantification is assumed to bind more strongly than all of the propositional connectives; that is, it is given higher precedence. For example, in the above example we wrote

$$\forall x \, (\, K(x) \Rightarrow Q(x) \,)$$

and not

$$\forall x \, K(x) \Rightarrow Q(x)$$

as the latter would be interpreted as

$$(\forall x\, K(x)) \Rightarrow Q(x)$$

which says *"if every dog has stayed in the kennel then x will have to go into quarantine."* This is certainly not what is intended; in particular, it is a predicate with a free variable x – appearing in $Q(x)$ – and is therefore not a proposition.

Example 4.5

The statement

"Nobody likes a sore loser"

is expressed as:

$$\forall x\, (S(x) \;\Rightarrow\; \forall y\, \neg L(y,x))$$

where $S(x) = $ *"x is a sore loser"*

$L(y,x) = $ *"y likes x"*.

The universe of discourse is (assumed to be) the collection of all people.

This proposition is saying the following is true of every person x: if x is a sore loser, then every person y does not like x.

Exercise 4.5   (Solution on page 424)

Using the predicates

$B(x) = $ *"x is a bee"*

$F(x) = $ *"x is a flower"*

$L(x,y) = $ *"x likes y"*

write each of the following statements in predicate logic.

1. All bees like all flowers.
2. Bees only like flowers.
3. Only bees like flowers.

4.2.2   Existential Quantification

When we want to express that a predicate $P(x)$ is true of at least *some* element x of the universe of discourse, we can write:

$$\exists x\, P(x)$$

which is pronounced as

"*there exists x such that P(x)*";

it is true if, and only if, the predicate $P(x)$ is true of *some* value of x. For example, instead of writing

$$Front(\text{Joel}) \lor Front(\text{Felix}) \lor Front(\text{Oskar}) \lor Front(\text{Amanda})$$

to express that $Front(x)$ is true of at least one of the four children, we can simply write

$$\exists x \, Front(x)$$

which says the same thing, that *someone* sits in the front seat (again, assuming the universe of discourse is the set of the four children).

Again, the variable x in $\exists x \, Front(x)$ is a bound variable, bound by the quantifier "$\exists x$"; and like universal quantification, existential quantification is assumed to bind more strongly than all of the propositional connectives.

Example 4.6

The statement

"*Someone didn't do the homework*"

is expressed as:

$$\exists x \, \neg H(x)$$

where $H(x) =$ "*x did the homework*".

The universe of discourse is again (assumed to be) the set of students who were assigned the homework to do.

This proposition states that $\neg H(x)$ holds of *some* student: perhaps no one did the homework (as expressed by the proposition given in Example 4.3); or perhaps several did the homework while several others didn't; or perhaps all but one person did the homework. This proposition doesn't distinguish between these possibilities; it merely notes that at least one element of the universe of discourse satisfies the predicate, that is, at least one person did not do the homework.

Example 4.7

The statement

"*If some dog that has stayed in the kennel has been in contact with a dog with rabies, then every dog that has stayed in the kennel will have to go into quarantine*"

is expressed as:

$$\exists x \left(K(x) \, \land \, \exists y \, (C(x,y) \land R(y)) \right) \;\Rightarrow\; \forall x \left(K(x) \Rightarrow Q(x) \right)$$

where $K(x) =$ "x has stayed in the kennel"

$R(x) =$ "x has rabies"

$C(x,y) =$ "x and y have been in contact"

$Q(x) =$ "x will have to go into quarantine".

Exercise 4.7   (Solution on page 424)

Assuming the universe of discourse is the set of human beings, consider the following predicates

$Male(x) =$ "x is male"

$Female(x) =$ "x is female"

$Parent(x,y) =$ "x is a parent of y"

$Father(x,y) =$ "x is the father of y"

$Mother(x,y) =$ "x is the mother of y"

$Sibling(x,y) =$ "x and y are siblings"

$Cousin(x,y) =$ "x and y are cousins"

Using these predicates, express the following properties in predicate logic.

1. Every human is either male or female, but no human is both.

2. Mothers are female parents.

3. Every human has exactly one mother and exactly one father.

4. Siblings have the same parents.

5. Cousins each have a parent who are siblings.

Exercise 4.8   (Solution on page 425)

Using the following predicates:

$Horse(h) =$ "h is a horse"

$Animal(a) =$ "a is an animal"

$Head(x,y) =$ "x is the head of y"

formalise the following argument in predicate logic:

All horses are animals.

Therefore, all horse heads are animal heads.

Explain why the argument is valid.

4.2.3   Bounded Quantifications

There are two forms of *bounded quantification* which we use for convenience. These restrict the range of the variables being quantified.

Firstly, to declare that the predicate $P(x)$ is true of every element of the set A, we write

$$\forall\, x{\in}A\ P(x)$$

which is pronounced as

    "for all values x in A, $P(x)$".

This is logically equivalent to

$$\forall x \left(x{\in}A \ \Rightarrow\ P(x) \right).$$

Similarly, to declare that the predicate $P(x)$ is true of some element of the set A, we write

$$\exists\, x{\in}A\ P(x)$$

which is pronounced as

    "there is some value x in A such that $P(x)$".

This is logically equivalent to

$$\exists x \left(x{\in}A \ \wedge\ P(x) \right).$$

One further useful restriction for the existential quantifier is declare that exactly one value x satisfies $P(x)$. This is written

$$\exists! x\ P(x)$$

which is pronounced as

    "there is exactly one value x such that $P(x)$".

This is logically equivalent to

$$\exists x \left(P(x)\ \wedge\ \neg\exists y\,(P(y) \wedge y \neq x) \right).$$

This says that there is a value x such that $P(x)$, but there is not a different value $y{\neq}x$ such that $P(y)$. For example, if the predicate *Front*(x) denotes that child x sits in the front seat of the car, where, again, the universe of discourse is the set of the four children, then

$$\exists! x\ Front(x)$$

states that exactly one of the children sits in the front seat.

Note that you can combine the last two constructions to declare that exactly one value from a set A satisfies $P(x)$: $\exists! x \in A\, P(x)$. Also note that x is of course bound by the quantifiers in each case.

Example 4.8

You may be aware that $\sqrt{2}$ is irrational: that it cannot be expressed as a fraction p/q. (We shall justify this claim in Example 5.6, page 139.) In fact, *any* nonnegative integer is either a perfect square, such as $25 = 5^2$, or its square root is irrational. We can express this fact as follows:

$$\forall n \in \mathbb{Z}\, (\, \exists k \in \mathbb{Z}\, (n = k^2) \quad \lor \quad \neg \exists q \in \mathbb{Q}\, (n = q^2)\,).$$

This says that for all integers n, either there exists another integer k such that $n = k^2$ (that is, n is a perfect square with square root k), or there does *not* exist a rational number q such that $n = q^2$ (that is, it does not have a rational square root).

Example 4.9

Recall the following puzzle from Exercise 1.16 (page 35). Joel, Felix and Oskar each write their name on a piece of paper, and then exchange the pieces of paper so that no one has the piece with their own name on it. They then hold these pieces of paper so that Amanda can't see what's on them, but tell her that each has the name of one of the others, and they challenge her to figure out who is holding each name. She is allowed to look at the name written on any one piece of paper. She decides to look at Joel's piece, and finds "Oskar" written on it.

Let Boys $= \{$ Joel, Felix, Oskar $\}$ be the set of three boys; and let Papers $= \{ J, F, O \}$ be the set of three pieces of paper with names written on them: J is the piece with "Joel" written on it; F is the piece with "Felix" written on it; and O is the piece with "Oskar" written on it. Furthermore, let $Holds(b, p)$ be the predicate which says that boy b holds the piece of paper p. Then we can formulate the conditions describe in this problem as follows:

1. Each boy holds precisely one piece of paper:

   $\forall b \in$ Boys $\exists! p \in$ Papers $Holds(b, p)$.

2. Each piece of paper is held by precisely one boy:

   $\forall p \in$ Papers $\exists! b \in$ Boys $Holds(b, p)$.

3. No piece of paper is being held by the boy whose name is on the paper:

$$\neg Holds(\text{Joel}, J) \ \wedge \ \neg Holds(\text{Felix}, F) \ \wedge \ \neg Holds(\text{Oskar}, O).$$

4. Joel's piece of paper has "Oskar" written on it:

$$Holds(\text{Joel}, O).$$

Exercise 4.9   (Solution on page 425)

Let $T(s, c)$ stand for the predicate *"student s takes course c."* Express the following statements in predicate logic.

1. *Alice and Bob take exactly one course together."*
2. *Alice and Bob take exactly two courses together."*

4.3   Rules for Quantification

If it is not the case that the predicate $P(x)$ is true for *all* values of x, then this must mean that $P(x)$ is not true for *some* value of x; that is,

$$\neg \forall x \, P(x) \ \Leftrightarrow \ \exists x \, \neg P(x).$$

Equally, if it is not the case that the predicate $P(x)$ is true for *some* value of x, then this must mean that $P(x)$ is not true for *all* values of x; that is,

$$\neg \exists x \, P(x) \ \Leftrightarrow \ \forall x \, \neg P(x).$$

These two laws coincide with De Morgan's Laws:

$$\neg (P \wedge Q) \ \Leftrightarrow \ \neg P \vee \neg Q$$
$$\neg (P \vee Q) \ \Leftrightarrow \ \neg P \wedge \neg Q$$

if we consider universal quantification as a (potentially) infinite conjunction, and existential quantification as a (potentially) infinite disjunction. Suppose that the universe of discourse is $\mathcal{U} = \{\, a, b, c, \ldots \,\}$. Then

$$\neg \forall x P(x) \ \Leftrightarrow \ \neg(P(a) \wedge P(b) \wedge P(c) \wedge \cdots)$$
$$\Leftrightarrow \ \neg P(a) \vee \neg P(b) \vee \neg P(c) \vee \cdots \quad (\textit{De Morgan's Law})$$
$$\Leftrightarrow \ \exists x \neg P(x);$$

and

$$\neg \exists x P(x) \ \Leftrightarrow \ \neg(P(a) \vee P(b) \vee P(c) \vee \cdots)$$
$$\Leftrightarrow \ \neg P(a) \wedge \neg P(b) \wedge \neg P(c) \wedge \cdots \quad (\textit{De Morgan's Law})$$
$$\Leftrightarrow \ \forall x \neg P(x).$$

Example 4.10

Recall from the Example in Section 4.2 that Joel, Felix, Oskar and Amanda must all brush their teeth before going to school in the morning; that is, that the proposition

$$\forall x \ Teeth(x)$$

is true, where – as before – we use $Teeth(x)$ to denote the statement that child x has brushed their teeth, and we continue to take the universe of discourse to consist of the set of four children in question:

Children = { Joel, Felix, Oskar, Amanda }.

On a particular day, it may be discovered that this statement is *not* true. For example, perhaps Joel, Oskar and Amanda have all brushed their teeth, but Felix has not. This is the reason that $\forall x \ Teeth(x)$ is false, i.e., that

$$\neg \forall x \ Teeth(x)$$

is true: that there is someone (namely Felix) who has *not* brushed their teeth:

$$\exists x \neg Teeth(x)$$

This is an example of the general law that

$$\neg \forall x \, P(x) \ \Leftrightarrow \ \exists x \neg P(x).$$

We have also earlier noted that, when driving to school, one of the children must sit in the front seat of the car: that is, that the statement

$$\exists x \, front(x)$$

must be true, where – as before – we use $front(x)$ to denote the statement that child x sits in the front seat. For this statement to be false, it would have to mean that *none* of the children are sitting in the front seat, or in other words that *all* of them are *not* sitting in the front seat:

$$\forall x \neg front(x).$$

This is an example of the general law that

$$\neg \exists x \, P(x) \ \Leftrightarrow \ \forall x \neg P(x).$$

Exercise 4.10   (Solution on page 425)

For each of the following statements, identify which of the options provided correctly expresses its negation. Translate each statement into predicate logic to confirm your choices.

1. Some people like mathematics.

   (a) Some people dislike mathematics.

   (b) Everybody dislikes mathematics.

   (c) Everybody likes mathematics.

2. All cats have fur and a tail.

   (a) No cat has fur and a tail.

   (b) Some cats are bald and tailless.

   (c) Some cats are bald or tailless.

3. Everyone who had not been vaccinated got sick.

   (a) Everyone who had been vaccinated did not get sick.

   (b) Some people who had been vaccinated got sick.

   (c) Some people who had not been vaccinated did not get sick.

Having established how quantifiers interact with negation, we next consider how they interact with conjunction and disjunction. Specifically, we may wonder which of the following is true:

1. $\forall x (P(x) \wedge Q(x)) \overset{?}{\Leftrightarrow} \forall x P(x) \wedge \forall x Q(x)$.

2. $\exists x (P(x) \wedge Q(x)) \overset{?}{\Leftrightarrow} \exists x P(x) \wedge \exists x Q(x)$.

3. $\forall x (P(x) \vee Q(x)) \overset{?}{\Leftrightarrow} \forall x P(x) \vee \forall x Q(x)$.

4. $\exists x (P(x) \vee Q(x)) \overset{?}{\Leftrightarrow} \exists x P(x) \vee \exists x Q(x)$.

We carefully consider each of these in turn.

1. This property is valid.

   If $P(x) \wedge Q(x)$ is true of every object x, then certainly $P(x)$ must be true of every object x and $Q(x)$ must be true of every object x.

   Equally, if $P(x)$ is true of every object x and $Q(x)$ is true of every object x, then $P(x) \wedge Q(x)$ must be true of every object x.

2. This property is *not* valid.

   If $P(x) \wedge Q(x)$ is true of some object x, then $P(x)$ must be true of that object x and $Q(x)$ must be true of that object x.

   However, $P(x)$ may be true of some object, and $Q(x)$ may be true of some *different* object, while $P(x) \wedge Q(x)$ may never be true of the same object x.

   For example, it is true that prime numbers and perfect squares exist:

   $$\exists x \, Prime(x) \wedge \exists x \, Square(x) \text{ is true.}$$

For instance $Prime(17)$ is true and $Square(25)$ is true. However, no number can be both prime *and* a perfect square at the same time:

$\exists x(Prime(x) \land Square(x))$ is false.

We have, however, established the weaker property:

2'. $\exists x(P(x) \land Q(x)) \Rightarrow \exists x P(x) \land \exists x Q(x)$.

3. This property is *not* valid.

If $P(x)$ is true for all objects x, then certainly $P(x) \lor Q(x)$ must be true of all objects x; Equally, if $Q(x)$ is true for all objects x, then $P(x) \lor Q(x)$ must be true of all objects x.

However, $P(x) \lor Q(x)$ may be true of all objects x without it being the case that $P(x)$ is true of all objects x, nor that $Q(x)$ is true of all objects x.

For example, it is true that all integers are either even or odd:

$\forall x\,(Even(x) \lor Odd(x))$ is true.

However, not every integer is even, and not every integer is odd:

$\forall x\,Even(x) \lor \forall x\,Odd(x)$ is false.

We have, however, established the weaker property:

3'. $\forall x(P(x) \lor Q(x)) \Leftarrow \forall x P(x) \lor \forall x Q(x)$.

4. This property is valid.

If $P(x) \lor Q(x)$ is true of some object x, then either $P(x)$ must be true of that object x or $Q(x)$ must be true of that object x.

Equally, if $P(x)$ is true of some object x or $Q(x)$ is true of some object x, then $P(x) \land Q(x)$ must be true of that object x.

As a final note, the following are clearly valid properties:

1. $\forall x \forall y\,P(x,y) \Leftrightarrow \forall y \forall x\,P(x,y)$;
2. $\exists x \exists y\,P(x,y) \Leftrightarrow \exists y \exists x\,P(x,y)$.

That is, we can rearrange the order in which universal quantifications are applied, as well as the order in which existential quantifications are applied. It is common practice to write these as $\forall x, y\,P(x,y)$ and $\exists x, y\,P(x,y)$, respectively. However, as we see in the following example, we cannot rearrange different quantifiers:

$\forall x \exists y\,P(x,y) \not\Leftrightarrow \exists y \forall x\,P(x,y)$.

Example 4.11

A certain mathematics textbook has an exercise which asks its reader to translate the following sentence into predicate logic:

"Every real number is smaller than some integer."

This informal English sentence can be interpreted in (at least) the following two different ways:

1. $\forall r \in \mathbb{R} \, \exists n \in \mathbb{Z} \, (r < n)$

   Given any real number r, we can find a larger integer n.

2. $\exists n \in \mathbb{Z} \, \forall r \in \mathbb{R} \, (r < n)$

   There is an integer which is larger than every real number.

The first of these statements is true – and is undoubtedly the interpretation intended by the author – while the second statement is blatantly false. The author of this mathematics textbook was trying to state a basic fact about numbers, but the ambiguity of English complicated this task.

4.4 Modelling in Predicate Logic

The language of predicate logic gives us tools on top of propositional logic and set theory with which to model scenarios. In this section we we present a few examples.

Example 4.12

Recall the Carrollean puzzle from Exercise 2.25, where we are given the three premises:

All babies are illogical.

Nobody is despised who can manage a crocodile.

Illogical persons are despised.

from which we are to deduce that no baby can manage a crocodile. Let us introduce the following predicates:

$B(x) = $ *"x is a baby"*

$I(x) = $ *"x is illogical"*

$D(x) = $ *"x is despised"*

$M(x) = $ *"x can manage a crocodile"*

Then the above three premises translate into the following propositions:

1. $\forall x(B(x) \Rightarrow I(x))$
2. $\forall x(M(x) \Rightarrow \neg D(x))$ or, equivalently, $\forall x(D(x) \Rightarrow \neg M(x))$
3. $\forall x(I(x) \Rightarrow D(x))$

and the conclusion translates into $\forall x(B(x) \Rightarrow \neg M(x))$.

However, for any x such that $B(x)$ is true (if x is a baby), by first premise, $I(x)$ is true (x is illogical); and thus by the third premise, $D(x)$ is true (x is despised); and therefore by the second premise, $\neg M(x)$ is true (x cannot manage a crocodile).

Hence the conclusion does indeed follow from the premises.

Exercise 4.12   (Solution on page 426)

Formalise the following two arguments in predicate logic:

1. *Everybody loves somebody.*

   Therefore somebody is loved by everybody.

2. *Somebody loves everybody.*

   Therefore everybody is loved by somebody.

In each case, discuss any ambiguities that you identify in the English statements, but use what you consider to be the intended interpretations.

Are these arguments valid?

Example 4.13

Figure 4.1 presents an example Sudoku puzzle which consists of a 9×9 grid with numbers entered into some of the squares. The objective is to completely fill in the grid so that each column, each row, and each of the nine 3×3 blocks contains the digits from 1 to 9 exactly once. Properly set, the initial numbers will allow for only one valid solution.

This is a classic logic-style puzzle, and as such is perfectly suited for modelling in predicate logic. If you struggle with solving the puzzle given in Figure 4.1, an Internet search engine will find any number of Web sites which will solve it for you; and the means by which these Web sites' software does this will inevitably work on the following formal representation (or something very similar).

We start by defining the universe of discourse to be the interval $I = [1..9]$ of integers from 1 to 9. This reflects the fact that there are:

- 9 rows, listed from top to bottom as row 1 through to row 9;

Figure 4.1: A Sudoku puzzle.

- 9 columns, listed from left to right as column 1 through to column 9;
- 9 blocks, listed from left to right and top to bottom as block 1 through to block 9;
- 9 values, 1 through 9, to be inserted into the squares.

We then define the following predicate:

$$V(i, j, k) = \text{"square } (i, j) \text{ holds the value } k\text{."}$$

That is, the number k is in the square located in row i and column j. Thus, for the example puzzle in Figure 4.1, the following propositions are true:

$$
\begin{array}{llll}
V(1,8,6) & V(1,9,7) \\
V(2,1,4) & V(2,5,9) \\
V(3,1,3) & V(3,4,2) & V(3,7,9) & V(3,8,8) \\
V(4,5,2) & V(4,6,3) & V(4,7,6) \\
V(5,2,2) & V(5,5,6) & V(5,8,5) \\
V(6,3,1) & V(6,4,7) & V(6,5,4) \\
V(7,2,3) & V(7,3,4) & V(7,6,7) & V(7,9,9) \\
V(8,5,1) & V(8,9,6) \\
V(9,1,9) & V(9,2,8)
\end{array}
$$

Next we define the following predicate:

$$B(i, j, b) = \text{"square } (i, j) \text{ is in block } b.\text{"}$$

This property is represented by the following nine propositions (one for each block):

$$B(i, j, 1) \Leftrightarrow (i, j) \in [1..3] \times [1..3]$$
$$B(i, j, 2) \Leftrightarrow (i, j) \in [1..3] \times [4..6]$$
$$B(i, j, 3) \Leftrightarrow (i, j) \in [1..3] \times [7..9]$$
$$B(i, j, 4) \Leftrightarrow (i, j) \in [4..6] \times [1..3]$$
$$B(i, j, 5) \Leftrightarrow (i, j) \in [4..6] \times [4..6]$$
$$B(i, j, 6) \Leftrightarrow (i, j) \in [4..6] \times [7..9]$$
$$B(i, j, 7) \Leftrightarrow (i, j) \in [7..9] \times [1..3]$$
$$B(i, j, 8) \Leftrightarrow (i, j) \in [7..9] \times [4..6]$$
$$B(i, j, 9) \Leftrightarrow (i, j) \in [7..9] \times [7..9]$$

Finally, we are ready to represent the properties satisfied by a valid solution to the puzzle.

1. Every square (i, j) holds exactly one value k: $\forall i \, \forall j \, \exists! k \, V(i, j, k)$.

2. Every row i contains every value k: $\forall i \, \forall k \, \exists j \, V(i, j, k)$.

3. Every column j contains every value k: $\forall j \, \forall k \, \exists i \, V(i, j, k)$.

4. Every block b contains every value k: $\forall b \, \forall k \, \exists i \, \exists j \, V(i, j, k) \wedge B(i, j, b)$.

All that is required now is to deduce truth values of the predicates $V(i, j, k)$ which satisfy these properties. This is a non-trivial and tedious task to do by hand, but is the sort of thing that computers can do very well (and very rapidly).

Exercise 4.13   (Solution on page 426)

Solve the Sudoku puzzle in Figure 4.1.

4.5   Additional Exercises

1. Let $V(x)$ stand for the predicate "x visits his parents every weekend", where the domain of discourse is the set of students in your class. Express each of the following quantifications in English:

   (a) $\exists x V(x)$
   (b) $\forall x V(x)$

(c) $\exists!x\neg V(x)$

(d) $\forall x\neg V(x)$

2. Using the predicates:

$$B(x) = \text{``}x \text{ is a bee''}$$
$$F(x) = \text{``}x \text{ is a flower''}$$
$$L(x,y) = \text{``}x \text{ likes } y\text{''}$$

write each of the following statements in predicate logic.

(a) All bees like some flowers.

(b) No bee likes only flowers.

(c) No bee hates (that is, does not like) all flowers.

3. Express the negation of each of the statements in the previous question, both in English as well as in predicate logic.

4. Let $T(s,c)$ stand for the predicate *"student s takes course c."* Express the following statements in predicate logic.

(a) *"Alice and Bob take all the same courses."*

(b) *"Alice and Bob do not take any courses together."*

5. Express the following properties in predicate logic, using only the usual operations of addition and multiplication as well as the less than relation $<$ between numbers.

(a) x is a divisor of y.

(b) x and y have no common divisors.

(c) x is a prime number.

(d) Every integer greater than one has a unique smallest prime divisor.

(e) *(Goldbach's Conjecture)* Every even integer greater than two can be written as the sum of two primes.

6. Express in English what each of the following propositions is saying about the set of real numbers $\mathbb{R}$, and determine whether they are true or false.

(a) $\forall x\,\exists y\,(\,x+y=x\,)$.

(b) $\exists y\,\forall x\,(\,x+y=x\,)$.

(c) $\forall x\,\exists y\,(\,x^2=y\,)$.

(d) $\forall y\,\exists x\,(\,x^2=y\,)$.

(e) $\forall x\,\forall y\,(\,x<y\;\vee\;y<x\,)$.

7. Express the following in predicate logic.

   (a) At least three items have property P.
   (b) At most 3 items have property P.
   (c) Exactly three items have property P.

8. A particular jazz standard recorded by Doris Day has the following title and lyrics:

   > *Everybody loves my baby*
   >
   > *But my baby don't love nobody but me*

   Express the above in predicate logic. What can you deduce from these two statements about who "my baby" is?

9. Samuel Goldwyn, on being told by a friend told him that he had named his son Sam, exclaimed, *"Why did you do that? Every Tom, Dick and Harry is named Sam!"* Assuming Goldwyn was right, and assuming he was restricting his attention to first names, how many Sams, Dicks and Harrys are there? Formulate your answer in predicate logic, including the assertion that every person has exactly one first name.

10. Lewis Carroll made the following argument.

    > *Everybody who is sane can do logic.*
    >
    > *No lunatics are fit to serve on a jury.*
    >
    > *None of your sons can do logic.*
    >
    > *Therefore, none of your sons is fit to serve on a jury.*

    Formulate the four claims in predicate logic. Do you consider this a valid argument?

11. Lewis Carroll also made the following three claims.

    > *No professor is ignorant.*
    >
    > *All ignorant people are vain.*
    >
    > *No professor is vain.*

    Formulate these three claims in predicate logic. Do any of them follow from the other two?

12. What is wrong with the following argument:

    > *A ham sandwich is better than nothing.*
    >
    > *Nothing is better than eternal happiness.*
    >
    > *Therefore, a ham sandwich is better than eternal happiness.*

Chapter 5

★ # Proof Strategies

You want proof? I'll give you proof!

- Sidney Harris.

So far, we have concentrated on developing formal languages for rigorously and unambiguously expressing properties of systems, namely the languages of propositional logic, predicate logic, sets and Boolean algebras. In the case of propositional logic we have used truth tables to determine the validity of logical arguments. We have also learned what it means for statements of predicate logic to be true or false, but we have not yet seen a procedure for determining truth or falsity. This is perhaps not too surprising, as predicates can range over infinite universes of discourse, hence infinitely many candidates potentially need to be inspected to test statements such as *"This program will terminate (with the correct result) at some point in time."*

A *proof* of a (true) statement is a demonstration of its validity which contains sufficient detail to convince someone that the statement is true. Statements which are provable are called *theorems*. We encountered formal proofs already in Chapter 3, where we derived the truth of various theorems of Boolean algebra; each such derivation ended with the symbol □ indicating that the truth of the theorem had been established.

Proofs allow us to reason formally about properties of systems, so that (ultimately) we can provide convincing and irrefutable evidence of their correctness. We have already explored some basic proof techniques, for instance reasoning with logical equivalences in propositional logic and reasoning equationally with Boolean algebras. However, thus far we have asked no more of our reader than to use common sense to follow our reasoning.

Proofs of theorems often require creativity and inspiration. Furthermore, there will always be many different ways to prove a given theorem, and any valid proof of a given theorem will be just as correct a proof as any other. However, some proofs will be more elegant and more easily grasped than others. The mathematician Paul Erdős often referred to "The Book" in which God keeps the most elegant proof of each mathematical theorem, and noted that "You don't have to believe in God, but you should believe in

F. Moller, G. Struth, *Modelling Computing Systems*,
Undergraduate Topics in Computer Science,
DOI 10.1007/978-1-84800-322-4_6, © Springer-Verlag London 2013

The Book."

Elegance aside, all formal proofs follow certain patterns that can be learned like the rules of chess. Different *proof strategies* can be applied, depending only on the form of the property being considered. Once these strategies are learned, proofs can be easily – even mechanically – constructed and checked. In this chapter, we develop such proof strategies that will allow us to verify (or indeed falsify) system properties from a systematic point of view, relieving us of the need for too much *Eureka!*-invoking inspiration.

5.1 A First Example

It is obvious – by drawing a Venn diagram – that the union $A \cup B$ of two sets A and B contains both A and B as subsets:

$$A \subseteq A \cup B \ \text{ and } \ B \subseteq A \cup B.$$

But $A \cup B$ is a very special superset of A and B: it consists of *precisely* the elements of A and the elements of B – no more and no less – and is therefore the *least* superset of both A and B. In other words, any set C which is a superset of both A and of B is also a superset of $A \cup B$:

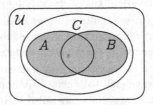

Although this fact should be intuitively clear, its validity deserves a formal proof such as the following.

Theorem 5.1

Let A, B and C be sets. Then

$$A \subseteq C \wedge B \subseteq C \ \Rightarrow \ A \cup B \subseteq C.$$

Proof: Assume that $A \subseteq C$ and $B \subseteq C$; we must show that $A \cup B \subseteq C$. To do this, we expand the definition of the set inclusion $A \cup B \subseteq C$:

every element of $A \cup B$ must also be in C.

So we pick an arbitrary element $x \in A \cup B$ and we show that $x \in C$. Noting that $x \in A \cup B$ is the same as $x \in A \vee x \in B$, we proceed by case analysis on whether $x \in A$ or $x \in B$.

1. If $x \in A$, then $x \in C$, since we assumed that $A \subseteq C$.

2. If $x \in B$, then again $x \in C$, since we assumed that $B \subseteq C$.

In each case, $x \in C$ follows from the assumptions.   □

At first sight you might find this proof perhaps more difficult to understand and less revealing than, for instance, a Venn diagram. However, in a few steps, it can be completely reduced to some basic proof strategies for predicate logic and some basic principles about sets. Everyone who has learned these strategies and principles can then easily check this proof, and, in fact, even machines can do that for you.

So let us take a quick initial look at some of the proof strategies that occur in this argument. They are based on logical principles of reasoning with propositional connectives and quantifiers. They deal with these connectives and quantifiers in two essentially different ways.

First, in order to prove the implication

$$A \subseteq C \wedge B \subseteq C \;\Rightarrow\; A \cup B \subseteq C,$$

we have *assumed* that $A \subseteq C \wedge B \subseteq C$ and proved that $A \cup B \subseteq C$ from this assumption. The underlying proof strategy allows us to prove an arbitrary implication $P \Rightarrow Q$ by assuming P and proving Q from this assumption. In a similar fashion, instead of proving

$$\forall x\,(x \in A \cup B \;\Rightarrow\; x \in C),$$

we have proved $x \in A \cup B \Rightarrow x \in C$ for an *arbitrary* x taken from the universe of discourse.

One way of understanding these proof strategies is that they decompose a proof goal, replacing it with a simpler one from which the original goal follows more or less automatically. These strategies narrow the distance between the assumptions and the goal from the goal side, hence in a bottom up way. Another way of understanding these strategies is to observe that they introduce a logical connective or quantifier into a proof. They can therefore be characterised as *introduction strategies* for connectives or quantifiers. The strategies mentioned above, for instance, introduce implication and universal quantification, respectively.

A second kind of strategy allows us to use complex assumptions or intermediate proof results (which can also be seen as assumptions) in proofs. In the above proof, for instance, we have used a strategy that allowed us to decompose the assumption $A \subseteq C \wedge B \subseteq C$ into two separate assumptions $A \subseteq C$ and $B \subseteq C$. Also, to prove $x \in C$ from the assumption $x \in A \vee x \in B$, we have used a case analysis strategy and proved $x \in C$ first from $x \in A$ and then from $x \in B$. The underlying proof strategy allows us to prove a goal R from a disjunction $P \vee Q$ by case analysis, that is, by proving R from the assumption P and from the assumption Q separately.

This second type of strategy can be understood as narrowing the distance between the assumptions and the goal from the assumptions side, hence in a top down way. They eliminate logical connectives or quantifiers and can therefore be characterised as *elimination strategies*.

When faced with the prospect of proving a theorem, a sensible approach would be to:

1. write out any assumptions, and previously-established facts that you suspect may be relevant, at the top of a page;

2. write out the statement which you wish to prove at the bottom of the page;

3. repeatedly apply elimination strategies to the statements at the top, and introduction strategies to the statements at the bottom, and look for how to make the logical argument meet in the middle.

With this in mind, we will present basic introduction and elimination strategies for each of the propositional connectives and quantifiers, and depict these as proof outlines with "holes" in the middle that need to be filled in. The justification behind each such proof outline will be made evident.

Exercise 5.2   (Solution on page 427)

Let A, B and C be sets. Prove the converse of Theorem 5.1, that

$$A \cup B \subseteq C \;\Rightarrow\; A \subseteq C \wedge B \subseteq C.$$

5.2   Proof Strategies for Implication

A proof of a theorem consists of a sequence of statements, each either being assumed or known to be true, or logically inferred from (i.e., implied by) earlier statements appearing in the proof. It is sensible, therefore, to start by considering proof strategies for implication.

In our introductory example we proved the theorem

$$A \subseteq C \wedge B \subseteq C \;\Rightarrow\; A \cup B \subseteq C.$$

by assuming that $A \subseteq C \wedge B \subseteq C$ and showing from this that $A \cup B \subseteq C$. This idea can be generalised to the following proof strategy for implication.

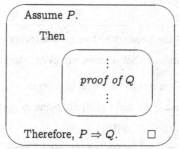

Assume P.

Then

$\vdots$

proof of Q

$\vdots$

Therefore, $P \Rightarrow Q$.   □

This is an introduction strategy for implication, as it gives a method for introducing a statement of the form $P \Rightarrow Q$ into a proof.

Example 5.2

Consider the fact that the average of two different numbers lies somewhere strictly between the two. For example, the average of the two numbers 13 and 25 is 19, which lies strictly between the two given numbers 13 and 25. This general fact is intuitively obvious. However, once it is rendered in precise mathematical terms, it becomes something that is nonetheless deserving of a proof.

As a mathematical statement, the above fact becomes:

If $a < b$ then $a < \frac{a+b}{2}$ and $\frac{a+b}{2} < b$.

More precisely, this statement is of the form

$P \Rightarrow Q$

where

$P = a < b$, and

$Q = a < \frac{a+b}{2} \wedge \frac{a+b}{2} < b$.

Here we prove one half of this result:

If $a < b$ then $\frac{a+b}{2} < b$.

Proof: Assume that $a < b$.

Then, by adding b to both sides, we get that $a+b < b+b$.

Thus, by dividing both sides by 2, we get that $\frac{a+b}{2} < \frac{b+b}{2}$.

Since $\frac{b+b}{2} = b$, we get that $\frac{a+b}{2} < b$.

Therefore, if $a < b$ then $\frac{a+b}{2} < b$.   □

The introduction strategy for implication is so fundamental that one usually just assumes P and proves Q without even mentioning that this yields a proof of $P \Rightarrow Q$.

Example 5.3

Prove that the product of two even integers is an even integer.

Proof: Assume that a and b are even integers.

An even integer is twice an integer.

Thus $a = 2p$ and $b = 2q$ for some integers p and q.

Hence $ab = (2p)(2q) = 4pq$

$\qquad\quad = 2k$ for the integer $k = 2pq$.

Therefore, ab is an even integer. □

Exercise 5.3 (Solution on page 427)

Prove that the product of two odd integers is an odd integer.

There is another introduction strategy for implication which may be more natural to apply on occasion. We can assume that Q is false and prove that, under this assumption, P must also be false. The form of such a proof would thus be as follows.

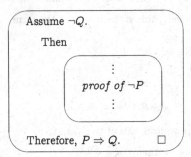

Assume $\neg Q$.

Then

$\vdots$

proof of $\neg P$

$\vdots$

Therefore, $P \Rightarrow Q$. □

A proof which employs this strategy is referred to as a *proof by contraposition*.

Example 5.4

Prove, by contraposition, the result from Example 5.2 that, for any two real numbers a and b, if $a < b$ then $\frac{a+b}{2} < b$.

Proof: Suppose that $\frac{a+b}{2} \geq b$.

Then, by multiplying both sides by 2, we get that $a + b \geq 2b$.

Thus, by subtracting b from both sides, we get that $a \geq b$.

Therefore, if $a < b$ then $\frac{a+b}{2} < b$.   □

Corresponding to the above two introduction strategies, there are two elimination strategies for implication which allow us to draw inferences from statements in a proof that involve implication. These strategies are as follows.

1. If $P \Rightarrow Q$ is true and P is true, then Q is true.

   A use of this proof strategy is referred to as *modus ponens*, and takes the following form.

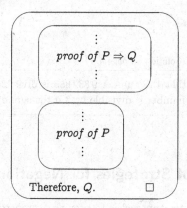

2. If $P \Rightarrow Q$ is true and $\neg Q$ is true, then $\neg P$ must be true.

   A use of this proof strategy is referred to as *modus tollens*, and takes the following form.

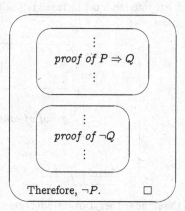

Indeed, we have already seen these proof principles in action, particularly extensively in the solution to the Amos Judd puzzle of Exercise 1.14 (page 33).

Example 5.5

Prove that if $a \in A$ and $A \subseteq B$ then $a \in B$.

Proof: Assume that $a \in A$, and that $A \subseteq B$.

By definition, $A \subseteq B$ means that $x \in A \Rightarrow x \in B$ for any x.

In particular, $a \in A \Rightarrow a \in B$.

Thus, by *modus ponens*, $a \in B$.   □

Exercise 5.5   (Solution on page 427)

Prove that the number $9\,839\,853$ is divisible by 3. (You may use the fact that a number is divisible by 3 if the sum of its digits is divisible by 3.)

5.3   Proof Strategies for Negation

The main approaches to proving a property of the form $\neg P$ is to assume that P is true and to infer from this a contradiction. By this, we mean that both some property Q and its negation $\neg Q$ can be inferred from our assumption P; as such a contradiction is impossible, the assumption from which it was inferred must be invalid. The form of such a proof would thus be as follows.

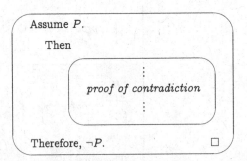

```
Assume P.
   Then
                  ⋮
          proof of contradiction
                  ⋮

   Therefore, ¬P.          □
```

This is the standard negation introduction strategy. The associated negation elimination strategy is nearly identical, allowing positive results to be proven by contradiction. It takes the following form.

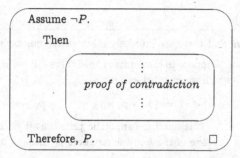

> Assume $\neg P$.
>
> Then
>
> > $\vdots$
> >
> > *proof of contradiction*
> >
> > $\vdots$
>
> Therefore, P. □

A proof which employs either of these strategies is referred to as a *proof by contradiction* or, more fancily, as *reductio ad absurdum*.

Our first example of a proof by contradiction is over 2000 years old and is attributed to the school of Pythagoras.

Example 5.6

Prove that $\sqrt{2}$ is irrational; that is, $\sqrt{2} \notin \mathbb{Q}$.

Proof. Suppose to the contrary that $\sqrt{2} \in \mathbb{Q}$; specifically, suppose that $\sqrt{2} = \frac{a}{b}$ where a and b are positive integers and $\frac{a}{b}$ is a fraction in lowest form; in particular, a and b are not both even.

Then squaring both sides gives us that $2 = \frac{a^2}{b^2}$, and then multiplying both sides by b^2 gives us that $2b^2 = a^2$.

Hence a must be even (since, by Exercise 5.3, if a were odd then a^2 would also be odd); that is, $a = 2c$ for some integer c.

As a and b are not both even, b must be odd.

But then $2b^2 = a^2 = (2c)^2 = 4c^2$, so $b^2 = 2c^2$, which means that b must be even, contradicting our earlier observation that b must be odd.

This must mean that our assumption that $\sqrt{2}$ is rational must be invalid; that is, $\sqrt{2}$ must in fact be irrational. □

Another famous example of a proof by contradiction that is also over 2000 years old, this time due to Euclid, is the following argument that there are infinitely many prime numbers. The proof relies on the *Fundamental Theorem of Arithmetic* – also proved by Euclid and which we prove in Exercise 9.9, page 235 – which states that every positive integer can be expressed as a product of prime numbers; in particular, every such number is divisible by some prime number.

Example 5.7

Prove that there are infinitely many prime numbers.

Proof. Suppose to the contrary that there are finitely many prime numbers, which we may list as $\{ p_1,\, p_2,\, p_3,\, \ldots,\, p_k \}$.

Let $n = (p_1 \times p_2 \times p_3 \times \cdots \times p_k) + 1$.

This number cannot be prime, as it is clearly larger than every one of the k prime numbers p_1 through p_k.

Thus, by the *Fundamental Theorem of Arithmetic*, some prime number p_i must divide evenly into n.

However this is impossible, as dividing n by p_i clearly leaves a remainder of 1, and hence p_i does not divide evenly into p.

Therefore, our assumption that there are finitely many prime numbers must be invalid; that is, there must in fact be infinitely many prime numbers. ☐

Example 5.8

Suppose that $A \cap C \subseteq B$ and that $a \in C$. Prove that $a \notin A \setminus B$.

As always, before blindly starting a proof, you should try to get a good impression in your mind as to what it is you are trying to prove. If possible, this is best done by drawing a picture, which in this case means a Venn diagram:

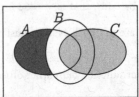

Here we have depicted three sets A, B and C which satisfy the premise of the proposition that we wish to prove: that $A \cap C \subseteq B$. From this we need to infer that any element $a \in C$ (i.e., which lies in the light gray area) will not be in $A \setminus B$ (i.e., cannot lie in the dark gray area). This seems obvious in the picture, but a rigorous argument is still demanded. Fortunately, now that we have a clear picture in our mind, a rigorous proof seems trivial.

Proof. Assume that the premises of the proposition are true, that $A \cap C \subseteq B$ and that $a \in C$. We shall show that assuming that $a \in A \setminus B$ leads to a contradiction.

Suppose that $a \in A \setminus B$; that is, that $a \in A$ but that $a \notin B$.

Since $a \in A$ and $a \in C$ (from the premise of the proposition), we have that $a \in A \cap C$.

But since $A \cap C \subseteq B$ (again from the premise of the proposition), from $a \in A \cap C$ we get that $a \in B$, contradicting $a \notin B$.

Therefore, we cannot have $a \in A \setminus B$; that is, we have $a \notin A \setminus B$. $\square$

As usual, there are various ways that this proposition can be proven, all of which being equally valid. The following is provided as an example.

A Different Proof. Assume that the premises of the proposition are true, that $A \cap C \subseteq B$ and that $a \in C$.

As $a \in A \setminus B$ if, and only if, $a \in A$ and $a \notin B$, we shall show that $a \notin A \setminus B$ by showing that we cannot have both $a \in A$ and $a \notin B$; that is, if we assume that $a \in A$ then we can deduce that $a \in B$.

Suppose then that $a \in A$.

Since $a \in C$ (from the premise of the proposition), we have that $a \in A \cap C$.

But then since $A \cap C \subseteq B$ (again from the premise of the proposition), we have that $a \in B$.

Therefore, we cannot have both $a \in A$ and $a \notin B$; that is to say, we must have $a \notin A \setminus B$. $\square$

Example 5.9

Assume that a and b are positive real numbers.

Prove that either $a \leq \sqrt{ab}$ or $b \leq \sqrt{ab}$.

Proof. Suppose to the contrary that $a > \sqrt{ab}$ and $b > \sqrt{ab}$.

Then $ab > \left(\sqrt{ab}\right)^2 = ab$, which is impossible.

Therefore, either $a \leq \sqrt{ab}$ or $b \leq \sqrt{ab}$. $\square$

Exercise 5.9   (Solution on page 427)

Prove that there is no such thing as the smallest positive rational number.

Exercise 5.10   (Solution on page 428)

Prove that every integer greater than 1 can be written as a product of prime numbers.

(Note that a prime number is the trivial product of one prime number.)

(5.4)   Proof Strategies for Conjunction and Equivalence

There is very little interesting or needed to say about dealing with conjunctions in proofs. To prove a property of the form $P \wedge Q$, we simply prove P and Q separately. The form of such a proof will look as follows.

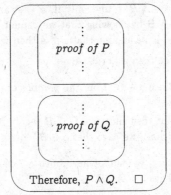

This is the basic introduction strategy for conjunction. The basic elimination strategy is equally straightforward: we may infer the truth of one of the conjuncts of an established conjunction. The form of such a proof will look like one of the following following.

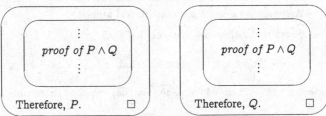

These will rarely be used in isolation, and their use inevitably comes naturally. As such, the following examples – while instructive – are somewhat contrived and superfluous.

Example 5.10

Prove that if $x \in A$ and $x \in B$ then $x \in A \cap B$.

Proof: Assume that $x \in A$ and that $x \in B$.

By the conjunction introduction strategy, we can infer from this that $x \in A \wedge x \in B$, which by definition means that $x \in A \cap B$.   □

Example 5.11

Prove that if $x \in A \cap B$ then $x \in A$ and $x \in B$.

Proof: Assume that $x \in A \cap B$

By definition this means that $x \in A \wedge x \in B$.

By the conjunction elimination strategy, we can infer from this both that $x \in A$ and that $x \in B$.   □

An equivalence $P \Leftrightarrow Q$ between properties P and Q simply represents the fact that each property implies the other; it is true if, and only if, $P \Rightarrow Q$ and $Q \Rightarrow P$. As such, proof strategies for equivalence are naturally based on those for conjunction. To prove $P \Leftrightarrow Q$, we simply prove $P \Rightarrow Q$ and $Q \Rightarrow P$ separately. The form of such a proof will look as follows.

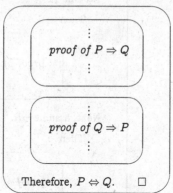

This is the basic introduction rule for equivalence. The basic elimination strategy is to infer the implication in one direction or the other from an established equivalence. The form of such a proof will look as follows.

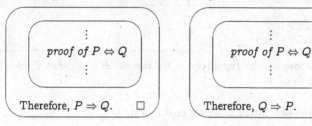

5.5   Proof Strategies for Disjunction

To prove that a disjunctive property $P \lor Q$, it suffices to prove one or the other of the disjuncts. The basic introduction strategy for disjunction is thus of the following forms.

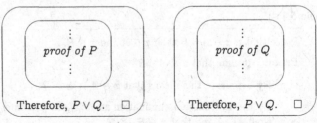

The above is rather weak, though. It might not be the case that one of P or Q always holds; rather, which holds might depend on some other factors. That is, it might be that P holds whenever some property R holds, and that Q holds when the property R does not hold. In this case the proof of $P \lor Q$ needs to be broken into cases. The relevant introduction strategy would then be of the following proof form.

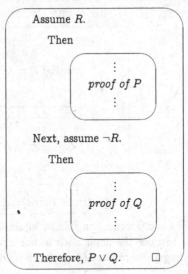

Example 5.12

Prove that for any integer n, the remainder of n^2 when divided by 4 is either 0 or 1.

Proof: Either n is even or it is odd.

- If n is even, then $n = 2k$ for some integer k, and

$$n^2 = (2k)^2 = 4k^2$$

which clearly has a remainder of 0 when divided by 4.

- If n is odd, then $n = 2k + 1$ for some integer k, and

$$n^2 = (2k+1)^2 = 4k^2 + 4k + 1 = 4(k^2 + k) + 1$$

which clearly has a remainder of 1 when divided by 4.

Thus the remainder of n^2 when divided by 4 is either 0 or 1.   □

A special case of the above strategy is to take the property R to be P itself. In this case, there would be no effort needed to infer P from the assumption P, so the form of the proof would be as follows.

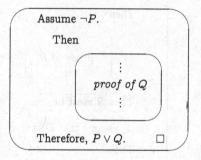

Assume $\neg P$.

Then

$\vdots$

proof of Q

$\vdots$

Therefore, $P \vee Q$.   □

Example 5.13

Prove that if x is a real number with $x^2 > x$ then either $x < 0$ or $x > 1$.

Proof: Assume as given that $x^2 > x$. Clearly this means that $x \neq 0$.

If it is *not* the case that $x < 0$, then $x > 0$, and we can divide each side of the given inequality $x^2 > x$ by x to deduce that $x > 1$.

Hence, either $x < 0$ or $x > 1$.   □

Exercise 5.13   (Solution on page 428)

Prove that if the product of two integers is even, then one of these two integers is itself even.

Exercise 5.14 (Solution on page 429)

Prove that if $A \subseteq B$ then either $x \notin A$ or $x \in B$.

The elimination strategy for disjunction is more interesting. If we have as given a property $P \vee Q$, we can prove that a further property R holds by breaking the proof into cases; that is, we show that $P \Rightarrow R$ and $Q \Rightarrow R$. This being the case, regardless of which of P or Q is true, R must be true. The form of this elimination strategy is thus as follows.

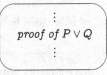
$$\vdots$$
$$\textit{proof of } P \vee Q$$
$$\vdots$$

Thus, either P is true, or Q is true.

Assume first that P is true.

Then

$$\vdots$$
$$\textit{proof of } R$$
$$\vdots$$

Thus R must be true.

Next, assume that Q is true.

Then

$$\vdots$$
$$\textit{proof of } R$$
$$\vdots$$

Thus, once again, R must be true.

Therefore, R is true (regardless of whichever of P or Q is true). $\square$

Example 5.14

Prove that $A \cap (B \cup C) \subseteq (A \cap B) \cup C$.

Proof. Let $x \in A \cap (B \cup C) = (A \cap B) \cup (A \cap C)$.

Then either $x \in A \cap B$, in which case $x \in (A \cap B) \cup C$;

or $x \in (A \cap C)$, in which case $x \in C$ so again $x \in (A \cap B) \cup C$. □

Example 5.15

Prove that if $|x - 3| > 3$ then $x^2 > 6x$.

Proof. If $|x - 3| > 3$ then either $x > 6$, in which case $x^2 > 6x$;

or $x < 0$, in which case $x^2 > 0 > 6x$. □

Exercise 5.15 (Solution on page 429)

Prove the triangle inequality: For real numbers a and b, $|a + b| \le |a| + |b|$.

Exercise 5.16 (Solution on page 429)

Prove that if n is an integer, then the final (units) digit of n^2 must be either 0, 1, 4, 5, 6 or 9; that is, n^2 cannot end with a 2, 3, 7 or 8.

Exercise 5.17 (Solution on page 430)

What is wrong with the following proof?

Fact: If $x + y = 12$ then $x \ne 7$ and $y \ne 8$.

Proof: Assume that the conclusion is false, that is, that it is *not* the case that $x \ne 7$ and $y \ne 8$.

Then $x = 7$ and $y = 8$,

Hence if $x + y = 12$ then $x \ne 7$ and $y \ne 8$. □

5.6 Proof Strategies for Quantifiers

5.6.1 Universal Quantification

A universal quantification $\forall x\, P(x)$ represents a potentially-infinite conjunction, asserting that $P(a)$ is true for *every* value a of the universe of discourse for the predicate P. As such, we look at how to generalise the proof strategies for conjunction.

To prove a property of the form $\forall x\, P(x)$, let a stand for an *arbitrary* object, and prove $P(a)$. The form of such a proof would thus be as follows.

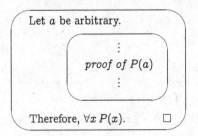

As long as we make no assumptions about a in the proof of $P(a)$, then this proof will be valid for whatever choice of a we make. That is, we will have shown that $P(x)$ must be true for every x (that is, for any and every choice of value a for x). It should be apparent how this introduction strategy generalises that for conjunction.

Note that we have already been tacitly using this strategy. For instance, in Example 5.9 we proved a result held for all positive real numbers a and b, by assuming as given arbitrary values for a and b. Usually this is fine – we generally don't have to think twice about taking arbitrary values as given. However, we do sometimes have to be more careful with introducing values.

We can next look to the elimination strategy for conjunction to derive a straightforward generalisation which tells us how to use a universal quantification within a proof. If we have ascertained that $\forall x\, P(x)$ is true and a is an element in the universe of discourse for the predicate P, then we can immediately infer that $P(a)$ is true. The form of such a proof will look as follows.

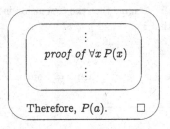

<hr />

Example 5.17

Prove that if $A \cap B = A$ then $A \subseteq B$.

Proof. Assume that $A \cap B = B$. We need to demonstrate that $A \subseteq B$, that is, that for any x, if $x \in A$ then $x \in B$:

$$\forall x\,(x \in A \;\Rightarrow\; x \in B).$$

To this end, let a be an arbitrary value.

To show that $a \in A \;\Rightarrow\; a \in B$, we assume that $a \in A$ and prove from this assumption that $a \in B$.

Assume then that $a \in A$.

Since $A \cap B = A$ (from the premise of the proposition), this means that $a \in A \cap B$.

But this means that $a \in A$ and $a \in B$; in particular, that $a \in B$.

Therefore, $\forall x \, (x \in A \;\Rightarrow\; x \in B)$; that is, $A \subseteq B$.   □

Example 5.18

Prove that $\forall x \left(P(x) \wedge Q(x) \right) \;\Leftrightarrow\; \forall x \, P(x) \;\wedge\; \forall x \, Q(x)$.

Proof. ($\Rightarrow$) Suppose $\forall x \left(P(x) \wedge Q(x) \right)$, and let a be an arbitrary value.

Then $P(a) \wedge Q(a)$, so $P(a)$ and $Q(a)$.

Since a is arbitrary, we can infer that $\forall x \, P(x)$ and $\forall x \, Q(x)$; that is, $\forall x \, P(x) \;\wedge\; \forall x \, Q(x)$.

($\Leftarrow$) Suppose $\forall x \, P(x) \;\wedge\; \forall x \, Q(x)$, and let a be an arbitrary value.

Then $P(a)$ and $Q(a)$, so $P(a) \wedge Q(a)$.

Since a is arbitrary, we can infer that $\forall x \left(P(x) \wedge Q(x) \right)$.   □

Exercise 5.18   (Solution on page 430)

Prove that if A and $B \setminus C$ are disjoint then $A \cap B \subseteq C$.

5.6.2   Existential Quantification

An existential quantification $\exists x \, P(x)$ represents a potentially-infinite disjunction, asserting that $P(a)$ is true for *some* value a of the universe of discourse for the predicate P. As such, we look at how to generalise the proof strategies for disjunction.

To prove a property of the form $\exists x \, P(x)$, we need only find a value a for which $P(a)$ holds, and prove $P(a)$. The form of such a proof would thus be as follows.

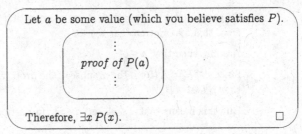

Let a be some value (which you believe satisfies P).

$$\vdots$$

proof of $P(a)$

$$\vdots$$

Therefore, $\exists x\, P(x)$.   □

Note the difference between this introduction strategy and the introduction strategy for $\forall x\, P(x)$. To prove $\forall x\, P(x)$ you need to prove that $P(a)$ holds for an *arbitrary* value a without making any assumptions about a. To prove $\exists x\, P(x)$ you need to prove that $P(a)$ holds for a single chosen value of a.

We next look to the elimination strategy for disjunction to derive a generalisation which tells us how to use an existential quantification within a proof. If we have ascertained that $\exists x\, P(x)$ is true, and if some property R holds under the assumption that $P(a)$ holds regardless of the specific value a of the universe of discourse, then we can infer that R is true. The form of such a proof will look as follows.

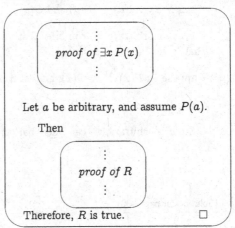

$$\vdots$$

proof of $\exists x\, P(x)$

$$\vdots$$

Let a be arbitrary, and assume $P(a)$.

Then

$$\vdots$$

proof of R

$$\vdots$$

Therefore, R is true.   □

<hr>

Example 5.19

Prove that, if $x \neq 1$, then $\frac{y-2}{y+1} = x$ for some y.

Proof. Let $y = \frac{x+2}{1-x}$ (noting that, since $x \neq 1$, $1-x \neq 0$, and so we are not inadvertently dividing by 0 in defining y).

Then $y-2 = \dfrac{x+2 - 2(1-x)}{1-x} = \dfrac{3x}{1-x}$,

and $y+1 = \dfrac{x+2 + 1-x}{1-x} = \dfrac{3}{1-x}$,

so $\frac{y-2}{y+1} = \left(\frac{3x}{1-x}\right) / \left(\frac{3}{1-x}\right) = \left(\frac{3x}{1-x}\right) \times \left(\frac{1-x}{3}\right) = x.$ $\square$

The difficulty with proving the existence of an object a for which a property P holds is: how do we find the particular value a? In the above example, why did we choose to take $y = \frac{x+2}{1-x}$? The answer in this case – as it typically will be – lies in working backwards. Since we wanted to find a value y such that $\frac{y-2}{y+1} = x$, we worked from this equation:

- by multiplying both sides by $(y+1)$ we get $y-2 = x(y+1) = xy + x$;

- by rearranging terms to get all (and only) terms involving y on one side (i.e.. by adding $2-xy$ to both sides) we get $x+2 = y-xy = y(1-x)$.

- Dividing each side by $(1-x)$ – noting that this will not be an illegal division by zero, since the premise stipulates that $x \neq 1$ – we arrive at the value we seek: $y = \frac{x+2}{1-x}$.

Exercise 5.19   (Solution on page 430)

Prove that for every real $x>0$ there is a real y such that $y(y+1) = x$.

Although typically the case, it isn't strictly necessary (nor sometimes even possible) to explicitly find the specific value x which witnesses the fact that $\exists x P(x)$; the mere fact that such a value exists is all that needs to be demonstrated.

Example 5.20   A Strange Proof of Existence

Fact: There are irrational numbers a and b such that a^b is rational.

Proof. We know from Example 5.6 that $\sqrt{2}$ is irrational.

Furthermore, either $\left(\sqrt{2}\right)^{\sqrt{2}}$ is rational or it is irrational.

- Suppose $\left(\sqrt{2}\right)^{\sqrt{2}}$ is rational. Let $a = b = \sqrt{2}$.

  Then a and b are irrational, and $a^b = \left(\sqrt{2}\right)^{\sqrt{2}}$ is rational.

- Suppose $\left(\sqrt{2}\right)^{\sqrt{2}}$ is irrational. Let $a = \left(\sqrt{2}\right)^{\sqrt{2}}$ and $b = \sqrt{2}$.

  Then a and b are irrational, and

$$a^b = \left(\left(\sqrt{2}\right)^{\sqrt{2}}\right)^{\sqrt{2}} = \left(\sqrt{2}\right)^{\left(\sqrt{2}\sqrt{2}\right)} = \left(\sqrt{2}\right)^2 = 2$$

is rational. $\square$

What is strange about this example is that we demonstrated the *existence* of two particular irrational numbers a and b which satisfy our conditions *without* discovering for certain what these particular numbers are!

Exercise 5.20   (Solution on page 431)

Prove that $\exists x \left(P(x) \vee Q(x) \right) \Leftrightarrow \exists x\, P(x) \vee \exists x\, Q(x)$.

(Hint: refer to the proof in Example 5.18.)

5.6.3   Uniqueness

There are two approaches to proving a property of the form $\exists ! x\, P(x)$, the first by proving existence and uniqueness separately, and the second by combining these two concerns.

1. First prove *existence*: $\exists x\, P(x)$

   and then *uniqueness*: $\forall y \forall z \left[\left(P(y) \wedge P(z) \right) \Rightarrow y = z \right]$.

2. Prove $\exists x \left[P(x) \wedge \forall y \left(P(y) \Rightarrow y = x \right) \right]$.

Either way, the proof strategies are derived from existing strategies.

Example 5.21

Prove that for every x there is a unique y such that $x^2 y = x - y$.

Proof. Let $y = \dfrac{x}{x^2 + 1}$.

$$\text{Then } x^2 y = \frac{x^3}{x^2 + 1} = \frac{x(x^2 + 1) - x}{x^2 + 1} = x - y.$$

Furthermore, if $x^2 z = x - z$, then $z(x^2 + 1) = x$,

so $z = \dfrac{x}{x^2 + 1} = y$.   □

Example 5.22

Suppose $\mathcal{F}$ is a family of sets. Prove that there is a unique set A that has the following two properties:

1. $\mathcal{F} \subseteq \mathcal{P}(A)$.
2. $\forall B \left(\mathcal{F} \subseteq \mathcal{P}(B) \Rightarrow A \subseteq B \right)$.

Proof. Let $A = \bigcup \mathcal{F}$.

1. Suppose $X \in \mathcal{F}$.

    Then $X \subseteq \bigcup \mathcal{F}$; that is, $X \subseteq A$.

    Hence $X \in \mathcal{P}(A)$.

2. Suppose B is *any* set satisfying $\mathcal{F} \subseteq \mathcal{P}(B)$.

    Let $a \in A$; that is, $a \in \bigcup \mathcal{F}$.

    Then $\exists X \in \mathcal{F}$ with $a \in X$.

    Thus $X \in \mathcal{P}(B)$, so $X \subseteq B$.

    Hence $a \in B$. $\square$

Exercise 5.22   (Solution on page 431)

Prove that there is a unique set A such that, for every set B, $A \cup B = B$.

5.7   Additional Exercises

1. Prove that, for any two real numbers $a, b \in \mathbb{R}$: if $a < b$ then $a < \frac{a+b}{2}$.

2. Assume that m and n are integers. Prove that if $m+n$ is even, then m and n are either both even or both odd.

3. Assume that n is an integer. Prove that if $3n + 2$ is an odd integer, then n must be an odd integer.

4. Prove that there is no even prime number greater than 2.

5. Prove that $\sqrt{3}$ is irrational.

6. Prove or disprove each of the following.

   (a) The sum of two rational numbers is rational.

   (b) The sum of two irrational numbers is irrational.

7. Assume that n is an integer. Prove that $n^2 \geq n$.

8. Prove that if n is an integer, then the final digit of n^4 must be either 0 or 1 or 5 or 6.

9. Prove that there are no integer solutions to the equation $x^2 + 2y^2 = 24$.

10. Prove the Distributivity Laws for sets:

   (a) $A \cap (B \cup C) = (A \cap B) \cup (A \cap C)$.

   (b) $A \cup (B \cap C) = (A \cup B) \cap (A \cup C)$.

11. Prove the following.

   (a) $P \Rightarrow (Q \Rightarrow P)$.
   (b) $(P \Rightarrow Q) \Rightarrow (P \vee R \Rightarrow Q \vee R)$.
   (c) $(P \Rightarrow Q) \Rightarrow (P \wedge R \Rightarrow Q \wedge R)$.

12. Prove the following.

   (a) $\forall x(\,P(x) \wedge Q(x)\,) \Leftrightarrow \forall x\, P(x) \wedge \forall x\, Q(x)$.
   (b) $\exists x(\,P(x) \vee Q(x)\,) \Leftrightarrow \exists x\, P(x) \vee \exists x\, Q(x)$.
   (c) $\forall x(\,P(x) \vee Q(x)\,) \Leftarrow \forall x\, P(x) \vee \forall x\, Q(x)$.
   (d) $\exists x(\,P(x) \wedge Q(x)\,) \Rightarrow \exists x\, P(x) \wedge \exists x\, Q(x)$.

13. Prove that the two approaches to proving $\exists! x\, P(x)$ from Section 5.6.3 are equivalent.

Chapter 6

Functions

Home computers are being called upon to perform many new functions, including the consumption of homework formerly eaten by the dog.

- Doug Larson.

We regularly want to associate to each value of one set A some particular value taken from another set B (which may be the same set A). Such a mapping of values in A to values in B is referred to as a *function*.

Functions arise everywhere in people's lives. For example, shoppers are ever calculating (or at least estimating) for themselves the cost of their basket of goods from the number and unit costs (plus relevant sales taxes). Functions are especially relevant to the computer scientist's world. Computer programs are written to turn input values into output values, and the design and implementation of Boolean circuits will inevitably start from a definition of the function of the circuit which describes its behaviour on each possible input. For this reason, it is necessary to take a careful look at what a function is and understand its definition and the various properties that it may enjoy.

6.1 Basic Definitions

A *function* f from a set A to a set B is an assignment of exactly one element of B to each element of A. We write $f : A \to B$ to denote that f is a function from A to B, and we write $f(a)$ to refer to the unique element of B assigned to the element a of A by the function f. Thus f maps each element a of A to an element $b = f(a)$ of B, which we will also denote by $f : a \mapsto b$. Figure 6.1 gives a pictorial representation of such a function.

Example 6.1

Each person in a class of twelve students is assigned a particular grade, in the form of an integer percentage between 0 and 100, which appears as

F. Moller, G. Struth, *Modelling Computing Systems*,
Undergraduate Topics in Computer Science,
DOI 10.1007/978-1-84800-322-4_7, © Springer-Verlag London 2013

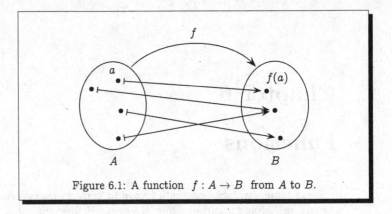

Figure 6.1: A function $f : A \to B$ from A to B.

follows on a list posted on a bulletin board:

Andrews	75	Evans	78	Parker	64
Archer	92	Fletcher	46	Smith	59
Collins	64	Greene	68	Taylor	100
Davies	88	Lewis	54	Williams	78

Here, each person in the set

Class $=$ { Andrews, Archer, Collins, Davies, Evans, Fletcher,
Greene, Lewis, Parker Smith, Taylor, Williams }

is assigned a value from the set

Marks $=$ { 0, 1, 2, 3, 4, ..., 100 }.

This describes a function

score : Class $\to$ Marks

in which, for example, score(Greene) $= 68$; the function score maps the value
Greene to the value 68, that is, score : Greene $\mapsto$ 68.

It is possible for a function $f : A \to B$ to assign the same value from B
to two different values of A. In the above example,

score(Collins) $=$ score(Parker) $= 64$.

However, only one value of B may be assigned to any value of A. In this
sense, a function $f : A \to B$ may be viewed as a machine into which you
input a value $x \in A$ and – depending only on that value – some value
$f(x) \in B$ will be output in response:

input $x \longrightarrow$ [f] $\longrightarrow f(x)$ output

If the same value for x is input on two separate occasions on the left, then the same value will be output on the right for $f(x)$ on both occasions.

If $f : A \to B$ is a function from A to B, we refer to A as the *domain* of f and B as its *codomain*. If $f(a) = b$ we refer to a as an *argument* of the function f, and to b as the *value* of the function f on argument a.

If the domain of the function f is the Cartesian product $A_1 \times A_2 \times \cdots \times A_n$, then we say that f has *arity* n or that f takes n arguments. A function which takes two arguments is called a *binary function*. Common binary functions are often written in *infix* form $x\, f\, y$ rather than $f(x,y)$. For example, we would naturally write $2 + 2 = 4$ rather than $+(2, 2) = 4$.

The *range* of the function $f : A \to B$, denoted range(f), is the subset of the codomain B consisting of all values that the function f can produce:

$$\text{range}(f) \;=\; \{\, f(a) \,:\, a \in A \,\}.$$

Given a subset $S \subseteq A$ of the domain of f, the *image* of S under f, denoted by $f(S)$, is the subset of the codomain B consisting of all values that the function f can produce: from arguments in S:

$$f(S) \;=\; \{\, f(a) \,:\, a \in S \,\}.$$

Thus, for example, range$(f) = f(A)$.

Given a subset $T \subseteq B$ of the codomain of f, the *preimage* of T under f, denoted by $f^{-1}(T)$, is the subset of the domain A consisting of all arguments of f which produce values in T:

$$f^{-1}(T) \;=\; \{\, a \in A \,:\, f(a) \in T \,\}.$$

Notice in particular that $f^{-1}(B) = A$, since every argument in A produces some value in B. We can also note that images and preimages allow us to view f and f^{-1} as functions between the powersets $\mathcal{P}(A)$ and $\mathcal{P}(B)$:

$$f : \mathcal{P}(A) \to \mathcal{P}(B) \qquad \text{and} \qquad f^{-1} : \mathcal{P}(B) \to \mathcal{P}(A).$$

Example 6.2

Consider the function $f : \{\, 1,\, 2,\, 3 \,\} \to \{\, a,\, b,\, c \,\}$ defined, as depicted below, by $f(1) = c$, $f(2) = a$ and $f(3) = c$.

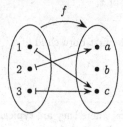

The domain of f is $\{\, 1,\, 2,\, 3 \,\}$.

The codomain of f is $\{\, a,\, b,\, c \,\}$.

The range of f is $\{\, a,\, c \,\}$.

$f(\{1,2\}) = \{\, a, c \,\}$ and $f(\{1,3\}) = \{\, c \,\}$.

$f^{-1}(\{b,c\}) = \{\, 1, 3 \,\}$ and $f^{-1}(\{c\}) = \{\, 1, 3 \,\}$.

Exercise 6.2 (Solution on page 431)

1. What is the range of the function score from Example 6.1?

2. If a score of 70 or higher is considered to be a first-class mark, express the set of students who have scored a first-class mark as a preimage of an appropriate set.

Example 6.3

Here are three example functions defined with respect to an arbitrary set A.

1. The *identity function* $\mathrm{id}_A : A \to A$ is the function which maps each element a of A to itself: $\mathrm{id}_A(x) = x$ for all $x \in A$.

2. the *cardinality function* $|\cdot| : \mathcal{P}_{\mathrm{fin}}(A) \to \mathbb{N}$ maps each finite subset of A to the number of elements in that subset: $|X| =$ the number of elements of X. (The cardinality of a set is simply the number of elements in the set.)

   Note that this function is only well-defined on finite sets. For example, there is no natural number n which denotes the number of elements in the set $\mathbb{N}$.

3. Given a subset $S \subseteq A$ of A, its *characteristic function* $\chi_S : A \to \mathbb{B}$ indicates whether or not an object is an element of S:

$$\chi_S(x) = \begin{cases} 1, & \text{if } x \in S; \\ 0, & \text{if } x \notin S. \end{cases}$$

Exercise 6.3 (Solution on page 432)

Indicate which of the following are functions from the set Humans of all humans to itself. For each that is not a function, indicate why it fails to be a function.

1. *Mother*(x) represents the mother of x.

2. *Parent*(x) represents the parent of x.

3. *Child*(x) represents the child of x.

4. *FirstBornChild*(x) represents the first-born child of x.

Example 6.4

Functions are common in mathematics, where they are typically given by a formula. For example, the function $f : \mathbb{R} \to \mathbb{R}$ defined by

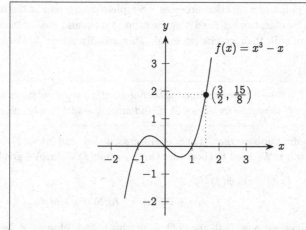

Figure 6.2: The graph of the function $f(x) = x^3 - x$.

$$f(x) = x^3 - x$$

takes a real value $x \in \mathbb{R}$ and returns another real value $f(x) \in \mathbb{R}$ which is computed from x by the formula $x^3 - x$. We can use this formula to calculate the value of $f(x)$ when $x = \frac{3}{2}$:

$$f\left(\frac{3}{2}\right) = \left(\frac{3}{2}\right)^3 - \frac{3}{2} = \frac{27}{8} - \frac{3}{2} = \frac{15}{8}.$$

Such functions are typically plotted as a *graph* on the xy-plane as in Figure 6.2, where we have indicated the point $\left(\frac{3}{2}, \frac{15}{8}\right)$ on the graph.

Motivated by the above example, we can represent a function $f : A \to B$ from A to B as a set of pairs over the Cartesian product $A \times B$. The *graph* of f, denoted graph(f), is the set of all pairs $(a, b) \in A \times B$ such that $b = f(a)$. Thus, for every $a \in A$ there is exactly one $b \in B$ such that $(a, b) \in$ graph(f), namely $b = f(a)$. As an example, for the score function from Example 6.1,

graph(score) $= \{$ (Andrews, 75), (Archer, 92), (Collins, 64),

(Davies, 88), (Evans, 78), (Fletcher, 46),

(Greene, 68), (Lewis, 54), (Parker, 64),

(Smith, 59), (Taylor, 100), (Williams, 78) $\}$;

and for $f(x) = x^3 - x$,

graph(f) $= \{ (x, x^3 - x) : x \in \mathbb{R} \}$.

The graph of a function provides a complete description of the function, in that two functions defined over the same domain and codomain are equal if, and only if, their graphs are equal. This is easily proven in the following.

Theorem 6.4

Let $f, g : A \to B$ be two functions defined on the same domain and codomain. Then $f(a) = g(a)$ for all $a \in A$ if, and only if, $graph(f) = graph(g)$.

Proof: Suppose that $f(a) = g(a)$ for all $a \in A$, and let $(a, b) \in A \times B$ be arbitrary. We need to show that $(a, b) \in graph(f) \Leftrightarrow (a, b) \in graph(g)$. But

$$(a, b) \in graph(f) \;\Leftrightarrow\; b = f(a)$$
$$\Leftrightarrow\; b = g(a) \;\Leftrightarrow\; (a, b) \in graph(g).$$

Suppose now that $graph(f) = graph(g)$, and let $a \in A$ be arbitrary. We need to show that $f(a) = g(a)$. But $(a, f(a)) \in graph(f)$, and since $graph(f) = graph(g)$, we have $(a, f(a)) \in graph(g)$, and hence $f(a) = g(a)$. □

Exercise 6.4   (Solution on page 432)

What is the graph of the function f from Example 6.2?

6.2   One-To-One and Onto Functions

A function $f : A \to B$ associates a single value $b \in B$ to each value $a \in A$, but the same value $b \in B$ may be associated to more than one value in A; that is, we may have two different values $a, a' \in A$ such that $f(a) = f(a')$. For example, given the function $f(x) = x^3 - x$ there are *three* values of x for which $f(x) = 0$, namely $x = -1$, $x = 0$ and $x = 1$.

If a function does *not* assign the same value to two different inputs, it is said to be *one-to-one* (*1-1*), or *injective*.

Definition 6.4

A function $f : A \to B$ is *one-to-one* (*1-1*), or *injective*, if, and only if, $f(a) = f(a')$ implies that $a = a'$ for all $a, a' \in A$. More formally:

$$\forall a, a' \in A \left(f(a) = f(a') \to a = a' \right).$$

In other words, there do not exist two different values in A which f maps to the same value in B:

$$\neg \exists\, a \in A\, \exists\, a' \in A \left(f(a) = f(a') \wedge a \neq a' \right).$$

Exercise 6.5   (Solution on page 432)

Indicate which of the following functions are one-to-one. For those that are not one-to-one, indicate the reason that they fail to be one-to-one.

1. The function score : Class $\rightarrow$ Marks from Example 6.1.
2. The function $f : \mathbb{R} \rightarrow \mathbb{R}$ defined by $f(x) = x^2$.
3. The function $f : \mathbb{N} \rightarrow \mathbb{N}$ defined by $f(x) = x^2$.

Definition 6.5

*A function $f : A \rightarrow B$ is **onto**, or **surjective**, if, and only if, its range is equal to its codomain, range$(f) = B$; that is, every value $b \in B$ is the image of some value $a \in A$:*

$$\forall\, b \in B\, \exists\, a \in A \left(f(a) = b \right).$$

Exercise 6.6   (Solution on page 432)

Indicate which of the following functions are onto. For those that are not onto, indicate the reason that they fail to be onto.

1. The function score : Class $\rightarrow$ Marks from Example 6.1.
2. The function $f : \mathbb{R} \rightarrow \mathbb{R}$ defined by $f(x) = x^2$.
3. The function $f : \mathbb{N} \rightarrow \mathbb{N}$ defined by $f(x) = x^2$.

A function $f : A \rightarrow B$ which is both one-to-one and onto is particularly special: it defines a perfect correspondence between the sets A and B, in that the function f pairs up the elements of A and B, with each element of one set paired to exactly one element of the other set. Such a function is referred to as a *bijection*.

Definition 6.6

*A function f is a **bijection** if it is both one-to-one and onto.*

Exercise 6.7   (Solution on page 432)

Indicate which of the functions f_1, f_2, f_3 or f_4 depicted by the following diagrams are one-to-one, which of them is onto, and which of them is both (i.e., a bijection).

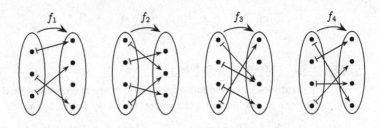

Let $f : A \to B$ be a bijection. Since f is onto, every element $b \in B$ is the image of some element $a \in A$; and since f is one-to-one, every element $b \in B$ is the image of a *unique* element $a \in A$. This suggest that we can turn the mapping around, to *invert* it, and associate a unique element of A with each element of B.

Definition 6.7

If f is a bijection, then the inverse function $f^{-1} : B \to A$ is the function that assigns to each element $b \in B$ the unique element $a \in A$ such that $f(a) = b$. That is, $f^{-1}(b) = a$ if, and only if, $f(a) = b$. This can be pictured as follows:

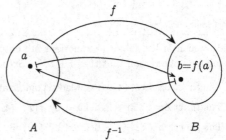

The function $f^{-1} : B \to A$ is also a bijection, and $(f^{-1})^{-1} = f$.

Example 6.7

The function $f : \mathbb{R} \to \mathbb{R}$ defined by $f(x) = 2x + 3$ is a bijection, with $f^{-1}(x) = (y - 3)/2$. For example,

$$f(5) = 2 \cdot 5 + 3 = 13$$

and

$$f^{-1}(13) = (13 - 3)/2 = 5.$$

More generally, if $f : A \to B$ is one-to-one, then f provides a bijection from A to range(f), and we can define the inverse function $f^{-1} : \text{range}(f) \to A$.

Example 6.8

Let $A = \{a, b, c, \ldots, z\}$ be the set consisting of the usual 26 characters of the alphabet. We can use a bijection $f : A \to A$ as the basis of a simple encryption scheme. For example, suppose we take the bijection f defined as follows:

a	b	c	d	e	f	g	h	i	j	k	l	m
↓	↓	↓	↓	↓	↓	↓	↓	↓	↓	↓	↓	↓
y	k	t	e	c	s	w	u	b	m	z	v	l

n	o	p	q	r	s	t	u	v	w	x	y	z
↓	↓	↓	↓	↓	↓	↓	↓	↓	↓	↓	↓	↓
q	g	d	p	o	j	f	r	h	n	a	x	i

To encode a message we apply the function f to each letter of the message. For example, the message

```
WE ATTACK AT DAWN
```

would be encoded as

```
NC YFFYTZ YF EYNQ
```

It is important that the function f is a bijection. No two letters can be mapped to the same letter, as otherwise it would be impossible to decode since different messages would give rise to the same encrypted text.

In order to decode messages that we receive which are encoded as above, we simply apply the inverse function f^{-1} to each of the letters of the encrypted text.

This encryption method is insecure; it is very easy to decode encrypted messages even if you don't know the function f with which they are encrypted. However, the idea of using a bijection f to encode messages, thus allowing such messages to be decoded with the inverse function f^{-1}, is fundamental.

Exercise 6.8   (Solution on page 433)

What is the inverse of the function f of Example 6.8?

6.3   Composing Functions

If we have a function $f : A \to B$ from A to B and another function $g : B \to C$ from B to C, we can:

- first apply the function f to some argument $a \in A$ to arrive at a value $b = f(a) \in B$;
- then use the value $b = f(a) \in B$ as an argument to the function g to arrive at a value $c = g(f(a)) \in C$.

Composing two function applications, one after the other, is a very common thing to do; it is commonly denoted by $g \circ f$, and can be pictured as follows:

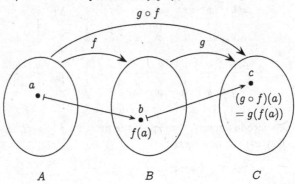

Definition 6.8

Given a function $f : A \to B$ from A to B and a function $g : B \to C$ from B to C, the *composition* of g and f is the function $g \circ f : A \to C$ from A to C defined by

$$(g \circ f)(x) = g(f(x)).$$

Note that the co-domain of the function f must be the same as the domain of the function g in order to form the composition. Also note carefully the order of the functions: the composition $g \circ f$ of the functions g and f first applies the function f to its input before applying the function g to the result. The reason for writing $g \circ f$ rather than $f \circ g$ is to coincide with the order in which the individual function applications appear: $(g \circ f)(x) = g(f(x))$.

Exercise 6.9   (Solution on page 433)

Consider the following two functions f and g from $\{1, 2, 3, 4\}$ to itself:

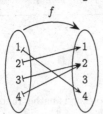

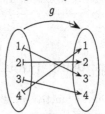

Find $f \circ g$ and $g \circ f$.

If $f : A \to A$ then we can compose f with itself. In this case we typically write f^2 for $f \circ f$, and more generally $f^{n+1} = f \circ f^n$. In other words,

$$f^n = \underbrace{f \circ f \circ \cdots \circ f}_{n \text{ times}}$$

As special cases we have $f^0 = \text{id}_A$ and $f^1 = f$, noting that $f \circ \text{id}_A = f$ (see exercise 8, page 177).

If we compose two one-to-one functions, we will arrive at yet another one-to-one function. The same is true of onto functions. These facts are demonstrated in the following two theorems.

Theorem 6.9

If $f : A \to B$ and $g : B \to C$ are both one-to-one, then so is $g \circ f : A \to C$.

Proof: Suppose $(g \circ f)(x) = (g \circ f)(y)$; that is, $g(f(x)) = g(f(y))$. What we need to demonstrate is that $x = y$.

Since g is one-to-one, $f(x) = f(y)$.

Hence, since f is one-to-one, $x = y$.   □

Theorem 6.10

If $f : A \to B$ and $g : B \to C$ are both onto, then so is $g \circ f : A \to C$.

Proof: Suppose $c \in C$.

What we need to demonstrate is that $c = (g \circ f)(a)$ for some $a \in A$.

Since g is onto, $c = g(b)$ for some $b \in B$.

Since f is onto, $b = f(a)$ for some $a \in A$.

Hence $c = g(f(a)) = (g \circ f)(a)$.   □

Exercise 6.10   (Solution on page 433)

Prove that if $f : A \to B$ and $g : B \to C$ are both bijections, then so is $g \circ f : A \to C$.

Exercise 6.11   (Solution on page 433)

Prove that if $f : A \to B$ is a bijection, then $f^{-1} \circ f = \text{id}_A$ and $f \circ f^{-1} = \text{id}_B$.

Exercise 6.12   (Solution on page 433)

Prove that function composition is associative: if $f : A \to B$, $g : B \to C$ and $h : C \to D$ then $h \circ (g \circ f) = (h \circ g) \circ f$.

★ 6.4   Comparing the Sizes of Sets

We can easily compare the sizes (the cardinalities) of two finite sets simply by counting their elements; the size of one is greater than the size of the other if it contains more elements, and the two sets are the same size if they contain the same number of elements.

Counting the number of elements in a finite set involves listing them in some arbitrary order, denoting one of them to be the first element, another to be the second element, and so on to the last element. For example, we would conclude that the set { Joel, Felix, Oskar, Amanda } has four elements by virtue of the fact that we could find a one-to-one and onto function (a bijection)

$$f : \{1, 2, 3, 4\} \to \{ \text{Joel, Felix, Oskar, Amanda} \}$$

which effectively lists the elements of the set. For example, the function f may list this set (alphabetically) as follows:

$$f : \quad 1 \mapsto \text{Amanda}$$
$$2 \mapsto \text{Felix}$$
$$3 \mapsto \text{Joel}$$
$$4 \mapsto \text{Oskar}$$

This bijection demonstrates that the two sets { Joel, Felix, Oskar, Amanda } and { 1, 2, 3, 4 } are the same size (i.e., have the same cardinality).

We can compare the sizes of any two sets by trying to find a bijection between them which would demonstrate that the two sets are the same size. If such a bijection doesn't exist, then one set must be bigger than the other. For example, if we try to find a bijection

$$f : \{ \text{Joel, Felix, Oskar, Amanda} \} \to \{ \text{cola, fanta, sprite} \}$$

we would quickly realise that this would be impossible, as no such function could be one-to-one: some element of the second set would have to be the image of more than one element of the first set since there are not enough elements in the second set to go around. If this function was aimed at providing each child with a drink, then it is clear that some drink would have to be shared.

For the same reason, no function

$$f : \{\,\text{cola, fanta, sprite}\,\} \to \{\,\text{Joel, Felix, Oskar, Amanda}\,\}$$

could be onto. If this function was aimed at distributing drinks to children, then it is clear that at least one child would not get a drink.

Given two arbitrary sets A and B, we would naturally consider B to be at least as large as A if we could find a *one-to-one* function $f : A \to B$, since f would associate each element of A with its own element of B, so intuitively there would have to be at least as many elements of B as there are of A. On the other hand, we would naturally consider A to be at least as large as B if we could find an *onto* function $f : A \to B$, since f would associate each element of B with at least one element of A which is not associated with any other element of B, so intuitively there would have to be at least as many elements of A as there are of B. Finally, we would naturally consider the two sets to be of the same size (cardinality) if we could find a bijection $f : A \to B$ giving a direct correspondence associating each element of one of the sets with its own element of the other set. We will denote that a set A is no bigger than, no smaller than, and the same size as B by $A \preceq B$, $A \succeq B$, and $A \cong B$, respectively, and summarise this discussion as follows.

Definition 6.12

- $A \preceq B$ if, and only if, there exists a one-to-one function $f : A \to B$.
- $A \succeq B$ if, and only if, there exists an onto function $f : A \to B$.
- $A \cong B$ if, and only if, there exists a bijection $f : A \to B$.

The following results show that these definitions make sense in terms of comparing sizes of sets. The first result says that one set is no bigger than a second if, and only if, the second is no smaller than the first. The second result says that two sets are the same size if, and only if, each is no larger than the other.

Theorem 6.12

$A \preceq B$ if, and only if, $B \succeq A$. That is, there exists a one-to-one function $f : A \to B$ if, and only if, there exists an onto function $g : B \to A$.

Proof: Suppose that $f : A \to B$ is one-to-one, and fix some element $a_0 \in A.$[†] We can define the function $g : B \to A$ as follows:

- if $b \in \text{range}(f)$ then $b = f(a)$ for a unique value $a \in A$, and we define $g(b)$ to be this unique value a;
- if $b \notin \text{range}(f)$ then we define $g(b)$ to be a_0.

[†] If no such a_0 exists, that is if $A = \emptyset$, then $\emptyset$ trivially represents the graph of a one-to-one function from A to B, as well as the graph of an onto function from B to A.

This function g is onto, as $a = g(f(a))$ for each element $a \in A$.

Suppose now that $g : B \to A$ is onto. For each value $a \in A$, fix some value b_a such that $g(b_a) = a$. Then the function $f : A \to B$ defined as $f(a) = b_a$ for each $a \in A$ is clearly one-to-one.   □

(Theorem 6.13)   Schröder-Bernstein Theorem

$A \cong B$ if, and only if, $A \preceq B$ and $B \preceq A$.

Proof: Suppose we have functions $f : A \to B$ and $g : B \to A$ which are both one-to-one; we wish to construct a bijection $h : A \to B$.

For any $a \in A$, consider the sequence generated from a by alternately applying g^{-1} and f^{-1} whenever possible:

$$a \;\mapsto\; g^{-1}(a) \;\mapsto\; f^{-1}(g^{-1}(a)) \;\mapsto\; g^{-1}\big(f^{-1}(g^{-1}(a))\big) \;\mapsto\; \cdots$$

This is possible since f and g are one-to-one, and hence $f^{-1} : \text{range}(f) \to A$ and $g^{-1} : \text{range}(g) \to B$ are well-define functions. However, this sequence may stop at some point, either at an element of A not in the range of g (and hence for which g^{-1} is not defined) or at an element of B not in the range of f (and hence for which f^{-1} is not defined).

We can then define our bijection $h : A \to B$ as follows:

$$h(a) \;=\; \begin{cases} g^{-1}(a), & \text{if the sequence generated by } a \\ & \text{ends at an element of } B; \\[2mm] f(a), & \text{otherwise.} \end{cases}$$

This is a well-defined function, since $g^{-1}(a)$ will be defined if the sequence generated by a ends at an element of B (in particular, not at a). It remains to demonstrate that this function is one-to-one and onto.

To demonstrate that h is one-to-one, let us assume that $h(x) = h(y)$, and show that we must have $x = y$.

- If the sequences generated by x and y both end at elements in B, then $h(x) = g^{-1}(x)$ and $h(y) = g^{-1}(y)$, so $g^{-1}(x) = g^{-1}(y)$, and hence $x = y$.

- If neither sequence generated by x and y ends at an element of B, then $h(x) = f(x)$ and $h(y) = f(y)$, so $f(x) = f(y)$, and hence $x = y$.

- If the sequence generated by x ends at an element of B, but not so for the sequence generated by y, then $h(x) = g^{-1}(x)$ and $h(y) = f(y)$, so $g^{-1}(x) = f(y)$. But then $y = f^{-1}(g^{-1}(x))$ would appear (as the third element) in the sequence generated by x, contradicting the assumption that its sequence ends differently to that generated by x.

- If the sequence generated by y ends at an element of B, but not so for the sequence generated by x, then $h(y) = g^{-1}(y)$ and $h(x) = f(x)$, so $g^{-1}(y) = f(x)$. But then $x = f^{-1}(g^{-1}(y))$ would appear (as the third element) in the sequence generated by y, contradicting the assumption that its sequence ends differently to that generated by y.

To demonstrate that h is onto, let us assume that $b \in B$, and show that we must have $b = h(a)$ for some $a \in A$.

- If the sequence generated by $g(b)$ ends at an element of B, then $h(g(b)) = g^{-1}(g(b)) = b$.

- If the sequence generated by $g(b)$ does not end at an element of B, then $f^{-1}(b)$ must be defined and appear (as the third element) in the sequence generated by $g(b)$, and hence $h(f^{-1}(b)) = f(f^{-1}(b)) = b$.   □

These definitions are unremarkable for finite sets, but reveal surprising relationships between infinite sets, as the following example demonstrates.

Example 6.13

The set $\mathbb{N}$ of nonnegative integers in some sense contains almost twice as many elements as the set $\mathbb{E} = \{0, 2, 4, \ldots\}$ of nonnegative even integers. However, the function $f : \mathbb{N} \to \mathbb{E}$ defined by $f(n) = 2n$ provides a bijection from $\mathbb{N}$ to $\mathbb{E}$, demonstrating that there are in fact the same "number" of even integers as there are integers. This bijection can be pictured as follows:

$$f : \quad \begin{array}{ccccccccccc} 0 & 1 & 2 & 3 & 4 & 5 & 6 & 7 & 8 & 9 & 10 \\ \updownarrow & \updownarrow & \updownarrow & \updownarrow & \updownarrow & \updownarrow & \updownarrow & \updownarrow & \updownarrow & \updownarrow & \updownarrow \\ 0 & 2 & 4 & 6 & 8 & 10 & 12 & 14 & 16 & 18 & 20 \end{array} \quad \cdots$$

The confusion arising from the above example is with the idea of the "number" of elements of an infinite set. There are in fact an infinite number of objects in each of the sets, and as such there is no problem with considering them to have the same cardinality.

Realising that the set of even integers is no smaller than the set of all integers, it may seem that one infinite set is as big as any other. In fact, some infinite sets are larger than others. To explore this idea, we start with the following definitions.

Definition 6.13

A set A is said to be *finite* if, and only if, there is a bijection

$$f : \{1, 2, 3, \ldots, n\} \to A$$

for some $n \in \mathbb{N}$. This function effectively lists all of the elements of A, and the value n is the cardinality of A: $|A| = n$.

A set A is said to be **countably infinite** if, and only if, there is a bijection

$$f : \mathbb{N} \to A.$$

This function lists the elements of A in an infinite list.

Finally, a set is said to be **countable** if, and only if, it is finite or countably infinite; and it is said to be **uncountable** if, and only if, it is not countable.

Example 6.14

The set of integers $\mathbb{Z}$ is countable. A bijection $f : \mathbb{N} \to \mathbb{Z}$ witnessing this fact can be defined as

$$f(n) = \begin{cases} \frac{n+1}{2} & \text{if } n \text{ is odd;} \\ -\frac{n}{2} & \text{if } n \text{ is even.} \end{cases}$$

This function would list the integers as follows:

$$f : \begin{array}{ccccccccccc} 0 & 1 & 2 & 3 & 4 & 5 & 6 & 7 & 8 & 9 & 10 \\ \downarrow & \downarrow & \downarrow & \downarrow & \downarrow & \downarrow & \downarrow & \downarrow & \downarrow & \downarrow & \downarrow \\ 0 & 1 & -1 & 2 & -2 & 3 & -3 & 4 & -4 & 5 & -5 \end{array} \cdots$$

Clearly this function is one-to-one and onto, as every integer will appear exactly once in this list.

Exercise 6.14   (Solution on page 433)

What is the inverse $f^{-1} : \mathbb{Z} \to \mathbb{N}$ of the bijection $f : \mathbb{N} \to \mathbb{Z}$ given in Example 6.14?

Exercise 6.15   (Solution on page 434)

Prove that $A \cong B$ for any two countable sets A and B.

That is, given bijections $f : \mathbb{N} \to A$ and $g : \mathbb{N} \to B$, show how to construct a bijection $h : A \to B$.

As an example of the difference between countable and uncountable sets, we shall see that there are far more numbers on the real number line than just the integers; that is, the set of real numbers $\mathbb{R}$ is uncountable. This may seem perfectly sensible, as these numbers fill the number line: between any two different real numbers, no matter how close they are to each other, you can find a third. The integers, on the other hand, are relatively few and far between.

While such an intuitive argument gives rise to a valid result in this case, the same intuition would lead you to believe that there are uncountably-many rational numbers, as between any two different rational numbers, no matter how close they are to each other, you can find a third. However, before we demonstrate that there are uncountably-many reals, we first demonstrate that this intuition about the rationals is faulty; the rationals are countable, and hence no more numerous that the integers.

Example 6.15

The set $\mathbb{Q}^+$ of positive rational numbers is countable. To see this, we need to find a bijection

$$f : \mathbb{N} \to \mathbb{Q}^+$$

which completely lists them. To this end, we first note that a positive rational is a number of the form $\frac{p}{q}$, where p and q are positive integers, and we can arrange these in an infinite number of infinite rows by:

- listing all the rationals with numerator $p = 1$ in the first row,
- listing all the rationals with numerator $p = 2$ in the second row,
- listing all the rationals with numerator $p = 3$ in the third row,

and so on as depicted in Figure 6.3. We can then zigzag diagonally through this arrangement as depicted in Figure 6.3, listing the rationals in the order in which they are encountered. However, we only list rationals that appear in lowest form, and ignore those (depicted crossed out in grey circles in Figure 6.3) that are not in lowest form; for example, we do not include $\frac{4}{6}$ in our listing as it will have already appeared earlier in our list as $\frac{2}{3}$.

The resulting listing provides the required bijection $f : \mathbb{N} \to \mathbb{Q}^+$:

$$f : \begin{array}{ccccccccccc} 0 & 1 & 2 & 3 & 4 & 5 & 6 & 7 & 8 & 9 & 10 \\ \downarrow & \downarrow & \downarrow & \downarrow & \downarrow & \downarrow & \downarrow & \downarrow & \downarrow & \downarrow & \downarrow \\ \frac{1}{1} & \frac{2}{1} & \frac{1}{2} & \frac{1}{3} & \frac{3}{1} & \frac{4}{1} & \frac{3}{2} & \frac{2}{3} & \frac{1}{4} & \frac{1}{5} & \frac{5}{1} \end{array} \cdots$$

This function is one-to-one and onto as only rationals $\frac{p}{q}$ in lowest form appear in the list and each of these is encountered once and only once while zigzagging through the arrangement.

Extending this result to show that the set $\mathbb{Q}$ of all the rational numbers is countable is straightforward.

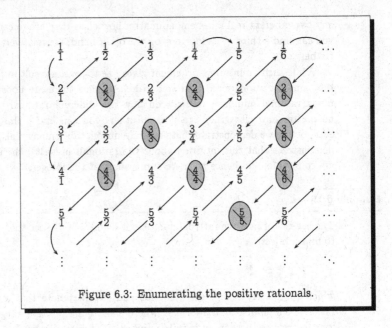

Figure 6.3: Enumerating the positive rationals.

Exercise 6.16   (Solution on page 434)

Prove that the set $\mathbb{Q}$ of all rationals is countable.

Example 6.16

The set $[0, 1]$ of nonnegative real numbers no greater than 1 is uncountable. To see this we must show that no bijection $f : \mathbb{N} \to [0, 1]$ exists. To this end, assume that f is such a function, and consider the listing of the real numbers that it gives:

$$
\begin{aligned}
f : \quad 0 &\mapsto 0 . \; \boxed{d_{00}} \; d_{01} \quad d_{02} \quad d_{03} \quad d_{04} \quad d_{05} \quad \cdots \\
1 &\mapsto 0 . \; d_{10} \quad \boxed{d_{11}} \; d_{12} \quad d_{13} \quad d_{14} \quad d_{15} \quad \cdots \\
2 &\mapsto 0 . \; d_{20} \quad d_{21} \quad \boxed{d_{22}} \; d_{23} \quad d_{24} \quad d_{25} \quad \cdots \\
3 &\mapsto 0 . \; d_{30} \quad d_{31} \quad d_{32} \quad \boxed{d_{33}} \; d_{34} \quad d_{35} \quad \cdots \\
4 &\mapsto 0 . \; d_{40} \quad d_{41} \quad d_{42} \quad d_{43} \quad \boxed{d_{44}} \; d_{45} \quad \cdots \\
5 &\mapsto 0 . \; d_{50} \quad d_{51} \quad d_{52} \quad d_{53} \quad d_{54} \quad \boxed{d_{55}} \cdots \\
&\qquad \vdots \qquad \vdots \quad \vdots \qquad \vdots \qquad \vdots \qquad \vdots \qquad \vdots \qquad \ddots
\end{aligned}
$$

Each number in this list, being a nonnegative real number no greater than 1, is given by an infinite decimal expansion with a leading 0. In particular, the value 0 appears as 0.00000··· and the value 1 appears as 0.99999···.

Consider now the real number

$$r = 0.r_1 r_2 r_3 r_4 r_5 \cdots$$

in which the ith decimal digit r_i is given by

$$r_i = (d_{ii} + 5) \bmod 10.$$

That is, the ith decimal digit of r is defined to differ by 5 from the ith decimal digit of $f(i)$.

Assuming that the function f above is indeed a bijection, and in particular onto, the value r must appear somewhere in the list; that is, we must have $r = f(n)$ for some $n \in \mathbb{N}$. However, for each n, r differs (by 5) from $f(n)$ in the nth decimal place, meaning that we cannot have $r = f(n)$.

An infinite set may thus be either countably infinite or uncountably infinite. In Exercise 6.15 we saw that any two countably-infinite sets are the same size, but the same is not true of two uncountable sets. The following exercise demonstrates that given any set, no matter how big, you can always construct an even bigger set by merely taking its powerset.

Exercise 6.17   (Solution on page 434)

Show that the powerset $\mathcal{P}(A)$ of any set A is strictly larger than A, by showing that no function $f : A \to \mathcal{P}(A)$ can be onto.

(Hint: Show that the set $B = \{x \in A \ : \ x \notin f(x)\}$ is different from $f(a)$ for all $a \in A$.)

★ **6.5**   **The Knaster-Tarski Theorem**

In this section, as an example in working with sets, we prove an important result on the existence of (greatest and least) fixed points of monotonic functions defined on the powerset of a given set. We also describe a procedure for calculating these fixed points.

Definition 6.17

Let S be a set, and let $f : \mathcal{P}(S) \to \mathcal{P}(S)$ be a function which maps subsets of S to subsets of S.

- f is *monotonic* if, and only if, $f(A) \subseteq f(B)$ whenever $A \subseteq B$.
- $A \subseteq S$ is a *fixed point* of f if, and only if, $f(A) = A$.

- $A \subseteq S$ is the *greatest fixed point* of f – denoted $gfp(f)$ – if, and only if, A is a fixed point (ie, $f(A) = A$) and A is larger than all other fixed points: if $f(B) = B$ then $B \subseteq A$.

- $A \subseteq S$ is the *least fixed point* of f – denoted $lfp(f)$ – if, and only if, A is a fixed point (ie, $f(A) = A$) and A is smaller than all other fixed points: if $f(B) = B$ then $A \subseteq B$.

Note that fixed points need not exist; and even if they do exist, then there is no guarantee that greatest and/or least fixed points exist.

Example 6.17

Let $S = \{0\}$ and define $f : \mathcal{P}(S) \to \mathcal{P}(S)$ by $f(\emptyset) = S$ and $f(S) = \emptyset$.

Clearly f does not have a fixed point.

Exercise 6.18 (Solution on page 434)

Define a function $f : \mathcal{P}(S) \to \mathcal{P}(S)$ over the set $S = \{1, 2\}$ which has two fixed points which are neither greatest nor least fixed points.

The following result, however, shows that both greatest and least fixed points exist for monotonic functions.

Theorem 6.18 Knaster-Tarski Theorem

If $f : \mathcal{P}(S) \to \mathcal{P}(S)$ is monotonic, then f has both greatest and least fixed points. Furthermore, these can be defined as follows:

- $gfp(f) \; = \; \bigcup\{A \subseteq S : A \subseteq f(A)\};$ and
- $lfp(f) \; = \; \bigcap\{A \subseteq S : f(A) \subseteq A\}.$

Proof: We will prove the result about the greatest fixed point $gfp(f)$ and leave the result about the least fixed point $lfp(f)$ as an exercise (Exercise 12, page 178).

To this end, let $G = \bigcup\{A \subseteq S : A \subseteq f(A)\}$ as in the Theorem. We first demonstrate that $G \subseteq f(G)$ by showing that given any $a \in G$ we must have that $a \in f(G)$.

Suppose $a \in G$. By the definition of G, this means that $a \in A$ for some $A \subseteq S$ such that $A \subseteq f(A)$. Hence $a \in f(A)$. Moreover, $A \subseteq G$ (as G is the union of all such sets), so by the monotonicity of f we have that $f(A) \subseteq f(G)$. Hence $a \in f(G)$ as required.

Next, we demonstrate the reverse inclusion, that $f(G) \subseteq G$.

Since we've shown that $G \subseteq f(G)$, by the monotonicity of f we have that $f(G) \subseteq f(f(G))$. This means that $f(G)$ is one of the sets in the family of sets whose union is G, and hence $f(G) \subseteq G$.

We've thus shown that G is a fixed point of f. It remains to show that it is the greatest fixed point. To this end, suppose that X is any fixed point of f. Since $X \subseteq f(X)$, X is one of the sets in the family of sets whose union is G, and hence $X \subseteq G$. □

Beyond knowing that greatest and least fixed points of f exist, we would like to know how to calculate them without having to calculate $f(A)$ for all subsets $A \subseteq S$. To do this, we can exploit the following observations.

Theorem 6.19

For all $n \in \mathbb{N}$,

1. $f^n(\emptyset) \subseteq f^{n+1}(\emptyset)$ and $f^n(\emptyset) \subseteq lfp(f)$;
2. $f^n(S) \supseteq f^{n+1}(S)$ and $f^n(S) \supseteq gfp(f)$.

From this we can deduce the following:

(a) $\bigcup_{n \in \mathbb{N}} f^n(\emptyset) \subseteq lfp(f)$ and $\bigcup_{n \in \mathbb{N}} f^n(S) \supseteq gfp(f)$;

(b) *If* $f^n(\emptyset) = f^{n+1}(\emptyset)$ *then* $lfp(f) = f^n(\emptyset)$;

(c) *If* $f^n(S) = f^{n+1}(S)$ *then* $gfp(f) = f^n(S)$;

(d) *If* $|S| = n$ *then* $lfp(f) = f^n(\emptyset)$ and $gfp(f) = f^n(S)$.

Proof: We prove only 1., by straightforward induction, and leave 2. and the corollaries (a)-(d) as exercises (Exercise 13, page 178).

For the base case, $f^0(\emptyset) = \emptyset$, so clearly $f^0(\emptyset) \subseteq f^1(\emptyset)$ and $f^0(\emptyset) \subseteq lfp(f)$.

For the induction case, assuming that $f^{n-1}(\emptyset) \subseteq f^n(\emptyset)$ and that $f^{n-1}(\emptyset) \subseteq lfp(f)$,

- $f^n(\emptyset) = f(f^{n-1}(\emptyset)) \subseteq f(f^n(\emptyset)) = f^{n+1}(\emptyset)$; and
- $f^n(\emptyset) = f(f^{n-1}(\emptyset)) \subseteq f(lfp(f)) = lfp(f)$. □

Thus, in order to calculate the least fixed point $lfp(f)$ of f, we can repeatedly apply f starting from the empty set $\emptyset$ until we arrive at a fixed point, which by above will be $lfp(f)$:

$$\emptyset = f^0(\emptyset) \subset f^1(\emptyset) \subset f^2(\emptyset) \subset \cdots \subset f^n(\emptyset) = f^{n+1}(\emptyset) = lfp(f).$$

A similar procedure, starting from S, will give us the greatest fixed point. This is guaranteed to work if the set S is finite; however, if S is infinite, we may generate infinite sequences of sets which approach yet never reach the fixed points.

Exercise 6.20 (Solution on page 434)

Let $f : \mathcal{P}(\mathbb{N}) \to \mathcal{P}(\mathbb{N})$ be defined by $f(S) = \{0\} \cup \{n{+}2 \,:\, n \in S\}$.

1. Prove that f is monotonic.

2. Show that $f^n(\emptyset) \subset f^{n+1}(\emptyset)$ and $f^n(\mathbb{N}) \supset f^{n+1}(\mathbb{N})$ for each $n \in \mathbb{N}$.

3. Determine the least and greatest fixed points $lfp(f)$ and $gfp(f)$.

6.6 Additional Exercises

1. Identify the domain, codomain, and range of the following functions.

   (a) the function that assigns to each nonnegative integer the least prime number greater than it.

   (b) the function that assigns to each pair of positive integers the maximum of these two values.

2. Let $A = \{1, 2, 3, 4\}$ and $B = \{a, b, c, d\}$, and let f_1, f_2 and f_3 be functions from A to B with the following graphs:

   $$\text{graph}(f_1) = \{(1,d), (2,a), (3,c), (4,c)\}$$
   $$\text{graph}(f_2) = \{(1,d), (2,c), (3,a), (4,b)\}$$
   $$\text{graph}(f_3) = \{(1,b), (2,c), (3,a), (4,d)\}$$

   Indicate which of these functions are one-to-one, which are onto, and which are bijections.

3. Give an example of a function from $\mathbb{N}$ to $\mathbb{N}$ that is

   (a) one-to-one but not onto.

   (b) onto but not one-to-one.

   (c) one-to-one and onto, but which does not map any value to itself.

   (d) neither one-to-one nor onto.

4. Find all functions from $X = \{a, b\}$ to $Y = \{1, 2, 3\}$. In each case, indicate whether or not the function is one-to-one, and whether or not it is onto.

5. Define the function $f : [0,1] \to (0,1)$ by: $f(0) = 1/2$; $f(1/n) = 1/(n{+}2)$ for all positive integers n; and $f(x) = x$ otherwise. Prove that f is a bijection.

6. Consider the following three functions f, g and h from $\{1, 2, 3\}$ to itself:

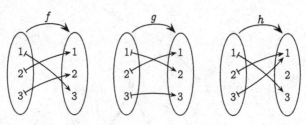

Find $f \circ g$, $g \circ f$, $f \circ h$, $h \circ f$, $g \circ h$, $h \circ g$.

7. Find $g \circ f$ and $f \circ g$, where $f(x) = x^2 + 1$ and $g(x) = x - 2$ are functions from $\mathbb{R}$ to $\mathbb{R}$.

8. Prove that for any function $f : A \to B$, $f = f \circ \mathrm{id}_A$ and $f = \mathrm{id}_B \circ f$, where $\mathrm{id}_X : X \to X$ is the identity function on X, that is, $f(x) = x$ for all $x \in X$.

9. Assuming that $f : A \to B$ and $g : B \to C$, prove or disprove the following.

   (a) If f and $g \circ f$ are both one-to-one, then g must also be one-to-one.

   (b) If g and $g \circ f$ are both one-to-one, then g must also be one-to-one.

   (c) If f and $g \circ f$ are both onto, then g must also be onto.

   (d) If g and $g \circ f$ are both onto, then g must also be onto.

10. Prove that if $A \subseteq B$ and B is countable, then A is countable.

11. In Example 6.15 we saw how to construct a function $f : \mathbb{N} \to \mathbb{Q}^+$ which listed all of the positive rational numbers by zigzagging through an infinite array of rational numbers. However, we had to disregard the rational numbers that we came across which were not in lowest terms. In this exercise we explore an alternative approach which avoids this complication.

   Consider the tree-like diagram in Figure 6.4 which is constructed by starting with $1/1$ at the top, and from each branching point labelled i/j drawing a left branch labelled $i/(i+j)$ and a right branch labelled $(i+j)/j$.

   Argue that every positive rational number appears exactly once in this tree, by arguing that each of the following is true.

   (a) Every node is labelled by a rational number in lowest form.

   (b) Every rational number appears somewhere in the tree.

   (c) No rational number appears twice in the tree.

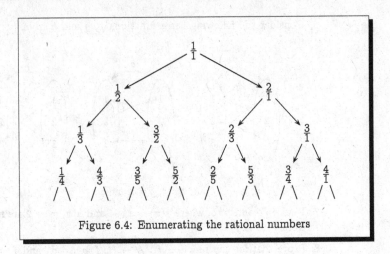

Figure 6.4: Enumerating the rational numbers

Thus, to list the rational numbers without repetition we need merely list the successive rows of the tree.

12. Prove the second part of Theorem 6.18 from page 174, that L as defined there is the least fixed point of f.

13. (a) Prove the second part of Theorem 6.19 (page 175).

  (b) Prove the four corollaries *(a)-(d)* to Theorem 6.19 (page 175).

Chapter 7

Relations

It is a melancholy truth that even great men have their poor relations.

— Charles Dickens, *Bleak House.*

In previous chapters we looked at grouping objects together into sets, as well as logics to reason about the elements in a set. We also studied functions $f : A \to B$ mapping elements in one set A to elements in another set B.

In this chapter we shall turn our attention towards more general *relationships* between elements of sets than simple mappings. Some everyday examples of such relationships are "parenthood" amongst the set of people *("A is a parent of B")* and "divisibility" amongst the set of integers *("x divides evenly into y")*. More generally, relationships can exist between elements of different sets, such as the "enrolment" relationship between the sets of students and courses *("student s takes course c")*. Relationships may even exist amongst elements of three or more sets, such as the "grade" relationship between students, courses and grades *("student s got a grade of g in course c")*.

7.1 Basic Definitions

We start by recalling that the *truth set* of a predicate such as

$$S(x, y, z) = \text{"student } x, \text{ in course } y, \text{ scored a grade of } z\text{"}$$

denotes a subset of a Cartesian product, in this case $S \times C \times G$, where S, C and G are the sets of students, courses, and grades, respectively. In this example, $S(s, c, g)$ is true if, and only if, s is a student who scored a grade of g in course c; and the truth set for this property is

$$Grades = \{ (s, c, g) : s \text{ is a student who} \\ \text{scored a grade of } g \text{ in course } c \}.$$

An *n-ary relation R* is just such a subset of *n*-tuples. In the above example, the set *Grades* is a ternary (that is, a 3-ary) relation over $S \times C \times G$:

F. Moller, G. Struth, *Modelling Computing Systems*,
Undergraduate Topics in Computer Science,
DOI 10.1007/978-1-84800-322-4_8, © Springer-Verlag London 2013

$Grades \subseteq S \times C \times G.$

The most obvious use of n-ary relations is in representing databases. For example, the above relation *Grades* might represent a particular University's database of students' course grades.

Example 7.1

The Internet Movie Database (IMDb) http://www.imdb.com is a Web site which contains a massive, and ever-increasing, online database of films and TV shows, associating with each of these its actors and production crew personnel (directors, writers, producers, etc), as well as many other attributes such as year of release and genre.

For example, the table in Figure 7.1 represents the database of James Bond films, recording their title, year of release, starring actor, and director. This is a fraction of the information, presented in tabular form, delivered by IMDb as a result of a search on the term "James Bond". It can be viewed as a relation

$BondFilms \subseteq$ Titles $\times$ N $\times$ Names $\times$ Names

over the sets

Titles $=$ film titles;

N $=$ natural numbers representing years;

Names $=$ names of people;

containing the 20 records (i.e., 4-tuples):

$r01 =$ (Dr. No , 1962 , Sean Connery , Terence Young)

$r02 =$ (Thunderball , 1965 , Sean Connery , Terence Young)

$\vdots$

$r20 =$ (Skyfall , 2012 , Daniel Craig , Sam Mendes).

The main use to which such a database is put is for being queried. As an example, we may wish to query the database to find out which James Bond films star Roger Moore. The answer to this query would be a particular set of records:

$Q = \{ r \in BondFilms : r \text{ stars Roger Moore} \}$

$= \{ r06, r07, r08, r10, r11 \}.$

Exercise 7.1   (Solution on page 435)

Express and answer the following queries about the above database of James Bond Films.

	Title	Year	Star	Director
r01	Dr. No	1962	Sean Connery	Terence Young
r02	Thunderball	1965	Sean Connery	Terence Young
r03	You Only Live Twice	1967	Sean Connery	Lewis Gilbert
r04	On Her Majesty's Secret Service	1969	George Lazenby	Peter R. Hunt
r05	Diamonds Are Forever	1971	Sean Connery	Guy Hamilton
r06	The Spy Who Loved Me	1977	Roger Moore	Lewis Gilbert
r07	Moonraker	1979	Roger Moore	Lewis Gilbert
r08	For Your Eyes Only	1981	Roger Moore	John Glen
r09	Never Say Never Again	1983	Sean Connery	Irvin Kershner
r10	Octopussy	1983	Roger Moore	John Glen
r11	A View to a Kill	1985	Roger Moore	John Glen
r12	The Living Daylights	1987	Timothy Dalton	John Glen
r13	Licence to Kill	1989	Timothy Dalton	John Glen
r14	Golden Eye	1995	Pierce Brosnan	Martin Campbell
r15	Tomorrow Never Dies	1997	Pierce Brosnan	Roger Spottiswoode
r16	The World Is Not Enough	1999	Pierce Brosnan	Michael Apted
r17	Die Another Day	2002	Pierce Brosnan	Lee Tamahori
r18	Casino Royale	2006	Daniel Craig	Martin Campbell
r19	Quantum of Solace	2008	Daniel Craig	Marc Forster
r20	Skyfall	2012	Daniel Craig	Sam Mendes

Figure 7.1: James Bond Films.

1. Which Bond films were directed by Lewis Gilbert?
2. Which Bond Films were released in the 1970s?

7.2   Binary Relations

Binary (that is, 2-ary) relations are the most common types of relations, and are of particular importance. Concepts such as

- order *("element a comes before element b")*,

- equivalence *("element a is the same as element b")*, and
- function *("input a results in output b")*

are all examples of binary relations, relating one thing a to another thing b. They are often written in *infix* style, so that we would write aRb rather than $(a, b) \in R$.

A binary relation $R \subseteq A \times B$ is thus just a set of ordered pairs, and is said to be a relation *from* the set A *to* the set B. The sets A and B are referred to as the *source* and *target*, respectively, of R.

A binary relation $R \subseteq A \times A$ from a set A to itself is said to be a relation *on* A. In this case, the relation is said to be *homogeneous*, whereas a relation $R \subseteq A \times B$ with $A \neq B$ is said to be *heterogeneous*.

Example 7.2

As an example of a binary relation on the natural numbers $\mathbb{N}$ we can take the usual *less-than-or-equal-to* relation $\leq \ \subseteq \mathbb{N} \times \mathbb{N}$:

$$\leq \ = \{(x, y) : x \leq y\}$$

$$= \{(0, 0), (0, 1), (1, 1), (0, 2), (1, 2), (2, 2), \dots\}.$$

As an example of a binary relation from the set H of humans to the natural numbers $\mathbb{N}$ we can take the relation $R \subseteq H \times \mathbb{N}$ given by:

$$R = \{(x, n) \in H \times \mathbb{N} : x \text{ has } n \text{ children}\}.$$

As an example of a binary relation from the set C of cities to the set N of countries (nations) we can take the relation $R \subseteq C \times N$ given by:

$$R = \{(c, n) \in C \times N : c \text{ is located in } n\}.$$

Example 7.3

Joel likes mint ice cream and coffee ice cream; Felix likes vanilla ice cream and cherry ice cream; Oskar likes vanilla ice cream and chocolate ice cream; and Amanda likes chocolate ice cream and mint ice cream. These properties can be related by the binary relation

$$Likes \subseteq \text{Children} \times \text{Flavours}$$

where

$$\text{Children} = \{\text{Joel, Felix, Oskar, Amanda}\} \quad \text{and}$$

$$\text{Flavours} = \{\text{Vanilla, Chocolate, Coffee, Cherry, Mint}\}$$

consisting of the following ordered pairs:

$$Likes \;\; = \;\; \{\, (\text{Joel}, \text{Mint}), \; (\text{Joel}, \text{Coffee}),$$
$$(\text{Felix}, \text{Vanilla}), \; (\text{Felix}, \text{Cherry}),$$
$$(\text{Oskar}, \text{Vanilla}), \; (\text{Oskar}, \text{Chocolate}),$$
$$(\text{Amanda}, \text{Chocolate}), \; (\text{Amanda}, \text{Mint}) \,\}.$$

Thus,

$$Likes \;\; = \;\; \{\, (c, f) \;\in\; \text{Children} \times \text{Flavours} \; :$$
$$\text{child } c \text{ likes ice cream flavour } f \,\}.$$

Put differently, this relation is the truth set of the predicate L defined by

$$L(c, f) \;=\; \text{child } c \text{ likes ice cream flavour } f.$$

Exercise 7.3   (Solution on page 435)

Referring to the database of James Bond films in Example 7.1, give the binary relation $StarsIn \subseteq \text{NAMES} \times \text{TITLES}$ defined by

$$StarsIn = \{\, (x, y) \; : \; x \text{ stars as James Bond in } y \,\}.$$

Binary relations can be visualised pictorially by drawing arrows connecting the related objects.

- A heterogeneous relation $R \subseteq A \times B$ from A to B would most naturally be depicted by drawing the two sets A and B side-by-side, and drawing an arrow from each element $a \in A$ in the first set to each of those elements $b \in B$ to which it is related; i.e., such that $(a, b) \in R$.

- A homogeneous relation $R \subseteq A \times A$ on A on the other hand might more naturally be depicted by simply laying out the elements of A in some natural fashion, and drawing an arrow from $a \in A$ to $b \in A$ whenever $(a, b) \in R$.

Example 7.4

The relation *Likes* of Example 7.3 is pictured as follows:

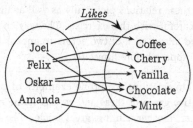

We have an arrow from a child $c \in$ Children to a flavour $f \in$ Flavours whenever $(c, f) \in$ *Likes.*

Example 7.5

The subset relation $\subseteq$ on the powerset of $\{a, b\}$:

$$\mathcal{P}(\{a, b\}) = \Big\{\emptyset, \{a\}, \{b\}, \{a, b\}\Big\}$$

is pictured as follows:

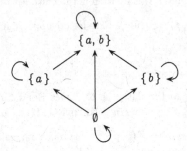

We have an arrow from one set A to another set B whenever $A \subseteq B$.

(Solution on page 435)

Referring to the database of James Bond films in Example 7.1, let

BONDACTORS $\subseteq$ NAMES

be the set of six actors who have played the role of James Bond, and define the two binary relations *Before* and *FirstBefore* on BONDACTORS as follows:

$$Before = \{(x, y) \ : \ x \text{ stars as James Bond in an earlier film}$$
$$\text{than one in which } y \text{ stars as James Bond}\};$$

$$FirstBefore = \{(x, y) \ : \ x \text{ starred as James Bond before } y \text{ did}\}.$$

Present these relations pictorially as well as list out their elements.

(Be careful with this exercise. The way that the binary relation *Before* is defined allows each of two actors to appear before the other, and for one actor to appear before himself!)

Kinship relations are prime examples of binary relations. We all have an intuitive grasp of these and we can name a wide range of relationships, e.g. father, mother, sibling, great uncle. The English language is not even particularly rich in this respect. In Swedish, for example, you don't just refer to your aunt or your uncle, but more specifically to your *farbror* (father's brother), your *morbror* (mother's brother) your *faster* (father's sister), or your *moster* (mother's sister).

Example 7.6

The Duck family consists of the parents Hortense and Quackmore Duck, and their two children Della and Donald. Hortense has a brother Scrooge, and Della has three sons: Huey, Louis and Dewey. Let us consider the set of these eight Ducks:

$$\text{DUCKS} = \big\{ \text{Quackmore, Hortense, Scrooge,}$$
$$\text{Donald, Della, Huey, Louis, Dewey} \big\}.$$

There are a variety of kinship relations defined over DUCKS × DUCKS, such as the following:

Father = $\big\{$ (Quackmore, Donald), (Quackmore, Della) $\big\}$.

Mother = $\big\{$ (Hortense, Donald), (Hortense, Della),

(Della, Huey), (Della, Louis), (Della, Dewey) $\big\}$.

Parent = $\big\{$ (Quackmore, Donald), (Quackmore, Della),

(Hortense, Donald), (Hortense, Della),

(Della, Huey), (Della, Louis), (Della, Dewey) $\big\}$.

Uncle = $\big\{$ (Scrooge, Donald), (Scrooge, Della),

(Donald, Huey), (Donald, Louis), (Donald, Dewey) $\big\}$.

Exercise 7.6   (Solution on page 436)

Define the kinship relations *Child*, *Brother*, *Sister* and *Sibling* on the Duck family of Example 7.6, and present the *Child* relation pictorially.

7.2.1   Functions as Binary Relations

We have defined a function $f : A \to B$ to be an assignment of exactly one element of B to each element of A, and noted in Theorem 6.4 that such a function is completely determined by its graph:

$$\text{graph}(f) = \big\{ (a, b) \in A \times B : b = f(a) \big\}.$$

The graph of the function f is a binary relation from A to B satisfying the following special property: every element $a \in A$ is related to exactly one element $b \in B$.

Conversely, any binary relation $R \subseteq A \times B$ which satisfies this property defines a function $f_R : A \to B$.

Theorem 7.6

A binary relation $R \subseteq A \times B$ is the graph of a function from A to B if, and only if,

$$\forall a \in A \, \exists ! \, b \in B \, \big((a,b) \in R \big) \tag{$\star$}$$

Proof: If the relation $R \subseteq A \times B$ satisfies the property $(\star)$, then we can define a function $f_R : A \to B$ by mapping each $a \in A$ to the unique $b \in B$ such that $(a,b) \in R$. Clearly, graph$(f_R) = R$, as given any $(a,b) \in A \times B$,

$$(a,b) \in \text{graph}(f_R) \iff f_R(a) = b \qquad \text{(by definition of graph}(f_R))$$
$$\iff (a,b) \in R \qquad \text{(by definition of } f_R).$$

Conversely, if $R = \text{graph}(f)$ for some function $f : A \to B$, then R must clearly satisfy $(\star)$, as the graph of any function must satisfy $(\star)$. $\qquad \square$

7.3 Operations on Binary Relations

We have defined binary relations as certain sets; specifically, a binary relation from A to B is a subset of $A \times B$. With this view in mind, there are various operations which we can apply to binary relations to extract information from them, or to build further binary relations, typical of the sort employed by database queries.

7.3.1 Boolean Operations

As binary relations are sets (of pairs), the usual set operations can be applied to these, often quite usefully. In the above Duck family Example 7.6, for instance, the *Parent* relation is defined simply as the union of the *Father* and *Mother* relations:

Parent = *Father* ∪ *Mother*.

This is intuitively clear, as x is a parent of y if, and only if, either x is the father of y, or x is the mother of y:

$(x, y) \in$ *Parent* if, and only if, $(x, y) \in$ *Father* or $(x, y) \in$ *Mother*.

We can also express the *Father* relation in terms of the *Parent* and *Mother* relations, noting that a father is someone who is a parent but not a mother:

$$Father = Parent \setminus Mother.$$

Note that in order to apply set operations to binary relations, the relations being operated on must be defined over the same sets (in this case, DUCKS × DUCKS). It would not make much sense, for example, to take the union *Father* ∪ *Before* of the relation *Father* ⊆ DUCKS × DUCKS from Example 7.6. and the relation *Before* ⊆ NAMES × NAMES from Exercise 7.5.

Exercise 7.7   (Solution on page 437)

Let R_1, R_2 and R_3 represent the *less-than* relation $<$, the *equality* relation $=$, and the *less-than-or-equal-to* relation $\leq$, respectively, all on the set $\mathbb{N}$ of natural numbers:

$$R_1 = \{(x,y) \in \mathbb{N}^2 : x < y\};$$
$$R_2 = \{(x,y) \in \mathbb{N}^2 : x = y\};$$
$$R_3 = \{(x,y) \in \mathbb{N}^2 : x \leq y\}.$$

What are the following relations?

1. $R_1 \cup R_2$
2. $R_3 \cap \overline{R_2}$
3. $R_3 \setminus R_1$

7.3.2   Inverting Relations

Given a binary relation, an obvious and natural thing to do is to turn it around, or invert it, and consider the converse relation. For example, the opposite, or inverse, of the *less-than-or-equal-to* relation $\leq$ is the *greater-than-or-equal-to* relation $\geq$ (as $x \leq y$ if, and only if, $y \geq x$); and the opposite, or inverse, of the *Parent* relation is the *Child* relation (as x is a parent of y if, and only if, y is a child of x).

Given a binary relation $R \subseteq A \times B$ from a set A to a set B, the *inverse* relation $R^{-1} \subseteq B \times A$ from B to A is defined as

$$R^{-1} = \{(b,a) : (a,b) \in R\}.$$

If we consider the pictorial representation of the relation R, we can derive the pictorial representation of R^{-1} simply by reversing the direction of all of the arrows, thus replacing each arrow from a to b where $(a,b) \in R$ by an arrow from b to a.

Example 7.7

The inverse of the relation *Likes* ⊆ Children × Flavours from Example 7.3 is the relation *Likes*⁻¹ ⊆ Flavours × Children of *"is liked by"*:

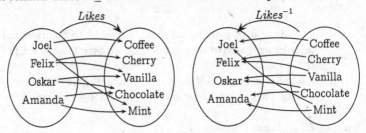

For example, (Joel, Mint) ∈ *Likes* indicates that Joel likes mint ice cream, while (Mint, Joel) ∈ *Likes*⁻¹ indicates that mint ice cream is liked by Joel.

Exercise 7.8   (Solution on page 437)

What is *Sibling*⁻¹, the inverse of the *Sibling* relation?

7.3.3   Composing Relations

As well as turn relations around, another natural operation is to combine, or compose, two relations by following one with another. Given relations $R \subseteq A \times B$ from A to B and $S \subseteq B \times C$ from B to C, the *composition* of S and R is the relation $S \circ R \subseteq A \times C$ from A to C defined as

$$S \circ R = \{(a,c) \in A \times C \ : \ \exists b \in B \text{ such that}$$
$$(a,b) \in R \text{ and } (b,c) \in S\}.$$

If we consider the pictorial representation of the relations R and S, we can derive the pictorial representation of $S \circ R$ simply by following an R-arrow by an S-arrow, as in the following example:

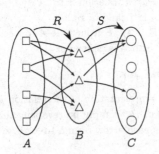
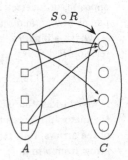

Note that the target of the relation R must be the same as the source of the relation S in order to form the composition. Also note carefully the order of the relations: the composition $S \circ R$ of the relations S and R first "applies" the relation R to its source before "applying" the relation S to the result. In this sense, the definition coincides with the composition of functions given in Definition 6.8.

Example 7.8

A grandfather is a father of a parent, and we can use this characterisation to define the *Grandfather* relation:

$$Grandfather \; = \; Parent \circ Father.$$

The order in which we write the two relations which are being composed is important. For example, a grandfather is a father of a parent, which is not the same thing as a parent of a father:

$$Father \circ Parent \; \neq \; Parent \circ Father.$$

Exercise 7.9   (Solution on page 437)

Define the relations *Uncle* and *Nephew* in terms of simpler relations, and derive these relations for the Duck family of Example 7.6.

7.3.4   The Domain and Range of a Relation

Given the relation $R \subseteq A \times B$ from A to B,

- the *domain* of R is the set

    $\mathrm{domain}(R) \; = \; \{\, a \in A \; : \; \exists\, b \in B \text{ such that } (a, b) \in R \,\};$

- the *range* of R is the set

    $\mathrm{range}(R) \; = \; \{\, b \in B \; : \; \exists\, a \in A \text{ such that } (a, b) \in R \,\}.$

That is to say, the domain of a relation consists of all elements of the source A of the relation which are related to something in the target B, and the range of a relation consists of all elements of the target B of the relation which are related to something in the source A.

Example 7.9

Consider the following relations on humans H:

$$Parent = \{(x,y) \; : \; x \text{ is a } parent \text{ of } y\}$$
$$Brother = \{(x,y) \; : \; x \text{ is a } brother \text{ of } y\}$$

Then

domain(*Parent*) = the set of parents (*not* all of *H*);

range(*Parent*) = the set of children (*all* of *H*);

domain(*Brother*) = the set of brothers (males with siblings);

range(*Brother*) = the set of humans with a brother.

Exercise 7.10   (Solution on page 438)

Prove that if $R \subseteq A \times B$ is the graph of a function $f : A \to B$, then domain(R) = A (i.e., the domain of f) and range(R) = range(f).

7.4 Properties of Binary Relations

There are various properties that a binary relation on a set A may or may not satisfy. Of particular interest are the properties of *reflexivity*, *symmetry* and *transitivity*, all of which we shall explore in this section.

7.4.1 Reflexive and Irreflexive Relations

The difference between the *less-than* relation < and the *less-than-or-equal-to* relation ≤ on numbers is that any number is *less-than-or-equal-to* itself (since it is equal to itself), but no number is *less-than* itself. For example, $2 \leq 2$ is true but $2 < 2$ is not true. This motivates our first property.

Definition 7.10

A relation R on a set A is *reflexive* if, and only if, every element of A is related to itself by R:

$$\forall x \in A \; (xRx).$$

The relation is *irreflexive* if, and only if, no element of A is related to itself:

$$\forall x \in A \; \neg(xRx).$$

Thus, for example, the *less-than-or-equal-to* relation ≤ is reflexive, while the *less-than* relation < is irreflexive. Note that irreflexive is not the same as non-reflexive: it is possible for a binary relation to relate some but not all elements to themselves, thus making the relation neither reflexive nor irreflexive.

Exercise 7.11   (Solution on page 438)

Is the relation *Before* from Exercise 7.5 reflexive, irreflexive, or neither? What about the relation *FirstBefore*?

7.4.2   Symmetric and Antisymmetric Relations

Equality between objects suggests – amongst other things – a certain symmetry between the objects, which is captured by the next property of interest.

Definition 7.11

A relation R on a set A is *symmetric* if, and only if, y is related to x whenever x is related to y:

$$\forall x, y \in A \, (xRy \Rightarrow yRx).$$

The relation is *antisymmetric* if, and only if, y is never related to x whenever x is related to y, except possibly for when $x = y$:

$$\forall x, y \in A \, \big((xRy \wedge yRx) \Rightarrow x = y\big).$$

Thus, for example, the relations $<$ and $\leq$ are both antisymmetric, while the relation $=$ is symmetric (as well as anytisymmetric).

Exercise 7.12   (Solution on page 438)

Is the relation *Before* from Exercise 7.5 symmetric, antisymmetric, or neither? What about the relation *FirstBefore*?

7.4.3   Transitive Relations

If one number is less than a second number which is itself less than a third number, then clearly the first number will also be less than the third number. This property of the *less-than* relation is embodied in the final property of interest.

Definition 7.12

A relation R on a set A is *transitive* if, and only if, x is related to z whenever x is related to some y which is related to z:

$$\forall x, y, z \in A \, \big((xRy \wedge yRz) \Rightarrow xRz\big).$$

Thus, for example, the relations $<$ and $\leq$ are both transitive.

Exercise 7.13   (Solution on page 438)

Is the relation *Before* from Exercise 7.5 transitive? What about the relation *FirstBefore*?

Example 7.13

Consider the sibling (brother or sister) relationship over people.

1. This is not reflexive, as you would not consider someone to be their own sibling. It is in fact irreflexive.

2. It is symmetric as anyone is obviously a sibling to each of their siblings. Clearly it is not antisymmetric.

3. Finally, it is not transitive, as this would imply that any person who has a sibling must be a sibling of themselves. Also, if we allow half-siblings, one person may be a sibling to a second person due to sharing a common father whilst having different mothers; and the second person may be a sibling to yet a third person due to sharing a common mother whilst having different fathers. In this scenario, the first and third children would not be siblings, as they do not share a common parent.

Exercise 7.14   (Solution on page 439)

Consider the relations *is-an-ancestor-of* and *is-married-to* defined over people. Indicate whether these are reflexive, irreflexive, symmetric, anti-symmetric, and/or transitive. Justify your answers.

7.4.4   Orderings Relations

Various common binary relations arrange the elements of their domain into some specific ordering. For example the *less-than-or-equal-to* relation $\leq$ orders the natural numbers into an increasing sequence: $0 \leq 1 \leq 2 \leq 3 \leq \cdots$. Note that this ordering is total in the sense that any two numbers a and b are related in one way or the other: either $a \leq b$ or $b \leq a$.

Whether or not a particular binary relation defined on a set orders the elements of that set depends on whether or not it satisfies certain of the properties defined above. Naturally, a *less-than-or-equal-to* relation should be:

- reflexive – any element should be *less-than-or-equal-to* itself;

- antisymmetric – if a is *less-than-or-equal-to* b and b is also *less-than-or-equal-to* a, then a and b should be equal.

- transitive – if a is *less-than-or-equal-to* b and b is *less-than-or-equal-to* c, then a should be *less-than-or-equal-to* c.

In fact, these three properties taken together indicate that a relation is an ordering relation as defined as follows.

Definition 7.15

A binary relation R on a set is a *partial order* if, and only if, it is reflexive, antisymmetric, and transitive. It is a *total order* if, and only if, it is a partial order in which any two elements are related in one way or the other:

$$\forall x, y \in A\, (xRy \vee yRx).$$

Example 7.15

- The *equality* relation $=$ on integers is a partial order, but it is not a total order.

- The *less-than-or-equal-to* relation $\leq$ on integers is a total order. However, the *less-than* relation $<$ on integers is not a (total or partial) order, as it is not reflexive.

- The *subset* relation $\subseteq$ on sets is a partial order but not a total order; for example, $\{1\} \not\subseteq \{2\}$ and $\{2\} \not\subseteq \{1\}$.

7.4.5   Equivalence Relations

A binary relation on a set may reflect a notion of *sameness* between elements of that set, defining when we might want to consider two elements of the set to be indistinguishable – that they are in some sense *equivalent*.

As with orderings, whether or not a particular relation over a set defines an *equivalence* between elements of that set depends on whether or not it satisfies certain of the properties defined above. Naturally, such a relation should be:

- reflexive – any element should be *the same as* itself;

- symmetric – if a is *the same as* b then b should be *the same as* a;

- transitive – if a is *the same as* b and b is *the same as* c, then a should be *the same as* c.

These three properties suffice to define a notion of *sameness*.

Definition 7.16

A binary relation R on a set is an *equivalence relation* if, and only if, it is reflexive, symmetric, and transitive.

Example 7.17

- The *equality* relation $=$ on integers is an equivalence.
- The *less-than-or-equal-to* relation $\leq$ on integers is not an equivalence relation, as it is not symmetric. Furthermore, the *less-than* relation $<$ on integers fails to be an equivalence relation for this same reason, as well as for not being reflexive.
- The *subset* relation $\subseteq$ on sets is not an equivalence relation, as it is not symmetric.

Example 7.18

Consider splitting up a set A of people into twelve groups depending on the month of their birthday; for example, one of the groups might consist of all those people in A whose birthday is in September. (There may actually be fewer than twelve groups, if there are months in which no one in A was born.) This naturally defines an equivalence relation R on A in which two people are related if, and only if, their birthdays are in the same month:

$$R = \{(x, y) : x \text{ and } y \text{ have birthdays in the same month}\}.$$

Clearly this relation is reflexive, symmetric and transitive.

Exercise 7.18   (Solution on page 439)

Which of the following binary relations on $\mathbb{N}$ are partial orders? Which are total orders? Which are equivalences? Explain your answers.

1. The identity relation $I = \{(n, n) : n \in \mathbb{N}\}$.
2. The universal relation $U = \{(m, n) : m, n \in \mathbb{N}\}$.
3. The parity relation $P = \{(m, n) : m = n \,(mod\,2)\}$.

Exercise 7.19   (Solution on page 439)

Consider a set S of students who are each taking some number of courses chosen from a set C of courses. Define the following binary relations on S:

$$R_1 = \{ (s_1, s_2) \; : \; s_1 \text{ and } s_2 \text{ take all the same courses} \}.$$

$$R_2 = \{ (s_1, s_2) \; : \; s_1 \text{ and } s_2 \text{ take } some \text{ course together} \}.$$

Are either of these an equivalence relation? Justify your answer.

7.4.6   Equivalence Classes and Partitions

Consider the equivalence relation R from Example 7.18 defined over some set A of people:

$$R = \{ (x, y) \; : \; x \text{ and } y \text{ have birthdays in the same month} \}.$$

We based this equivalence relation on a *partitioning* of the set A into disjoint sets. This idea is formalised in the following.

(Definition 7.20)

A partition of a set A is a collection $\{ A_i \; : \; i \in I \}$ of disjoint non-empty subsets of A which together contain all of A. That is:

1. $A_i \cap A_j = \emptyset$ whenever $i \neq j$; and

2. $\bigcup_{i \in I} A_i = A$.

*The subsets A_i are called the **blocks** of the partition. We say that one partition is a refinement of a second partition if, and only if, every block of the first is a subset of some block of the second.*

(Example 7.20)

We can refine the relation R from Example 7.18 by splitting the people of A not just according to the month of their birth, but according to sex as well, thus creating (up to) 24 groups; for example, one of the groups might consist of all females in A whose birthday is in September. This new partition of A is clearly a refinement of the original coarser partition defined only by birth month.

(Exercise 7.21)   (Solution on page 439)

What is the finest partition of a set A, in the sense that it cannot be refined into a different partition? What is the coarsest (i.e., least fine) partition?

Any partition of a set A naturally defines an equivalence relation, in just the way the partition of Example 7.18 gave rise to the equivalence relation R; two elements of A will be deemed equivalent if, and only if, they appear in the same block of the partition. Just as clearly, any equivalence relation partitions the elements over which it is defined into disjoint non-empty subsets, called *equivalence classes*.

Definition 7.21

Given an equivalence relation R on a set A, the *equivalence class* of an element a of A with respect to R, denoted $[a]_R$, is the set of elements of A which are related to a by R:

$$[a]_R = \{x \in A : aRx\}.$$

Theorem 7.22

The collection of equivalence classes $\{[a]_R : a \in A\}$ of an equivalence relation R is a partition of A.

Proof: To prove this we need to show the following:

1. Each $[a]_R$ is non-empty.

   This is true since $a \in [a]_R$.

2. The union of the equivalence classes is A.

   This is true since each $a \in A$ is in the equivalence class $[a]_R$.

3. The equivalence classes are disjoint; in other words, two non-disjoint equivalence classes must be equal.

   To see this, let us assume that $[a]_R$ and $[b]_R$ are not disjoint, that they contain a common element x; that is, aRx and bRx, which by symmetry means also that xRa, and thus by transitivity that bRa. Then

   $$y \in [a]_R \Leftrightarrow aRy$$
   $$\Leftrightarrow bRy \text{ (by transitivity, since } bRa \text{ and } aRy)$$
   $$\Leftrightarrow y \in [b]_R.$$

   Thus we must have that $[a]_R = [b]_R$.   □

Exercise 7.23   (Solution on page 440)

What are the equivalence relations defined by the finest and coarsest partitions of a set A identified in Exercise 7.21?

Exercise 7.24   (Solution on page 440)

Let the relation R on the set $A = \{1, 2, 3, \ldots, 29\}$ of positive integers less than 30 be defined by:

$(x, y) \in R$ if, and only if, x and y have the same prime factors.

For example, $(12, 18) \in R$ since $12 = 2 \times 2 \times 3$ and $18 = 2 \times 3 \times 3$ have the same prime factors 2 and 3. Clearly this is an equivalence relation.

How many equivalence classes does R partition A into? List each of these equivalence classes.

7.5   Additional Exercises

1. Consider the following family members of Don Vito Corleone and his wife Carmella have four children: Santino, Federico, Michael and Constanzia. Santino is married to Sandra and they have four children: Santino Jr, Francesca, Kathryn and Frank. Michael is married to Kay and they have two children: Anthony and Mary. Constanzia is married to Carlo and they have two children: Victor and Michael Francis. Federico is not married and has no children.

   (a) List out the set CORLEONES of all persons mentioned above.
   (b) List out the relations *Father*, *Mother*, *Husband* and *Sibling*.
   (c) Define the relation *Father* in terms of *Mother* and *Husband*.
   (d) Define the relations *Parent*, *Wife* and *Spouse* in terms of the above relations, and list these out.
   (e) Define the relations *Father-In-Law*, *Mother-In-Law* and *Cousin* in terms of the above relations, and list these out.

2. Indicate which of the following relations defined over the integers $\mathbb{Z}$ are reflexive, which are irreflexive, which are symmetric, which are antisymmetric, and which are transitive. Justify your answers.

   (a) $R_1 = \{(a, b) : a = b \text{ or } a = -b\}$.
   (b) $R_2 = \{(a, b) : a = b-1\}$.
   (c) $R_3 = \{(a, b) : a+b \leq 10\}$.
   (d) $R_4 = \{(a, b) : a < 2b\}$.

3. Indicate which of the following relations defined over the positive integers are reflexive, which are irreflexive, which are symmetric, which are antisymmetric, and which are transitive. Justify your answers.

   (a) The *divisibility* relation $a \mid b$ which holds if, and only if, a divides evenly into b.
   (b) The *relatively prime* relation which holds between a and b if, and only if, their greatest common divisor is 1.

(c) The relation which holds between a and b if, and only if, their difference (i.e., the larger minus the smaller) is divisibly by 3.

4. What does a symmetric and transitive relation look like? Is it true that any binary relation which is symmetric and transitive must also be reflexive? Justify your answer.

5. Suppose R and S are symmetric relations on a set A. Which of the following must be a symmetric relation? Justify your answers.

   (a) $R \cup S$.   (b) $R \cap S$.   (c) $R \circ S$.   (d) $\overline{R}$.   (e) R^{-1}

6. Suppose R and S are transitive relations on a set A. Which of the following must be a transitive relation? Justify your answers.

   (a) $R \cup S$.   (b) $R \cap S$.   (c) $R \circ S$.   (d) $\overline{R}$.   (e) R^{-1}

7. Match the property of the binary relation R on A listed on the left to a characterisation of that property on the right:

1. reflexive	(a) $R \circ R \subseteq R$
2. irreflexive	(b) $\mathrm{id}_A \cap R = \emptyset$
3. symmetric	(c) $R = R^{-1}$
4. antisymmetric	(d) $\mathrm{id}_A \subseteq R$
5. transitive	(e) $R \cap R^{-1} \subseteq \mathrm{id}_A$

8. The *reflexive closure* of a relation R over a set A is the smallest reflexive relation that contains R. Similarly, the *symmetric closure* of a relation R over a set A is the smallest symmetric relation that contains R, and the *transitive closure* of a relation R over a set A is the smallest transitive relation that contains R.

   Compute the reflexive, symmetric and transitive closures of the binary relation $R = \{(0,1), (1,2), (3,4), (4,3)\}$ over the set $A = \{0, 1, 2, 3, 4\}$.

9. Prove that $R \cup \{(a,a) : a \in R\}$ is the reflexive closure of R.

10. Prove that $R \cup R^{-1}$ is the symmetric closure of R.

11. Prove that

$$\{(a_1, a_n) : \exists a_2, a_3, \ldots, a_{n-1} \text{ such that } (a_i, a_{i+1}) \in R$$
$$\text{for each } i = 1, 2, \ldots, n-1\}$$

    is the transitive closure of R.

12. Let us say that two real numbers x and y are approximately equal, and write $x \approx y$, if, and only if, they differ by no more than $1/1000$. Thus, the relation $\approx$ on $\mathbb{R}$ is defined as follows:

$$\approx \; = \; \{(x,y) : |x - y| < 1/1000\}.$$

Intuitively this ought to be an equivalence relation. Explain why this relation is – or is not – reflexive, symmetric and transitive.

13. Consider the relation $\leq$ defined on a Boolean algebra B as follows: for all $x, y \in B$, $x \leq y$ if, and only if, $x + y = y$.

   (a) Prove that $\leq$ is a partial order.

   (b) What does $\leq$ correspond to in the Boolean algebra of sets?

   (c) What does $\leq$ correspond to in the Boolean algebra of propositions?

14. Assuming that R is an equivalence relation on A, show directly from the definitions that the following statements about two elements a and b of A are equivalent:

   (a) aRb (b) $[a]_R = [b]_R$ (c) $[a]_R \cap [b]_R \neq \emptyset$

Chapter 8

Inductive and Recursive Definitions

Great fleas have little fleas,
Upon their backs to bite 'em,
And little fleas have lesser fleas,
And so ad infinitum.

- Augustus De Morgan.

Most of the objects under study within Computer Science are defined *inductively*: that is, they are defined in terms of smaller instances of themselves. Numbers, lists, binary trees, and even computer programs themselves, are all built up from smaller objects of the same type. For example, two computer programs stuck together, typically with a semicolon between them, so that the second is executed once the first completes its task is nothing more than a program defined in terms of two smaller programs. Also, functions defined over such objects are typically given by inductive definitions, whereby the value of the function on an inductively-defined object is defined by the value of the function on smaller objects. More generally, a *recursive* definition allows a function to be defined in terms of its value on arbitrary objects, not necessarily smaller objects, and can be meaningfully employed.

Understanding inductively-defined objects, and the functions defined on them, will naturally rely on understanding the inductive nature of such definitions. In this chapter, we explore such inductive definitions and recursively-defined functions.

8.1 Inductively-Defined Sets

As we saw, we can define finite sets by simply listing their elements, such as

F. Moller, G. Struth, *Modelling Computing Systems*,
Undergraduate Topics in Computer Science,
DOI 10.1007/978-1-84800-322-4_9, © Springer-Verlag London 2013

$$\text{BINARYDIGITS} = \{0, 1\}$$

$$\text{DECIMALDIGITS} = \{0, 1, 2, 3, 4, 5, 6, 7, 8, 9\}$$

$$\text{LETTERS} = \{\, a, b, c, d, e, f, g, h, i, j, k, l, m,$$
$$n, o, p, q, r, s, t, u, v, w, x, y, z \,\}$$

$$\text{Children} = \{\text{Joel, Felix, Oskar, Amanda}\}$$

However, for infinite sets we have had to resort to using some (implicit or explicit) rule for generating their members. For example, the set of natural numbers

$$\mathbb{N} = \{0, 1, 2, 3, \dots\}$$

which we defined (informally) in Chapter 2 relies on our ability as intelligent beings to extract the implicit rule hinted at by the ellipses which says that adding one to any element of this set gives the "next" element in the set. However, this approach to defining sets is fraught with complications.

1. How can we expect a non-intelligent entity (such as a computer) to be able to understand such a definition? At the very least we would somehow have to make explicit the rule for generating the elements of the set.

2. How can we even be certain of the implicit rule underlying the defining equation? For example, the author of the above definition may intend $\mathbb{N}$ to represent the decimal digits (and thus end at the digit 9), or the roots (i.e., solutions) of the equation $x^4 - 6x^3 + 11x^2 - 6x = 0$ (in which case $\mathbb{N}$ would contain only the four values listed).

3. The order in which we list the elements of a set is irrelevant, so what sense does it make to refer to the "next" element in a set?

4. How can we determine when some object, $\sqrt{9}$ say, is in the set we are defining while another object, $\sqrt{10}$ say, is not?

One easy way of defining an infinite collection of objects is to provide a method for generating new elements from existing ones. This idea is encompassed by the following definition.

Definition 8.1

An *inductive definition* of a set has three components.

1. *The basis clause, which establishes that certain objects are in the set. These elements constitute the "building blocks" for constructing further elements in the set.*

2. *The inductive clause, which defines the ways in which elements of a set can be used to produce further elements which are also in the set.*

3. The *extremal clause*, which asserts that no object is an element of the set being defined unless its membership can be established from the first two clauses. In other words, the set being defined is the *smallest* set which satisfies the first two clauses.

Example 8.1

We can represent precisely the set $\mathbb{N}$ of natural numbers by way of the following inductive definition.

1. $0 \in \mathbb{N}$.

2. $(n+1) \in \mathbb{N}$ whenever $n \in \mathbb{N}$.

   In other words, $n \in \mathbb{N} \Rightarrow (n+1) \in \mathbb{N}$.

3. Nothing else is in $\mathbb{N}$. That is, nothing is in $\mathbb{N}$ unless it can be constructed from the first two clauses.

   In other words, $\mathbb{N}$ is the *smallest* set satisfying the first two clauses.

The basis clause declares the number 0 as a basic element of the set $\mathbb{N}$; and the inductive clause says that given a natural number n, we can produce another natural number $n+1$ by adding 1 to the given number n. In this way we can conclude that $\sqrt{9} = 3$ is an element of $\mathbb{N}$, since 0 is an element of $\mathbb{N}$ (by the basis clause), and hence $0+1 = 1$ is an element (by the inductive clause), and hence $1+1 = 2$ is an element (again by the inductive clause), and thus finally $2+1 = 3$ is an element (by a further use of the inductive clause).

The extremal clause tells us that an element of $\mathbb{N}$ has to be either 0 (from the basis clause) or the successor of another element of $\mathbb{N}$ (from the inductive clause). We could not infer that $\sqrt{10} \approx 3.16$ is an element of $\mathbb{N}$, as there is no way to construct $\sqrt{10}$ from these basis and inductive clauses: $\sqrt{10}$ is clearly not 0; and no matter how many times we add 1 to 0 we will never generate the value $\sqrt{10}$. Hence we must conclude that $\sqrt{10}$ is *not* an element of $\mathbb{N}$ as defined.

Alternatively, we can easily see that the set $\{0, 1, 2, 3, 4, \ldots\}$ satisfies clauses (1) and (2) of the definition. Therefore, since $\mathbb{N}$ is being defined to be the *smallest* set satisfying these clauses, $\mathbb{N}$ must be a subset of this; since this set does not contain $\sqrt{10}$, $\sqrt{10} \notin \mathbb{N}$.

Exercise 8.1 (Solution on page 440)

Explain, using this inductive definition of $\mathbb{N}$, why $4 \in \mathbb{N}$ while $4.5 \notin \mathbb{N}$.

Example 8.2

We can inductively define the set

$$\text{ODD} = \{1, 3, 5, 7, \ldots\}$$

of odd natural numbers as the smallest set satisfying the following:

1. $1 \in \text{ODD}$.
2. If $n \in \text{ODD}$ then $(n+2) \in \text{ODD}$.

Note that in this example, we incorporated the extremal clause into the preamble of the definition, by defining the set to be the smallest set satisfying the basis and inductive clauses; being the smallest such set, only those elements which must be in the set due to the basis and inductive clauses are actually members. We could have instead included the extremal clause; however, the above is a common useful abbreviated form.

Exercise 8.2   (Solution on page 441)

The set $\mathbb{N}$ satisfies the two clauses in the definition of ODD; that is, it contains 1, and it contains $(n+2)$ whenever it contains n. Why does this not imply that $\text{ODD} = \mathbb{N}$?

Exercise 8.3   (Solution on page 441)

Give an inductive definition for the set POWERS-OF-2 of powers of 2,

$$\text{POWERS-OF-2} = \{1, 2, 4, 8, 16, 32, 64, \ldots\}.$$

Example 8.3

Given a finite set S, we can define the powerset $\mathcal{P}(S)$ of S inductively as the smallest set satisfying the following:

1. $\emptyset \in \mathcal{P}(S)$.
2. If $X \in \mathcal{P}(S)$ and $a \in S$ then $X \cup \{a\} \in \mathcal{P}(S)$.

For example, if $S = \{1, 2, 3\}$, then by the basis clause $\emptyset \in \mathcal{P}(S)$, and by one application of the inductive clause we get that the following sets are in $\mathcal{P}(S)$:

$$\emptyset \cup \{1\} = \{1\} \qquad \emptyset \cup \{2\} = \{2\} \qquad \emptyset \cup \{3\} = \{3\}$$

This application reveals that all of the singleton sets $\{1\}$, $\{2\}$ and $\{3\}$ are in $\mathcal{P}(S)$. A second application of the inductive clause tells us that the following sets are in $\mathcal{P}(S)$:

$$\emptyset \cup \{1\} = \{1\} \qquad \emptyset \cup \{2\} = \{2\} \qquad \emptyset \cup \{3\} = \{3\}$$

$$\{1\} \cup \{1\} = \{1\} \qquad \{1\} \cup \{2\} = \{1, 2\} \quad \{1\} \cup \{3\} = \{1, 3\}$$

$$\{2\} \cup \{1\} = \{1, 2\} \quad \{2\} \cup \{2\} = \{2\} \qquad \{2\} \cup \{3\} = \{2, 3\}$$

$$\{3\} \cup \{1\} = \{1, 3\} \quad \{3\} \cup \{2\} = \{2, 3\} \quad \{3\} \cup \{3\} = \{3\}$$

This second application reveals that all of the two-element sets $\{1, 2\}$, $\{1, 3\}$ and $\{2, 3\}$ are also in $\mathcal{P}(S)$. A third application of the inductive clause would reveal that, apart from the above sets, the three-element set $S = \{1, 2, 3\}$ itself is in $\mathcal{P}(S)$. Further applications of the inductive clause would generate no new elements.

Exercise 8.4 (Solution on page 441)

Why can the above definition not be applied to infinite sets? (Hint: Why would this definition not provide ODD $\in \mathcal{P}(\mathbb{N})$, where ODD is as defined in Example 8.2?)

8.2 Inductively-Defined Syntactic Sets

The elements of the set $\mathbb{N}$ of natural numbers as defined above are *semantic* values, not *syntactic* objects. To understand the distinction clearly, if we define the set

Children = { Joel, Felix, Oskar, Amanda }

we have to make clear whether we mean the set of four names, or the collection of people which make up four specific children. Each name in the list is merely a syntactic object unless we assign some meaning or semantic content to it.

In the same way, we have that $\sqrt{9}$ is an element of $\mathbb{N}$, as $\sqrt{9} = 3$ and 3 is an element of $\mathbb{N}$. The set $\mathbb{N}$ represents the collection of values making up the natural numbers, not some arbitrary representation of them such as decimal numbers (sequences of decimal digits) or binary numbers (sequences of binary digits).

To define sets of such syntactic objects, we first introduce some terminology. An *alphabet* is a finite set of *symbols* or *characters*. A finite sequence of characters from an alphabet A is called a *string* or *word* over A. The *length* of a word $w = a_1 a_2 a_3 \cdots a_n$, where $n \in \mathbb{N}$ and $a_i \in A$ for each $1 \le i \le n$, is given by the number n of (occurrences of) characters in w. We shall use the special symbol ε (which cannot be a character of the alphabet A) to denote the *empty word*, that is, the only word of length 0. Note that $\varepsilon w = w \varepsilon = w$ for any word w.

Finally, we shall use A^* to denote the set of all words over A, and A^+ to denote the set of non-empty words over A. We can define these two sets inductively as follows.

Definition 8.4

The set A^* of words over alphabet A is the smallest set satisfying the following:

1. $\varepsilon \in A^*$; and
2. if $w \in A^*$ and $a \in A$ then $aw \in A^*$.

The set A^+ of non-empty words over alphabet A is the smallest set satisfying the following:

1. $a \in A^+$ for each $a \in A$; and
2. if $w \in A^+$ and $a \in A$ then $aw \in A^+$.

Example 8.4

If $A = \{a, b\}$, then A^* is the set consisting of all sequences of a's and b's, including the empty sequence containing no characters:

$$A^* = \{\varepsilon, a, b, aa, ab, ba, bb, aaa, aab, aba, abb, \dots\}.$$

This is since:

- by the first (basis) clause, $\varepsilon \in A^*$;
- by the second (inductive) clause, adding either an a or a b to the front of any word in A^* gives a word in A^*, and as we know $\varepsilon \in A^*$, this means that $\{a, b\} \subseteq A^*$;
- but then by the second (inductive) clause, since we now know that $\{\varepsilon, a, b\} \subseteq A^*$, we can infer that $\{a, b, aa, ab, ba, bb\} \subseteq A^*$;
- by a third application of the second (inductive) clause, we can now infer that

$$\{\ a, b, aa, ab, ba, bb,$$
$$aaa, aab, aba, abb, baa, bab, bba, bbb\} \subseteq A^*;$$

and each new application of the second (inductive) clause adds more new strings to the set.

Similarly, A^+ is the set consisting of all non-empty sequences of a's and b's:

$$A^+ = \{a, b, aa, ab, ba, bb, aaa, aab, aba, abb, \dots\}.$$

We could have defined the sets A^* and A^+ in various other equivalent ways. For example, we could have used wa instead of (or as well as) aw in each of the second (inductive) clauses; or we could have provided just one inductive definition and defined the second set directly in terms of the first, by observing that $A^* = A^+ \cup \{\varepsilon\}$ and $A^+ = A^* \setminus \{\varepsilon\}$.

We can now define the sets of decimal and binary numbers as the sets of non-empty words over decimal, respectively binary, digits.

$$\text{DECIMALNUMBERS} = \text{DECIMALDIGITS}^+$$

$$\text{BINARYNUMBERS} = \text{BINARYDIGITS}^+$$

Exercise 8.5   (Solution on page 441)

Give an inductive definition of PosDecimalNumbers, the set of positive decimal numbers. Such numbers should not have leading zeros; that is, $35 \in \text{PosDecimalNumbers}$ but $035 \notin \text{PosDecimalNumbers}$.

8.3   Backus-Naur Form

A common style of presenting an inductive definition of a set of syntactic objects is the so-called *Backus-Naur Form (BNF)*, in which the syntactic forms are presented equationally. For example, the set A^* of words over A is given by the BNF equation

$$w \ ::= \ \varepsilon \mid aw$$

and the set A^+ of non-empty words over A is given by the BNF equation

$$w \ ::= \ a \mid aw$$

where in both cases a is taken to range over the alphabet A. In this way, BNF provides a short-hand form of writing out inductive definitions.

As another example, the natural numbers $\mathbb{N}$ were defined in terms of zero 0 and the *successor function* $s(n) = n+1$. These elements can be specified by the BNF equation

$$n \ ::= \ 0 \mid s(n).$$

Hence, for example, the number 4 is formally defined as $s(s(s(s(0))))$.

Inductive definitions of sets of syntactic expressions are very common in Computer Science. Indeed we have seen several already, such as the set of propositional formulæ, which we can now define formally as follows.

Example 8.5

The set of propositional formulæ can be defined inductively as the smallest set satisfying the following:

1. true and false are propositional formulæ, as is every propositional variable P.

2. If p and q are propositional formulæ then so are $\neg p$, $p \vee q$, $p \wedge q$, $p \Rightarrow q$ and $p \Leftrightarrow q$.

More succinctly, the following is a BNF equation for propositional formulæ.

$$p, q ::= \text{true} \mid \text{false} \mid P \mid \neg p \mid p \vee q \mid p \wedge q \mid p \Rightarrow q \mid p \Leftrightarrow q$$

Here, P is taken to range over the set of propositional variables.

Exercise 8.6 (Solution on page 441)

Give an inductive definition of the set of formulæ of predicate logic.

BNF notation was invented in 1959 by John Backus (and later simplified by Peter Naur) to define the syntax of the ALGOL programming language. It then became a common feature of the appendix to programming language reference books. This is due to the fact that the set of programs which can be written in a given programming language can be defined inductively from the constructs of the language.

Example 8.6

The following BNF equation describes a very simple programming language.

$$p ::= x := e \mid p_1 ; p_2 \mid \text{if } b \text{ then } p_1 \text{ else } p_2 \mid \text{while } b \text{ do } p$$

For readability, this is typically rendered in list fashion as follows:

$$
\begin{aligned}
p ::= \quad & x := e \\
\mid \quad & p_1 ; p_2 \\
\mid \quad & \text{if } b \text{ then } p_1 \text{ else } p_2 \\
\mid \quad & \text{while } b \text{ do } p
\end{aligned}
$$

In the above, x is taken to range over program variables; and e and b range over integer expressions and Boolean expressions, respectively, which themselves will similarly be defined inductively. Thus a program in this programming language is either

- an assignment statement "$x := e$" which evaluates the integer expression e and assigns this value to the variable x; or

- the sequential composition "$p_1 ; p_2$" of two (smaller) programs p_1 and p_2, which first executes the program p_1, and then executes the program p_2 if and when program p_1 has terminated; or

- a conditional statement "if b then p_1 else p_2" involving a Boolean test b and two (smaller) programs p_1 and p_2, which first evaluates the Boolean expression b, and then either executes the program p_1 if b evaluated to true, or executes the program p_2 if b evaluated to false; or

- a while loop "while b do p" involving a Boolean test b and a (smaller) program p, which repeatedly executes the program p for as long as the Boolean test b is true; that is, it first evaluates the Boolean expression b, and then either terminates if b evaluated to false, or executes the program p and repeats itself (starting with re-evaluating the Boolean expression b) if b evaluated to true.

We shall include one further minor – yet essential – piece of syntax in this language: we will allow ourselves to add braces around any program, thus writing $\{p\}$, in order to avoid ambiguity. This is illustrated in the following example.

The following is a program in this language for computing the sum of the first n positive integers: $s = 1 + 2 + 3 + \cdots + n$.

```
i := 0;
s := 0;
while i < n do
  { i := i + 1;
    s := s + i }
```

This 5-line program consists of two smaller programs combined with the sequential composition symbol:

```
i := 0 ;
s := 0;
while i < n do
  { i := i + 1;
    s := s + i }
```

The first of these programs is a simple assignment statement, while the second program is itself built up from two even smaller programs combined with the sequential composition symbol:

```
s := 0 ;
while i < n do
  { i := i + 1;
    s := s + i }
```

Again, the first of these programs is a simple assignment statement, while the second program is a while loop, the body of which is a program consisting of two simple assignment statements combined with the sequential composition symbol. The whole program thus breaks down as follows:

```
i := 0 ;
s := 0 ;
while i < n do
   { i := i + 1 ;
     s := s + i   }
```

It is possible to interpret this program differently, namely as two programs combined with the sequential composition symbol, the first being itself two simple assignment statements composed together sequentially, and the second being the while loop. The break down would then look as follows.

```
i := 0 ;
s := 0        ;
while i < n do
   { i := i + 1 ;
     s := s + i   }
```

This particular ambiguity is harmless. However, the potential for dangerous ambiguity is why the program includes braces around the body of the while loop. Without these, it would be possible (and moreover likely) that the program would be interpreted wrongly as follows.

```
i := 0 ;
s := 0        ;
while i < n do
     i := i + 1      ;
s := s + i
```

This program – or rather this interpretation of the program – would return the incorrect result $s = n$, as the while loop would do nothing but increment the counter i until it reached this value.

8.4 Inductively-Defined Data Types

Most data types used in computer programming languages are inductively defined, either by the compiler (the integers, for example) or by the programmer. For example, a list of natural numbers can be defined by the following BNF equation.

$$L \; ::= \; [\,] \; \mid \; n : L$$

In this definition, n ranges over natural numbers, and the colon symbol ":" represents the operation of adding an element to the front of a list. Thus, a list is either the *empty list* $[\,]$ (the list containing no items), or a list obtained by adding a natural number n to the head of a (smaller) list L. For example, the list $[1, 2, 3]$ is built up inductively starting from $[\,]$ as $1 : 2 : 3 : [\,]$. For clarity this could be written using parentheses as $1 : (2 : (3 : [\,]))$.

Of course, we could choose any other type of data to form a list over; e.g., a list of names is defined as above but by letting n range over names rather than numbers.

As a further example, the binary tree is a widely used data structure, and can be defined inductively as follows.

Example 8.7

We may inductively define *binary trees* using the following BNF equation.

$$t \; ::= \; \star \; \mid \; N(t_1, t_2)$$

That is, a tree is either a *leaf* $\star$ or an *internal node* $N(t_1, t_2)$ with two subtrees t_1 and t_2. For example, the tree

$$N(N(\star, \star), N(N(\star, N(\star, \star)), \star))$$

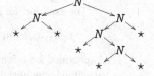

may be represented by the picture shown.

This binary tree definition only provides the *structure* of the data structure, but you typically want to store data in data structures. For example, a *dictionary* might be represented by a binary tree with names stored in the (internal) nodes, with the intention that all names stored in the left subtree precede (alphabetically) the name stored in the parent node, and all names stored in the right subtree follow the name stored in the parent node. For example, valid dictionaries for storing the list of names

{Joel, Felix, Oskar, Amanda}

may be given by either of the following trees:

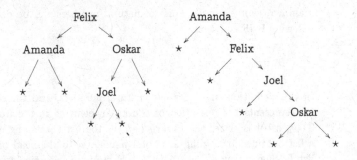

Exercise 8.7 (Solution on page 441)

Give an inductive definition for the the dictionary data structure outlined above. Note that the data structure would only define the syntactic structure; the fact that the names are stored in proper lexicographic order is a semantic issue which will not be reflected in the definition.

8.5 Inductively-Defined Functions

We can exploit the inductive definition of a set to provide convenient definitions for functions on that set. The function is defined by specifying its values on the basic elements of the set, and then specifying its values on the inductively-defined elements in terms of its previously-defined values.

For example, an infinite sequence

$$a_0, a_1, a_2, a_3, a_4, a_5, \ldots$$

is provided by a function whose domain is $\mathbb{N}$, and can often be defined by specifying the initial value a_0 and each subsequent value a_n in terms of the values a_k for $k < n$.

Example 8.8

The *factorial function* $n!$ is defined to be the product of the integers from 1 to n:

$$n! = 1 \times 2 \times 3 \times \cdots \times n.$$

More formally, it can be defined inductively as follows.

$$0! = 1; \quad and$$
$$n! = n \times (n-1)! \quad (for\ n > 0).$$

Thus, for example,

$$5! = 5 \times 4!$$
$$= 5 \times (4 \times 3!)$$
$$= 5 \times 4 \times (3 \times 2!)$$
$$= 5 \times 4 \times 3 \times (2 \times 1!)$$
$$= 5 \times 4 \times 3 \times 2 \times (1 \times 0!)$$
$$= 5 \times 4 \times 3 \times 2 \times 1 \times 1$$
$$= 120$$

Exercise 8.8   (Solution on page 442)

Compute the first few values of the sequence s_n defined inductively by:

$$s_0 = 0$$
$$s_n = s_{n-1} + 2n - 1$$

Can you recognise this sequence as a function of n?

Example 8.9

The *harmonic numbers* H_n are informally defined by

$$H_n = \tfrac{1}{1} + \tfrac{1}{2} + \tfrac{1}{3} + \cdots + \tfrac{1}{n}$$

and can be defined inductively as follows.

$$H_0 = 0; \quad and$$
$$H_n = H_{n-1} + \tfrac{1}{n} \quad (for \ n > 0).$$

Exercise 8.9   (Solution on page 442)

Compute the harmonic number H_6 from its inductive definition.

Example 8.10

The *Fibonacci numbers* are defined inductively as follows.

$$f_0 = 0;$$
$$f_1 = 1; \quad and$$
$$f_n = f_{n-1} + f_{n-2} \quad (for \ n > 1).$$

That is, each number in this sequence is obtained by adding together the previous two numbers in the sequence. The first few Fibonacci numbers are

$$0, 1, 1, 2, 3, 5, 8, 13, 21, 34, 55, 89, 144, 233, \ldots.$$

This sequence derives its name from the Italian mathematician Leonardo of Pisa, more commonly known by his nickname Fibonacci. Fibonacci was instrumental in spreading the use of the modern Hindu-Arabic numeral system to Europe, as an alternative to Roman numerals, through his book on arithmetic *Liber Abaci (The Book of Calculation)*, which was published in the early 13th century. The Fibonacci numbers appear in the solution of the following problem posed in this book.

Exercise 8.10   (Solution on page 442)

Suppose you have a pair of new-born rabbits at the start of month 1, and that each pair of rabbits produces a new pair of rabbits after 2 months and each month thereafter. How many pairs of rabbits will you have at the start of the nth month? (Work out the first few months and look for a pattern.)

It is worth looking more carefully at the above inductive definitions of sequences. As the natural numbers $\mathbb{N}$ are defined inductively in terms of zero 0 and the *successor function* $s(n) = n+1$, functions over them are naturally defined inductively. The above sequences are simple examples, but induction can be used to define more complicated functions than just sequences.

Example 8.11

By resorting to the inductive definition of the natural numbers

$$n \ ::= \ 0 \ | \ s(n).$$

as given on page 207, we can inductively define the function

$$\text{add} : \mathbb{N} \times \mathbb{N} \to \mathbb{N}$$

which adds two numbers as follows:

$$\text{add}(m, 0) = m; \quad and$$
$$\text{add}(m, s(n)) = s(\text{add}(m, n)).$$

The first clause merely states that $m+0 = m$; and the second, inductive, clause is the precise way of writing what we would more naturally write as:

$$\text{add}(m, n+1) = \text{add}(m, n) + 1.$$

Thus, for example,

$$add(3,2) = add(3,1) + 1$$
$$= add(3,0) + 1 + 1$$
$$= 3 + 1 + 1$$
$$= 5.$$

Exercise 8.11   (Solution on page 443)

Give an inductive definition of the function

$$mult : \mathbb{N} \times \mathbb{N} \to \mathbb{N}$$

which multiplies two numbers, in terms of zero and the successor function, as well as the function add defined above.

We can, of course, define functions inductively over any inductively-defined set. The inductive function definitions will naturally follow the structure of the inductive definitions of the domain.

Example 8.12

The length of a word $w \in A^*$ can be defined inductively as follows.

$$length(\varepsilon) = 0$$
$$length(aw) = 1 + length(w) \quad \textit{(for } a \in A\textit{)}.$$

The length of a list of natural numbers can be defined inductively as follows.

$$length([\,]) = 0$$
$$length(n : L) = 1 + length(L) \quad \textit{(for } n \in \mathbb{N}\textit{)}.$$

The height of a binary tree can be defined inductively as follows.

$$height(\star) = 0$$
$$height(N(t_1, t_2)) = 1 + \max\left(height(t_1), height(t_2)\right).$$

Exercise 8.12   (Solution on page 443)

Give an inductive definition of the function $sum(L)$ which computes the sum of a list L of numbers. Use it to verify that $sum([6,2,5]) = 13$.

Exercise 8.13   (Solution on page 443)

The *append* function $L_1 \mathbin{+\!\!+} L_2$ joins two lists L_1 and L_2 together. For example, $[1,2] \mathbin{+\!\!+} [3,5,7] = [1,2,3,5,7]$. Give an inductive definition of the append function.

Exercise 8.14   (Solution on page 443)

Referring to the inductive definition for formulæ of predicate logic given for Exercise 8.6 (page 208), give an inductive definition for a function which takes a formula of predicate logic and returns the set of variables which appear free in that formula.

8.6   Recursive Functions

In each of the functions defined in the previous section, the value of the function on a given argument is defined either directly, or in terms of its values on smaller arguments. In particular, for functions defined over $\mathbb{N}$ the value of the function on the argument 0 is defined directly, as there are no natural numbers $k < 0$.

Such inductively-defined functions are examples of *recursive functions*, which merely means that the value of a function applied to a given argument is expressed in terms of the value of that function applied to other – not necessarily smaller – arguments. Such definitions may not be well-founded, though. For example, it would not make sense to define a function by $f(n) = f(n+1) + 1$; in this case, we'd be forever lost trying to compute $f(0) = f(1) + 1 = f(2) + 2 = f(3) + 3 = \cdots$.

Example 8.14

McCarthy's 91-function $f : \mathbb{N} \to \mathbb{N}$ is defined as follows.

$$f(n) = \begin{cases} n - 10, & \text{if } n > 100; \\ f(f(n+11)), & \text{if } n \leq 100. \end{cases}$$

This function is recursively defined, but not inductively defined. Because of this, it is difficult even to see that this definition is well-founded – that is, that it even defines a value for each argument. In actual fact, $f(n) = 91$ for each $n \leq 100$, and $f(n) = n - 10$ for each $n > 100$.

Exercise 8.15   (Solution on page 443)

Prove that McCarthy's 91-function does indeed satisfy $f(n) = 91$ for each $n \leq 100$, and $f(n) = n - 10$ for each $n > 100$.

Example 8.15

Consider the following function $f : \mathbb{N} \to \mathbb{N}$.

$$f(n) = \begin{cases} 1, & \text{if } n \leq 1; \\ f(n/2), & \text{if } n > 1 \text{ even}; \\ f(3n+1), & \text{if } n > 1 \text{ odd}. \end{cases}$$

We can attempt to calculate the first few values of f:

$f(0) = 1$

$f(1) = 1$

$f(2) = f(1) = 1$

$f(3) = f(10) = f(5) = f(16) = f(8) = f(4) = f(2) = f(1) = 1$

$f(4) = f(2) = f(1) = 1$

$f(5) = f(16) = f(8) = f(4) = f(2) = f(1) = 1$

$f(6) = f(3) = f(10) = f(5) = f(16) = f(8) = f(4) = f(2) = f(1) = 1$

$f(7) = f(22) = f(11) = f(34) = f(17) = f(52)$

$\quad = f(26) = f(13) = f(40) = f(20) = f(10)$

$\quad = f(5) = f(16) = f(8) = f(4) = f(2) = f(1) = 1$

$f(8) = f(4) = f(2) = f(1) = 1$

We quickly realise that the value of the function must be 1 – if it has a value: the only value it could have on some input n is

$$f(n) = \cdots = f(1) = 1.$$

Indeed, this function seems to be well-defined: we don't seem to get into any cycles like

$$f(n) = \cdots = f(n);$$

and we always seem eventually to "bottom out" at $f(1) = 1$, although the route to this is rather chaotic: it took 6 unrollings of the function definition to compute $f(5)$, 9 unrollings to compute $f(6)$, and 17 unrollings to compute $f(7)$. It takes 11 unrollings to compute $f(26)$ (as can be seen in the calculation of $f(7)$ above, but it takes no fewer than 112 unrollings to compute $f(27)$, including computing $f(9232)$ along the way which itself requires only 35 unrollings.

It is unknown whether or not this function is in fact well defined, that is, that every sequence n, $f(n)$, $f^2(n)$, $f^3(n)$, ... eventually arrives at 1, although it has been confirmed for all numbers up to

$$n = 5.764 \times 10^{18} = 5,764,000,000,000,000,000.$$

The *Collatz conjecture* is the unproven claim that this sequence *does* converge to 1 regardless of the starting value n.

8.7   ## Recursive Procedures

As the data manipulated by computer programs is typically defined inductively, it should come as no surprise that programs typically manipulate this data recursively. That is, programs are written to run on some input data by recursively calling themselves to run on (generally smaller) input – unless the input data is so trivial that the program can immediately solve the problem at hand.

Example 8.16   Insertion Sort

Consider the problem of sorting a list of integers into increasing order. One method for doing this, called *Insertion Sort*, works as follows:

1. If the list only has only one element in it, then there is nothing to do: the list is clearly already sorted.

2. Otherwise, put the top card to one side and sort the remaining cards.

3. Insert the reserved card into the correct position in the sorted list.

This breaks the problem of sorting a list of numbers down to that of sorting a smaller list. But the trick is that this procedure is applied *recursively* in Step 2: the smaller list is itself sorted by the same procedure of putting one card to the side and (recursively) sorting the remaining cards – again with the above procedure – before inserting the reserved card into the resulting sorted list.

This procedure is based on the following function defined inductively over lists of numbers:

$$isort\,[\,] \quad = [\,]$$
$$isort\,(a:L) = (insert\,a)\,(isort\,L)$$

The definition of the auxiliary function $(insert\,a)$, which inserts the number a into a sorted list, is left as an exercise.

Exercise 8.16   (Solution on page 444)

Define the function $(insert\,a)$ inductively over (sorted) lists of numbers. Your definition should look as follows:

$$(insert\,a)\,[\,] \quad = \cdots$$
$$(insert\,a)\,(b:L) = \cdots (insert\,a)\,L \cdots$$

You can use the insertion sort procedure to sort a deck of 52 cards into some fixed order, say Ace through King, with all of the Clubs first, followed by the Diamonds, then Hearts, and finally the Spades. To sort the cards, you put the top one down onto a table and sort the remaining 51 cards; to do this, you put the top one down onto the table and sort the remaining 50 cards; continuing in this way, you will eventually find yourself with one card in your hand and 51 cards on the table, which you pick up one-by-one and insert into the correct place into the cards you are holding in your hand.

By doing this, the essence of recursion is hidden; the procedure could simply start with all 52 cards on the table, and picking them up and inserting them one-by-one into your hand, as many bridge players are accustomed to doing. The following example, however, gives a good example of the power of recursion in providing a sorting procedure which works much faster in practice than insertion sort.

Example 8.17)   Merge Sort

Another method for sorting a list of numbers, called *Merge Sort*, works as follows:

1. If the list only has only one element in it, then there is nothing to do: the list is clearly already sorted.

2. Otherwise, divide the list into two equal-sized lists (plus-or-minus one number, if the list consists of an odd number of integers).

3. Sort each of the two shorter lists.

4. Merge the two sorted lists together to produce the desired sorted list.

This breaks the problem of sorting a list of numbers down to that of sorting two smaller lists. But the trick is that this procedure is applied *recursively* in Step 2: the two half-sized lists are each sorted by the same procedure of dividing them into equal-sized lists and (recursively) sorting them – again with the above procedure – before merging them together.

This procedure can be elegantly demonstrated by having a group of people sort a deck of cards. Everyone in the group is to carry out the following procedure if they are handed a pile of cards:

1. If there is only one card in the pile that they are handed, then hand the pile right back to the person who gave it to you.

2. Otherwise, split the pile into two equal-sized piles and pass these smaller piles to two other people who are not holding any cards.

3. Take each of the two piles back when they are handed back to you. You will discover that – as if by magic – these two piles are each sorted.

4. Merge these two sorted piles into one sorted pile, and hand this sorted pile back to the person who gave it to you.

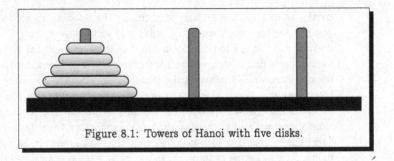

Figure 8.1: Towers of Hanoi with five disks.

Exercise 8.17   (Solution on page 444)

Figure 8.1 depicts the puzzle of the *Towers of Hanoi* in which we have three pegs and a number of discs of varying diameter; each disc has a hole in its centre allowing it to be positioned on the pegs. Starting with all of the discs on the first peg in increasing size with the largest on the bottom and the smallest on the top, the puzzle is to move all of the discs to a different peg by moving discs one at a time from peg to peg without ever placing any disc on top of a smaller disc.

Describe a recursive procedure for solving this puzzle. How many individual disc moves would your procedure take on the five-disc puzzle in Figure 8.1?

8.8 Additional Exercises

1. Consider the two quotes given at the start of this chapter and the next chapter. Only one of these properly underlies the principle of inductive definitions. Why is this? (Hint: Consider what the base case may be in each quote.)

2. Consider the set $X \subseteq \mathbb{N}$ defined as follows.

   (a) $0 \in X$.
   (b) if $n \in X$ then $(n+3) \in X$ and $(n+7) \in X$.
   (c) Nothing is in X unless its membership can be established from the above.

   Give three elements of $\mathbb{N}$ which are elements of X, and three elements of $\mathbb{N}$ which are not elements of X, explaining for each one why it is or is not an element.

   Can you give a complete description of the set X?

3. Describe the set P defined as the smallest set satisfying the following:

   (a) $\{\varepsilon, 0, 1\} \subseteq P$.
   (b) if $w \in P$ then $\{0w0, 1w1\} \subseteq P$.

   Give three elements of $\{0, 1\}^*$ which are elements of P, and three elements of $\{0, 1\}^*$ which are not elements of P, explaining for each one why it is or is not an element.

4. Give an inductive definition of the function $nodecount(t)$ which computes the number of internal nodes in the binary tree t, where the definition of a binary tree is as given in Example 8.7. Us this function to verify that

   $$nodecount\Big(N(N(\star, \star), N(N(\star, N(\star, \star)), \star))\Big) = 5.$$

5. Give a BNF equation for (a fragment of) your favourite programming language.

6. Given an inductive definition of the function $listnames(d)$ which takes a dictionary of names d, as defined by Exercise 8.7, and produces a list of names in alphabetic order (assuming the names are properly arranged alphabetically in the dictionary).

7. Give an inductive definition for a function which takes a formula of predicate logic and returns an equivalent formula in which negation symbols appear only applied to predicates.

8. Give an inductive definition of the function rev which takes a list and returns its reverse. Thus, for example, $rev([1, 2, 4]) = [4, 2, 1]$.

   Use your definition to compute $rev([1, 2, 4])$.

9. Male bees hatch from unfertilised eggs, and so have a mother but no father. Female bees hatch from fertilised eggs, and so have both a mother and a father. The family tree of a male bee can be seen in Figure 8.2 How many ancestors does a male bee have in the tenth generation back? How many of these ancestors are male?

10. Give an inductive definition of the function $msort$ upon which merge sort is based. You will want to define auxiliary functions $split$ which splits a list into two equal-size lists, and $merge$ which merges two sorted lists into one list.

11. Ackermann's Function is defined inductively as follows. For $n \geq 0$,

    $$A(0, n) = n + 1;$$

    and for $m, n \geq 1$,

    $$A(m, 0) = A(m-1, 1) \quad and$$
    $$A(m, n) = A(m-1, A(m, n-1)).$$

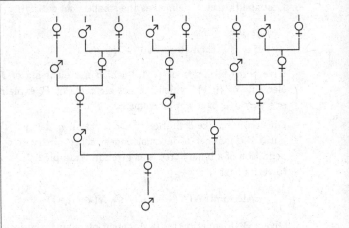

Figure 8.2: A family tree of male (♂) and female (♀) bees.

This is an extremely fast growing function. For example, that value of $A(4, 2)$ has $19,729$ decimal digits; and the value of $A(4, 3)$ is already well beyond astronomical.

(a) Work out the first few values of $A(1, n)$ to convince yourself that
$A(1, n) = n+2$.

(b) Work out the first few values of $A(2, n)$ to convince yourself that
$A(2, n) = 2n+3$.

(c) Work out the first few values of $A(3, n)$ to convince yourself that
$A(3, n) = 2^{n+3} - 3$.

(d) Work out the value of $A(4, 1)$.

Chapter 9

Proofs by Induction

In the middle of a cloudy thing is another cloudy thing, and within that another cloudy thing, inside which is yet another cloudy thing...

... and in that is yet another cloudy thing, inside which is something perfectly clear and definite.

- Ancient Sufi saying.

One of the most common forms of reasoning used within the subject of Computer Science is *inductive reasoning*. This is due to the fact, explored in the previous chapter, that Computer Science deals heavily with manipulating inductively-defined objects. Reasoning about such objects will naturally rely on exploiting the inductive nature of their definitions.

In Section 5.6 we explored the general technique for proving a property of the form $\forall x P(x)$, namely, to allow x to stand for an arbitrary value of the domain and to prove that $P(x)$ holds without making any assumptions about the value of x. Such a general approach is typically too weak to prove facts about natural numbers; we would like to be able to exploit the inductively-defined structure of natural numbers to arrive at our result. Such is the role of induction proofs.

9.1 Convincing but Inconclusive Evidence

Consider the following claim that the sum of the first n positive integers is $\frac{n(n+1)}{2}$:

Claim: For all $n \geq 0$, $1 + 2 + 3 + \cdots + n = \frac{n(n+1)}{2}$.

Note that the sum of the first zero natural numbers, which above is awkwardly written as $1 + 2 + 3 + \cdots + 0$, is naturally 0.

We can easily confirm this claim for various values of n:

F. Moller, G. Struth, *Modelling Computing Systems*,
Undergraduate Topics in Computer Science,
DOI 10.1007/978-1-84800-322-4_10, © Springer-Verlag London 2013

$$0 = \frac{0(1)}{2}, \quad \text{so the claim is true when } n = 0.$$

$$1 \qquad = 1 = \frac{1(2)}{2}, \quad \text{so the claim is true when } n = 1.$$

$$1 + 2 \qquad = 3 = \frac{2(3)}{2}, \quad \text{so the claim is true when } n = 2.$$

$$1 + 2 + 3 \qquad = 6 = \frac{3(4)}{2}, \quad \text{so the claim is true when } n = 3.$$

$$1 + 2 + 3 + 4 \qquad = 10 = \frac{4(5)}{2}, \quad \text{so the claim is true when } n = 4.$$

$$1 + 2 + 3 + 4 + 5 = 15 = \frac{5(6)}{2}, \quad \text{so the claim is true when } n = 5.$$

Each instance of the claim which we verify to be true seems to lend support to the validity of the claim. However, no (finite) amount of checking of individual cases can confirm the validity of the claim for all values of n.

Now consider each of the following claims.

- *Fermat's Last Theorem* claims that for no integer $n > 2$ does there exist a trio of positive integers x, y and z such that $x^n + y^n = z^n$. This claim went unproven for 350 years until Andrew Wiles' celebrated proof in the 1990s. By then, the conjecture was confirmed with the help of vast computer resources for all values of n up to 4 million. However, even if computers could have confirmed the truth of this conjecture for all values of n up to ten zillion, there would still be no reason why the conjecture should be true for ten zillion and one.

  Pierre de Fermat, after whom Fermat's Last Theorem is named, famously wrote the following about this Theorem in the margin of a textbook on arithmetic: *"Cuius rei demonstrationem mirabilem sane detexi. Hanc marginis exiguitas non caperet."* ("I have a truly marvellous proof of this proposition which this margin is too narrow to contain.") It is universally believed that whatever argument he may have had in mind could not have been valid. This is partly due to the fact that no proof was ever found amongst his papers, and partly due to the extreme complexity of the only known proof by Wiles – which can be understood in its entirety by only a small number of mathematicians worldwide. It also partly due to the fact that Fermat believed many things which ultimately turned out to be false, such as the next example.

- *Fermat numbers* are integers of the form $F_n = 2^{2^n} + 1$. They are so called on account of the fact that Pierre de Fermat wrote, in a letter to Marin Mersenne on 25 December 1640, that: *"If I can determine the basic reason why*

  $$3, \quad 5, \quad 17, \quad 257, \quad 65,537, \quad \ldots$$

  are prime numbers, I feel that I would find very interesting results." Based on the properties of the first few numbers of this form,

Fermat believed that they were all necessarily prime. Indeed the first few Fermat numbers listed by Fermat are prime:

$$F_0 = 2^{2^0} + 1 = 2^1 + 1 = 3$$
$$F_1 = 2^{2^1} + 1 = 2^2 + 1 = 5$$
$$F_2 = 2^{2^2} + 1 = 2^4 + 1 = 17$$
$$F_3 = 2^{2^3} + 1 = 2^8 + 1 = 257$$
$$F_4 = 2^{2^4} + 1 = 2^{16} + 1 = 65,537$$

Unfortunately for Fermat, his conjecture fails with the very next Fermat number:

$$F_5 = 2^{2^5} + 1 = 2^{32} + 1 = 4,294,967,297.$$

Fermat can be forgiven for not recognising this monstrosity to be a composite number. It was the great mathematician Leonhard Euler who first discovered in 1732 that this number can be factored as

$$641 \times 6,700,417.$$

Indeed, it is unknown whether *any* further Fermat numbers are prime (though it is known that a vast many are not).

- *Goldbach's conjecture*, which states that every even number greater than 2 can be expressed as the sum of two prime numbers, has been confirmed, again with the help of vast computer resources, for all even numbers up to 10^{18} (i.e., $1,000,000,000,000,000,000$). But as far as anyone knows, there might be a yet larger even number which is not the sum of two primes. It worked out well for Fermat's Last Theorem, but this gives no reason for hope, as demonstrated by the next two examples.

- In 1919, the Hungarian mathematician George Pólya conjectured that most (i.e., more than 50%) of the natural numbers less than any given number have an *odd* number of prime factors. For example, every prime number has an odd number of prime factors, namely one, as does $12 = 2 \times 2 \times 3$ (three prime factors), while $14 = 2 \times 7$ has an even number (two) of prime factors. By the mid 1950's empirical evidence for Pólya's conjecture seemed clear: the conjecture was verified for all numbers up to 800,000. However, contrary to this ever-growing evidence, Pólya's conjecture was disproved in 1958 when C. Brian Haselgrove showed that it had to be false for some value around 2×10^{361} (that is, a 2 followed by 361 zeros). It has since been shown to fail already for $n = 906,150,257$.

- Consider the following claim:

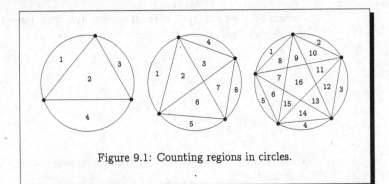

Figure 9.1: Counting regions in circles.

For all $n \geq 1$, $991n^2 + 1$ is *not* a perfect square;

that is, $\sqrt{991n^2+1}$ is not an integer.

We could confirm the validity of this claim for as many values of n as we have patience, but we could never conclude on the basis of the validity of a large number of cases that the claim is valid for all values of n. The claim is in fact false; however, the first value of n for which the claim fails is

$$n = 12,055,735,790,331,359,447,442,538,767.$$

We cannot be content with the mere experience of witnessing various instances of when a claim is true to lend reckless support to its universal truth. We cannot confidently lend any credence to Collatz's conjecture of Example 8.15 despite the comfort offered by the knowledge it holds for all values up to 2.22×10^{18}. Similarly, and more worrisome, a train may run perfectly for arbitrarily long – several years even – before a fault in its software control system contributes to a devastating crash.

Exercise 9.1   (Solution on page 444)

Some number of spots are placed randomly around the circumference of a circle, and every spot is connected to every other spot by a straight line. Assuming that no three lines intersect at a point inside the circle, we would like to know into how many regions is the circle divided?

For example, given 1, 2, 3, 4, or 5 spots, the circle is divided into 1, 2, 4, 8, or 16 regions, respectively; the final three of these are depicted in Figure 9.1.

How many regions are created by connecting six spots?

9.2 A Primary School Induction Argument

Suppose you wish to check that the formula

$$1 + 2 + 3 + \cdots + n = \frac{n(n+1)}{2}$$

is true for the first 30 values of n, and you ask a classroom of 30 ten-year-olds to check this formula, each child checking if it is true for some value of n. For example, the 17th child will check that

$$1 + 2 + 3 + \cdots + 17 = \frac{17 \times 18}{2}.$$

You watch each child working diligently on their individual problems and, as expected the first few children, working on confirming the formula for small values of n, are quick to report their success. Those working on larger values of n are taking longer. For example, it is taking a long while for the 17th child to add up the first 17 numbers to find they add up to 153, and then to compute $\frac{17 \times 18}{2} = 153$ to discover that the claim is true for $n=17$. Some children are reporting failure before checking their work and finding errors in their calculations before ultimately reporting success.

Alone in the crowd is the 28th child, a little girl who is sitting quietly reading a novel instead of working away on her calculations. You ask her if she is done, and she says yes. You ask her if the formula is true for $n=28$ and she says she doesn't know – yet. Confused, you look at her sheet of paper and see the following calculation:

$$1 + 2 + 3 + \cdots + 28 = \underbrace{1 + 2 + 3 + \cdots + 27}_{} + 28$$

$$= \frac{27 \times 28}{2} + 28$$

$$= 28 \times \left(\frac{27}{2} + 1 \right)$$

$$= 28 \times \left(\frac{29}{2} \right)$$

$$= \frac{28 \times 29}{2}$$

As you look over this calculation, the boy at the next desk announces that he has finished adding up the first 27 numbers and that they add up to $378 = \frac{27 \times 28}{2}$ as expected: the formula is true for $n=27$. The little girl immediately responds to this by announcing that the formula is true for $n=28$.

What this precocious little girl realised was that she could leave most of the hard work of adding up the first 28 numbers to her friend beside her, the little boy who is busily adding up the first 27 numbers. Once he has done that, all she needs to do is add 28 to his total. Knowing *what* the first

27 numbers are *supposed* to add up to, namely $\frac{27 \times 28}{2}$, she doesn't wait for him to do his job, but rather goes to work under the assumption that her friend will confirm this expectation. This is the calculation that she carried out.

Having carried out this calculation, can she say that the first 28 numbers add up to $\frac{28 \times 29}{2}$? Not right away, as she made the assumption that the first 27 numbers add up to $\frac{27 \times 28}{2}$; once her friend, the 27th child, confirms this assumption, she can (and does) announce boldly that the formula is true for $n=28$.

There is nothing special about the number 28, just something special about this little girl. If she had the problem of checking the formula for any other number, she would have done the same thing. She was no doubt quietly wondering why her friend beside her was busily adding up all the first 27 numbers; and indeed why her other friend on her other side was busily adding up the first 29 numbers.

Exercise 9.2 (Solution on page 445)

What calculation would this little girl do if she was the 27th child?

Exercise 9.3 (Solution on page 445)

When he was ten years old, the great mathematician Carl Friedrich Gauss was reportedly set the problem of adding up the first 100 numbers. His teacher's intention was to keep the class busy and quiet for some time, but Gauss solved the problem almost immediately. What clever trick did young Gauss employ?

9.3 The Induction Argument

Just as we can inductively define functions over inductively-defined domains, we can exploit the structure of an inductive definition to reason about the objects it defines. For example, *mathematical induction* allows you to prove that a property $P(n)$ of natural numbers $n \in \mathbb{N}$ holds for all natural numbers if:

1. (Base Case) it holds for the value 0, that is, $P(0)$;   and

2. (Induction Step) it holds for the value $k+1$ whenever it holds for k; that is,

$$P(k) \Rightarrow P(k+1).$$

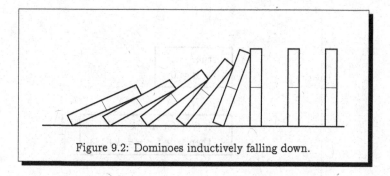

Figure 9.2: Dominoes inductively falling down.

$P(k)$ is referred to as the *inductive hypothesis*, from which we want to deduce $P(k+1)$.

Clause 2 can be equally expressed as follows

2′. it holds for the value $k > 0$ whenever it holds for $k-1$; that is,

$$P(k-1) \Rightarrow P(k).$$

The little girl discussed above did precisely this type of reasoning in showing that the property $P(n)$, which states that the first n numbers add up to $\frac{n(n+1)}{2}$, holds for the value 28 assuming it holds for the value 27.

As an analogy, imagine a (possibly infinite) string of dominoes standing side-by-side as in Figure 9.2. If we can prove that the first domino falls (i.e., gets pushed over), and that if one domino falls, the next domino will fall (i.e., gets pushed over by the preceding domino), then this is enough to conclude that *all* of the dominoes will fall over.

We can think of induction as a method of extending our knowledge of the truth: we establish the claim for the first relevant value (typically 0). Next we show that if the claim is true for some value k then it must also be true for the next value $k+1$. The important thing to note here is that k is not given a specific value although it might have some conditions imposed on it (in this case $k \geq 0$). Now since we know the claim to be true for 0, it must also be true for 1; but then it must also be true for 2; but then it must also be true for 3; and continuing in this fashion, we realise that the claim must be true for any value $n \in \mathbb{N}$. In this way we are viewing induction proofs as a form of bootstrapping argument, as depicted in Figure 9.3.

Alternatively, we can think of induction as a proof by contradiction: if the claim is false – that is, if the property does *not* hold for *all* values of $n \in \mathbb{N}$ – then it must fail for some *smallest* value $n \geq 0$; that is, the claim holds for all values less than n but not for n itself. The question then is: what can n be? It cannot be 0, as the base case established that the claim

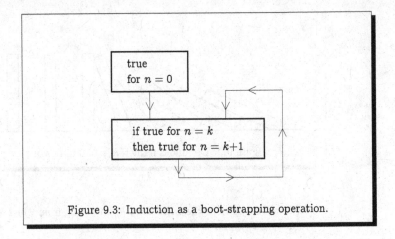

Figure 9.3: Induction as a boot-strapping operation.

holds for $n{=}0$. But then by the induction step, n cannot be 1 either; and hence not 2 either; and hence not 3 either; and hence not 4 either. We can carry on this reasoning indefinitely to show that n cannot be any value; for example, n cannot be $1,594$ since, being the smallest value for which the claim is false, the claim would be true for $1,593$, and thus by the induction step it must also be true for $1,594$. Continuing in this fashion, we realise our contradiction: the claim cannot actually fail for any value $n \in \mathbb{N}$.

Following this extensive discussion, we can finally offer the first formal proof by induction, as a model on which to base all other induction proofs.

Example 9.3

Fact: For all $n \geq 0$,

$$1 + 2 + 3 + \cdots + n \; = \; \frac{n(n+1)}{2}.$$

Proof: By induction on n.

Base Case: We note that

$$1 + 2 + 3 + \cdots + 0 \; = \; 0 \; = \; \frac{0(0+1)}{2}.$$

Induction Step: We assume that, for *some* k,

$$1 + 2 + 3 + \cdots + k \; = \; \frac{k(k+1)}{2},$$

and from this assumption (the inductive hypothesis) we prove that

$$1 + 2 + 3 + \cdots + k + (k+1) \; = \; \frac{(k+1)(k+2)}{2}.$$

That is, we demonstrate that if the statement of the theorem is true when $n = k$, then it must also be true when $n = k+1$.

By the inductive hypothesis we can rewrite the left-hand side of this equation that we want to prove true as

$$\frac{k(k+1)}{2} + (k+1).$$

We can then take out the common factor $(k+1)$ from these two terms, giving us

$$(k+1)\left(\frac{k}{2} + 1\right),$$

which is the same as

$$(k+1)\left(\frac{k+2}{2}\right),$$

or in other words,

$$\frac{(k+1)(k+2)}{2},$$

which is the right-hand side that we desire.

In other words, we carried out the following equational derivation:

$$1 + 2 + 3 + \cdots + k + (k+1)$$

$$= \frac{k(k+1)}{2} + (k+1) \quad \textit{(by the inductive hypothesis)}$$

$$= (k+1)\left(\frac{k}{2} + 1\right)$$

$$= (k+1)\left(\frac{k+2}{2}\right)$$

$$= \frac{(k+1)(k+2)}{2}.$$

At this point you should reflect on what the little girl in the Primary School problem from Section 9.2 did, and relate it to the induction step of the above argument. If her reasoning is clear, the following formulæ should be straightforward to verify.

Exercise 9.4 (Solution on page 446).

Show, by induction on n, that the following formulæ are true for all $n \geq 0$.

1. $1^2 + 2^2 + 3^2 + \cdots + n^2 = \dfrac{n(n+1)(2n+1)}{6}$.
2. $1 + 3 + 5 + \cdots + (2n-1) = n^2$.

3. $1 \cdot 2 + 2 \cdot 3 + 3 \cdot 4 + \cdots + n(n+1) = \dfrac{n(n+1)(n+2)}{3}$.

Exercise 9.5 (Solution on page 447)

Show, by induction on n, that for all $n \geq 0$:

$$F_0 \times F_1 \times \cdots \times F_n = F_{n+1} - 2$$

where $F_n = 2^{2^n} + 1$ are the Fermat numbers.

Induction is a very common technique for establishing mathematical formulæ such as the following.

Exercise 9.6 (Solution on page 448)

Show, by induction on n, that for any real number $r \neq 1$,

$$1 + r + r^2 + r^3 + \cdots + r^n = \frac{1 - r^{n+1}}{1 - r}$$

for all $n \geq 0$.

Note that if $-1 < r < 1$ then r^{n+1} approaches 0 as n approaches infinity; hence, as a corollary to the above, we can deduce that for any real r with $|r| < 1$,

$$1 + r + r^2 + r^3 + \cdots = \frac{1}{1 - r}.$$

So far we have used induction merely to prove simple formulæ. However, induction is more general than this, and the base case can be some value or values other than 0, as the next examples demonstrate.

Example 9.6

Fact: Any amount of postage of at least 8 pence can be made up from just 3-pence and 5-pence stamps.

Proof: By induction on n.

Base Case: A 3-pence stamp and a 5-pence stamp make up 8 pence.

Induction Step: Assume that we have a collection of such stamps adding up to a total of $n \geq 8$ pence.

- if there is a 5-pence stamp in this collection, remove it and replace it with two 3-pence stamps;
- If there are no 5-pence stamps, then there must be (at least) three 3-pence stamps in the collection; remove these and replace them with two 5-pence stamps.

In either case, we arrive at a collection of stamps adding up to $(n+1)$ pence. □

Example 9.7

Fact: The sum of the interior angles of a convex polygon with n sides is equal to $(n-2)180°$ for all $n \geq 3$. (A polygon is *convex* if every line joining two points of the polygon lies within the polygon.)

Proof: By induction on n.

Base Case: The sum of the interior angles of any triangle is 180°.

Induction Step: We assume that the theorem is true for some value $k \geq 3$: that the sum of the interior angles of any convex polygon with k sides is equal to $(k-2)180°$.

From this inductive hypothesis, we demonstrate that it must also be true for $k+1$: that the sum of the interior angles of any convex polygon with $k+1$ sides is equal to $(k-1)180°$.

Any $(k+1)$-gon can be decomposed into a triangle and a k-gon by connecting two non-adjacent vertices, as depicted in the diagram.

The sum of the interior angles of this $(k+1)$-gon is then the sum of the interior angles of the triangle, 180°, added to the sum of the interior angles of the k-gon which, by induction, is

$$180° + (k-2)180° = (k-1)180°.$$

□

Exercise 9.7   (Solution on page 449)

Suppose we draw n circles ($n \geq 1$) so that any two intersect at two points but no three intersect at any point. Prove, by induction on n, that these circles divide the plane into $n^2 - n + 2$ regions. Deduce from this that we cannot draw a Venn diagram for four or more sets with circles representing sets.

Induction is of immense importance in Computer Science where a great many of the objects under study are inductively defined. It is imperative that a Computer Scientist be comfortable with inductive reasoning in order to be successful with designing and understanding computing systems.

The following provides an example of reasoning inductively about a simple program.

Exercise 9.8   (Solution on page 449)

Consider the following piece of recursive program code:

```
function f(n)
    if n=0 then return 0
    else return f(n−1) + 2n − 1
```

This program code computes the following inductively-defined function:

$$f(n) = \begin{cases} 0, & \text{if } n=0 \\ f(n-1) + 2n - 1, & \text{if } n>0. \end{cases}$$

Show, by induction on n, that $f(n) = n^2$ for all $n \geq 0$.

9.4 Strong Induction

In a proof by induction we demonstrate that some property holds of some number based on the assumption that the property holds of the the previous number. Occasionally we may want to assume that the property holds of other smaller numbers, not just the previous number. An alternative form of induction which permits this is *strong induction* which allows you to prove that a property $P(n)$ of natural numbers holds of all natural numbers by demonstrating the following:

- $P(n)$ holds for n whenever it holds for all $k<n$; that is,

$$\left(\forall k<n : P(k)\right) \;\Rightarrow\; P(n).$$

You may well wonder at this point: what happened to the base case? In the case of $n=0$, the assumption that $P(k)$ holds for all values $k<n$ is vacuous, since there are no such values, and hence this one clause incorporates the base case of demonstrating that $P(0)$ holds under no assumption.

Example 9.8

Let

$$f(n) = \begin{cases} 0, & \text{if } n=0; \\ 2 \cdot f(n/2), & \text{if } n>0 \text{ even}; \\ f(n-1) + 1, & \text{if } n \text{ odd}. \end{cases}$$

Fact: $f(n) = n$ for every $n \geq 0$.

Proof: By (strong) induction on n, arguing by cases on the "structure" of n.

$\underline{n=0}$: $f(0) = 0$.

$\underline{n > 0 \text{ even}}$: $f(n) = 2 \cdot f(n/2)$

$$= 2 \cdot (n/2) = n.   \quad (By\ induction)$$

$\underline{n\ odd}$: $f(n) = f(n-1) + 1$

$$= (n-1) + 1 = n. \quad (By\ induction)$$

$\square$

Exercise 9.9   (Solution on page 450)

Prove, by strong induction, that every integer $n > 1$ is either prime or a product of primes.

This result, attributed first to Euclid over 2000 years ago, is referred to as the *Fundamental Theorem of Arithmetic*.

9.5   Induction Proofs from Inductive Definitions

We showed earlier how to define functions inductively, e.g., the Harmonic numbers H_n (Example 8.9) and the Fibonacci numbers (Example 8.10). Induction proofs are naturally used to reason about such inductively-defined functions, as evidenced by the following examples.

Example 9.9

Fact: For all $n \geq 0$,

$$H_1 + H_2 + H_3 + \cdots + H_n \;=\; (n+1)H_n \,-\, n.$$

Proof: By induction on n.

Base Case $(n = 0)$:

$$H_1 + H_2 + H_3 + \cdots + H_0 \;=\; 0 \;=\; (0+1)H_0 - 0.$$

Induction Step: $(n > 0)$:

$$H_1 + H_2 + H_3 + \cdots + H_n$$

$$= (H_1 + H_2 + H_3 + \cdots H_{n-1}) + H_n$$

$$= nH_{n-1} - (n-1) + H_n \qquad \textit{(by inductive hypothesis)}$$

$$= n(H_n - \tfrac{1}{n}) - (n-1) + H_n \qquad \textit{(since } H_n = H_{n-1} + \tfrac{1}{n}\textit{)}$$

$$= (n+1)H_n - n. \qquad\qquad \square$$

Exercise 9.10 (Solution on page 450)

Prove that for all $m \geq 1$ and all $n \geq m$, $\ H_n - H_m \geq \frac{n-m}{n}$.

Do this by assuming $m \geq 1$ and proving the result by induction on n.

Example 9.10

Fact: $f_0 + f_1 + f_2 + \cdots + f_n = f_{n+2} - 1$ for all $n \geq 0$.

Proof: By induction on n.

Base Case $(n = 0)$:

$$f_0 + f_1 + f_2 + \cdots + f_0 = f_0 = 0 = 1 - 1 = f_2 - 1.$$

Induction Step $(n > 0)$:

$$f_0 + f_1 + f_2 + \cdots + f_n + f_{n+1}$$

$$= (f_{n+2} - 1) + f_{n+1} \qquad \textit{(by the inductive hypothesis)}$$

$$= (f_{n+1} + f_{n+2}) - 1 = f_{n+3} - 1 \qquad\qquad \square$$

Exercise 9.11 (Solution on page 450)

Show, by induction on n, that

$$(f_0)^2 + (f_1)^2 + (f_2)^2 + \cdots (f_n)^2 = f_n f_{n+1}$$

for all $n \geq 0$.

We have seen that the base case may be some value n other than 0. There are also instances in which more than one base case is required. A simple example of this is provided by the following.

Example 9.11

Fact: For all $m \geq 2$ and for all $n \geq 1$, $\quad f_{n+m-2} = f_n f_{m-1} + f_{n-1} f_{m-2}$.

Proof: We assume that $m \geq 2$ is fixed, and we prove the result by induction on n.

Base Case $(n = 1)$: $f_{1+m-2} = f_{m-1} = f_1 f_{m-1} + f_0 f_{m-2}$.

Base Case $(n = 2)$: $f_{2+m-2} = f_m = f_{m-1} + f_{m-2} = f_2 f_{m-1} + f_1 f_{m-2}$.

Induction Step: $(n > 2)$:

$$f_{n+m-2} = f_{(n-1)+m-2} + f_{(n-2)+m-2}$$
$$= (f_{n-1} f_{m-1} + f_{n-2} f_{m-2}) + (f_{n-2} f_{m-1} + f_{n-3} f_{m-2})$$
$$\text{(by inductive hypothesis, twice)}$$
$$= (f_{n-1} + f_{n-2}) f_{m-1} + (f_{n-2} + f_{n-3}) f_{m-2}$$
$$= f_n f_{m-1} + f_{n-1} f_{m-2} \qquad\qquad \square$$

The above proof required two base cases, as the inductive hypothesis is invoked twice for the two values $n-1$ and $n-2$. If in the above proof we only do the base case for $n = 1$, and in the induction step we try to cater for all cases of $n > 1$ (in particular, $n = 2$), then the second invocation of the inductive hypothesis would be invalid in the particular instance where $n = 2$.

★ **9.6   Fun with Fibonacci Numbers**

In this section we explore three extended induction arguments involving Fibonacci numbers.

9.6.1   A Fibonacci Number Test

Suppose we are given an arbitrary positive integer x and asked whether or not it is a Fibonacci number. For example, how might we determine whether or not the number 517 is a Fibonacci number? The only apparent way is to use the inductive definition to compute successive Fibonacci numbers until we reach (or – more likely – exceed) 517. This is, however, not necessary; we can instead use the following simple test:

A positive integer x is a Fibonacci number if, and only if,

$5x^2 \pm 4$ is a perfect square.

For example, $x=3$ is a Fibonacci number, and $5 \cdot 3^2 + 4 = 49 = 7^2$; and $x=5$ is a Fibonacci number, and $5 \cdot 5^2 - 4 = 121 = 11^2$. However, $x=4$ is not a Fibonacci number, and neither $5 \cdot 4^2 - 4 = 76$ nor $5 \cdot 4^2 + 4 = 84$ is a perfect square.

For our less-modest example $x = 517$ above, a few calculator keystrokes tells us that $5 \cdot 517^2 - 4 = 1336441$, and pressing the square root button gives us 1156.0454, so $5x^2 - 4$ is clearly not a perfect square; and $5 \cdot 517^2 + 4 = 1336449$, and pressing the square root button gives us 1156.0489, so $5x^2 + 4$ is also not a perfect square. Therefore, this test tells us that $x = 517$ is not a Fibonacci number. On the other hand, testing the value $x=610$, a few calculator keystrokes tells us that $5 \cdot 610^2 - 4 = 1860496$, and pressing the square root button gives us 1364; in this case $5x^2 - 4$ is a perfect square, meaning that the value $x=610$ is a Fibonacci number (indeed $f_{15} = 610$).

The following two exercises provide the basis for the argument that this test is valid.

Exercise 9.12   (Solution on page 451)

Show, by induction on n, that for all $n \geq 0$ the pair $(x, y) = (f_n, f_{n+1})$ satisfies the equation

$$y^2 - xy - x^2 = \pm 1.$$

Exercise 9.13   (Solution on page 451)

Show, by induction on $x+y$, that if the pair (x, y) of positive integers satisfies the equation

$$y^2 - xy - x^2 = \pm 1$$

then $(x, y) = (f_n, f_{n+1})$ for some $n \geq 0$. (Hint: For the induction step, show that the "smaller" positive integer pair $(y-x, x)$ also provides a solution.)

Theorem 9.13   Fibonacci Test

A positive integer x is a Fibonacci number if, and only if, $5x^2 \pm 4$ is a perfect square.

Proof: We start by recalling the *quadratic formula* which states that the quadratic equation

$$ay^2 + by + c = 0$$

is solved by the following values of y:

$$y = \frac{-b \pm \sqrt{b^2 - 4ac}}{2a}.$$

In particular, for a given positive value of x, the quadratic equation

$$y^2 - xy - x^2 = \pm 1$$

is solved by the following positive value of y:

$$y = \frac{x + \sqrt{x^2 + 4(x^2 \pm 1)}}{2} = \frac{x + \sqrt{5x^2 \pm 4}}{2}.$$

By Exercise 9.12, if $x = f_n$, then the value of y given by this formula must be f_{n+1}, from which we can deduce that $5x^2 \pm 4$ must be a perfect square.

Conversely, if $5x^2 \pm 4$ is a perfect square for some positive integer x, then the value of y given by this formula, like x, must be a positive integer, in which case Exercise 9.13 tells us that x (as well as y) must be a Fibonacci number.   □

9.6.2   A Carrollean Paradox

The following result is known as Cassini's Identity.

Exercise 9.14   (Solution on page 452)

Show, by induction on n, that $f_{n+1}^2 - f_n f_{n+2} = (-1)^n$ for all $n \geq 0$.

Cassini's Identity forms the basis of a famous puzzle devised by Lewis Carroll. The puzzle is described in the following exercise.

Exercise 9.15   (Solution on page 452)

Take a square whose sides are 8 units long, cut it into four sections (two triangles and two quadrilaterals), and rearrange these four sections into a rectangle whose sides are 5 units and 13 units long as shown here:

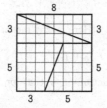
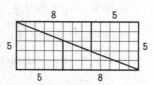

The area of the 8×8 square is 64 square units, but the area of the 5×13 rectangle is 65 square units! Where does the extra square unit come from?

This same phenomenon occurs with any square whose sides are of length taken from the Fibonacci numbers. For example consider the following 13×13 square cut up and rearranged into an 8×21 rectangle:

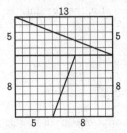 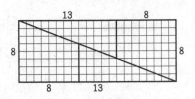

In this case, the area of the square is 169 square units, but the area of the rectangle is 168 square units, so this time we *lose* one square unit. Where did it go?

9.6.3   Fibonacci Decompositions

The Unique Prime Factorisation Theorem states that any positive integer n has a unique decomposition into the product of prime numbers. For example, the number $n=364$ decomposes uniquely into the product of primes as follows:

$$364 \ = \ 2 \cdot 2 \cdot 7 \cdot 13.$$

The proof of this theorem is carried out by induction on n, and can be found in any but the most basic algebra reference book. Here, we present a similar decomposition result which we shall find useful later.

Example 9.15   Zeckendorf's Theorem

Fact: Every integer $N \geq 0$ can be expressed uniquely as

$$N \ = \ f_{k_1} + f_{k_2} + f_{k_3} + \cdots + f_{k_n}$$

where $0 \ll k_1 \ll k_2 \ll k_3 \ll \cdots \ll k_n$.   (Here, $i \ll j$ means that $i \leq j-2$.)

For example, $100 \ = \ 3+8+89 \ = \ f_4 + f_6 + f_{11}$.

Proof: First we demonstrate, by induction on n, that for all $n \geq 1$,

$$f_{k_1} + f_{k_2} + f_{k_3} + \cdots + f_{k_n} \ < \ f_{k_n+1}$$

whenever $0 \ll k_1 \ll k_2 \ll k_3 \ll \cdots \ll k_n$.

Base Case $(n=1)$: $f_{k_1} < f_{k_1+1}$ as $k_1 \geq 2$ since $0 \ll k_1$.

Induction Step $(n>1)$:

$$f_{k_1} + f_{k_2} + f_{k_3} + \cdots + f_{k_{n-1}} + f_{k_n}$$

$$< f_{k_{n-1}+1} + f_{k_n} \qquad \text{(by inductive hypothesis)}$$

$$\leq f_{k_{n-1}} + f_{k_n} \qquad \text{(since } k_{n-1} \ll k_n, \text{ so } k_{n-1}+1 \leq k_n-1\text{)}$$

$$= f_{k_n+1} \qquad\qquad\qquad \text{(by definition)}$$

Thus if $N = f_{k_1} + f_{k_2} + f_{k_3} + \cdots + f_{k_n}$ where $0 \ll k_1 \ll k_2 \ll \cdots \ll k_n$ then we must have that $f_{k_n} \leq N < f_{k_n+1}$.

The main result then follows by induction on $N \geq 0$.

Base Case ($N = 0$): Trivially $0 = f_{k_1} + f_{k_2} + f_{k_3} + \cdots + f_{k_0}$.

Induction Step ($N > 0$): Let k be such that $f_k \leq N < f_{k+1}$. Then

$$(N - f_k) < f_{k+1} - f_k = f_{k-1}.$$

If N is to be represented as required, then by the above result, f_k must be one (indeed the largest) of the summands.

But then by the inductive hypothesis, $(N - f_k) \geq 0$ can be expressed uniquely as

$$(N - f_k) = f_{k_1} + f_{k_2} + f_{k_3} + \cdots + f_{k_n}$$

where $0 \ll k_1 \ll k_2 \ll k_3 \ll \cdots \ll k_n$.

Furthermore, since $f_{k_n} \leq (N-f_k) < f_{k-1}$, we must have that $k_n < k-1$, i.e. that $k_n \ll k$.

Taking $k_{n+1} = k$, we thus get that N is expressed uniquely in the required form as

$$N = f_{k_1} + f_{k_2} + f_{k_3} + \cdots + f_{k_n} + f_{k_{n+1}}. \qquad \square$$

9.7   When Inductions Go Wrong

We give here a few examples illustrating common mis-applications and mis-conceptions of induction.

Example 9.16

Let $T : \mathbb{Z} \to \mathbb{Z}$ be the function which is defined by:

$$T(n) = \begin{cases} n+6, & \text{if } n \leq 0; \\ T(T(n-7)), & \text{otherwise.} \end{cases}$$

We can show that $T(n) = 6$ for all $n \geq 0$. To do this, it is tempting to use induction on n as follows.

Base Case $(n = 0)$: $T(0) = 0 + 6 = 6$.

Induction Step $(n > 0)$: $T(n) = T(T(n-7))$

$$= T(6) \qquad \textit{(by inductive hypothesis)}$$

$$= 6 \qquad \textit{(by inductive hypothesis)}$$

There are two errors in the above argument. First of all if $n < 7$ then $n-7 < 0$ and the first inductive hypothesis cannot be applied. Secondly the claim that $T(6) = 6$ certainly doesn't follow from the inductive hypothesis unless $n > 6$. These observations show that to make the induction work we need to verify a *range* of base cases, namely, $T(n) = 6$ for $0 \leq n \leq 6$.

Although the claim is true in the above Example, the argument presented demonstrates how easy it is to make illegitimate arguments to back up a claim. On the other hand, the following exercise demonstrates a blatantly false claim to be true through a seemingly innocuous induction argument.

Example 9.17 Sorites Paradox

Consider the following "proof" that sandpiles do not exist.

Claim: For each $n \geq 0$, n grains of sand do not make a sandpile.

Proof: By induction on n.

Base Case $(n = 0)$:

   If there is no sand, then there can be no sandpile.

Induction Step: $(n > 0)$:

   Suppose we have $(n+1)$ grains of sand which constitute a sandpile. Clearly taking away a single grain of sand from a sandpile will still leave us with a sandpile. However, we will only have n grains of sand left, which by induction does not constitute a sandpile. Hence our $n+1$ grains of sand cannot constitute a sandpile. $\square$

This is known as the *sorites paradox* or the *heap paradox*. The name comes from the Greek word *soros* (σωρός) meaning "heap". It relies on the vagueness of words such as "heap" and "pile" and has many variations, each of which being a precise and accurate application of valid logical principles to arrive at a nonsensical conclusion.

- A man with only 1 hair is clearly bald.

- If a man with only 1 hair is bald,
  then a man with only 2 hairs is bald.

- If a man with only 2 hairs is bald,
  then a man with only 3 hairs is bald.

  ⋮

- If a man with only 9,999 hairs is bald,
  then a man with only 10,000 hairs is bald.

Each of these observations is precise and valid, yet chaining them all together allows us to conclude that a man with 10,000 hairs on his head is bald, a wholly nonsense claim.

In reasoning about systems, it is imperative that we use great care to employ only concepts that are as rigorously defined and precise as the logical means we use to analyse them.

The sorites paradox provides a playground for philosophers wanting to debate the validity of inductive arguments, but relies heavily on ill-defined terms removed from the rigour of mathematics. However, in the next exercise we provide a subtle error hidden in an otherwise air-tight inductive argument which leads to a clearly false conclusion. Can you uncover this error?

Exercise 9.17   (Solution on page 452)

What is wrong with the following "proof" that all people are the same age?

We show, by induction on n, that for every collection S of $n \geq 0$ people, all people in S are the same age.

Base Case $(n = 0)$: Trivially the claim holds when S consists of 0 people.

Base Case $(n = 1)$: Trivially the claim holds when S consists of 1 person.

Inductive Step $(n > 1)$: Assuming that the claim holds for all collections of size less than n, we show that it holds for any collection of size n. Let S be a collection of n people. Let S' and S'' be two overlapping collections of people which together make up S: $S = S' \cup S''$. By the inductive assumption, all people in S' are the same age, and all people in S'' are the same age. As S' and S'' overlap, all people in S must be the same age.

9.8   Examples of Induction in Computer Science

The following example is typical of the type of analysis which arises in the study of algorithms.

Example 9.18

Consider the following recursive algorithm $\text{MINMAX}(A, p, q)$ for calculating (x, y) where x and y are, respectively, the minimum and maximum values appearing in the array $A[1 \cdots n]$ between the indices p and q, inclusively (the intention is to initially call the algorithm with $\text{MINMAX}(A, 1, n)$).

```
MINMAX(A, p, q)
    1   if p = q then return (A[p], A[p])
    2   else if p = q−1 then
    3       if A[p] < A[q] then return (A[p], A[q])
    4       else return (A[q], A[p])
    5   else
    6       (minL, maxL) := MINMAX(A, p, p+1)
    7       (minR, maxR) := MINMAX(A, p+2, q)
    8       return ( min(minL, minR), max(maxL, maxR) )
```

We are interested in calculating the number of comparisons which this algorithm makes, as an indication of how long it takes to execute (a comparison is made in line 3, and two are made in line 8 through the use of the functions min and max). A simple analysis gives us that the number $T(n)$ of comparisons made by a call to $\text{MINMAX}(A, p, q)$ with $n = q−p+1$ is as follows:

1. if $n = 1$, that is, if $p = q$, then the algorithm terminates on line 1 without making any comparisons. Thus $T(0) = 0$.

2. if $n = 2$, that is, if $p = q−1$, then the algorithm terminates on line 3 after making one comparison. Thus $T(2) = 1$.

3. if $n > 2$ then the algorithm makes

   (a) $T(2)$ comparisons on line 6;   followed by
   (b) $T(n−2)$ comparisons on line 7;   followed by
   (c) 2 comparisons on line 8

   before terminating. Thus $T(n) = T(2) + T(n−2) + 2$ for all $n > 2$.

The inductive definition of $T(n)$ is thus summarised as follows.

$$T(1) = 0$$
$$T(2) = 1$$
$$T(n) = T(2) + T(n−2) + 2 \qquad \text{(for } n > 2\text{)}$$

Fact $T(n) = \left\lceil \frac{3n}{2} \right\rceil - 2$ (where $\lceil x \rceil$ is x rounded up to the nearest integer.)

Proof: By induction on n.

Base Case $(n \leq 2)$:   Clearly the result is true when $n{=}1$ or $n{=}2$.

Induction Step $(n > 2)$:   Suppose the result is true for all values $k \leq n$ for some $n \geq 2$. In particular,

$$T(n{-}2) \;=\; \left\lceil \frac{3(n-2)}{2} \right\rceil - 2 \;=\; \left\lceil \frac{3n}{2} \right\rceil - 5.$$

Thus

$$\begin{aligned}
T(n) &= T(2) + T(n{-}2) + 2 \\
&= 1 + \left(\left\lceil \tfrac{3n}{2} \right\rceil - 5 \right) + 2 \quad \text{(by inductive hypothesis)} \\
&= \left\lceil \tfrac{3n}{2} \right\rceil - 2. \qquad\qquad \square
\end{aligned}$$

The next two examples describe the technique of *structural induction*, which is arguably the most important variant of induction within computing.

Example 9.19

Let A be an alphabet containing (at least) two distinct characters a and b.

Fact $aw \neq wb$ for all words $w \in A^*$.

Proof: By induction on $length(w)$.

Base Case $(length(w) = 0)$:   In this case, we must have that $w = \varepsilon$, so

$$aw \;=\; a \neq b \;=\; wb.$$

Induction Step $(length(w) > 0)$:   We consider two subcases, depending on whether w begins with the character a or with some other character c.

$\underline{w = au}$:   Since $length(u) = length(w){-}1 < length(w)$, $au \neq ub$ by the inductive hypothesis. Hence

$$aw \;=\; aau \neq aub \;=\; wb.$$

$\underline{w = cu \text{ (where } c \neq a)}$: $aw = acu \neq cub = wb.$ $\square$

The above is an example of a proof based on *structural induction*: the inductive hypothesis assumes that the claim holds for all smaller structures (in this case, for all shorter words), and uses this assumption to establish that the claim holds for the structure in question. For this reason, such a proof is typically referred to as a proof by induction on the structure of words, and would more naturally be presented as follows.

Proof: By induction on the structure of words (that is, we prove the result for a word w under the inductive hypothesis that it is true for all smaller words), arguing by cases on the structure of w (that is, we consider in turn three possible forms of w, namely ε, au and cu where $c \neq a$).

$\underline{w = \varepsilon}$: $aw = a \neq b = wb$.

$\underline{w = au}$:   By induction (since u is smaller than w), $au \neq ub$, so

$$aw = aau \neq aub = wb.$$

$\underline{w = cu \ \ (\text{where } c \neq a)}$: $aw = acu \neq cub = wb$.

We give one further example, without the excessive explanations.

Example 9.20

Fact: Every binary tree t has exactly one more leaf than internal node.

Proof: By induction on the structure of t, arguing by cases on the structure of t.

$\underline{t = \star}$:   The tree $\star$ has 1 leaf and 0 internal nodes.

$\underline{t = N(t_1, t_2)}$:   By induction, t_i (for $i = 1, 2$) must have n_i nodes and $n_i + 1$ leaves, for some n_1, n_2. But then $N(t_1, t_2)$ must have $n_1 + n_2 + 1$ nodes and $(n_1 + 1) + (n_2 + 1) = (n_1 + n_2 + 1) + 1$ leaves.   □

Exercise 9.20   (Solution on page 453)

Prove by induction that $length(L_1 + \!\!+ L_2) = length(L_1) + length(L_2)$ for all lists L_1 and L_2, using the inductive definition of the length of a list from Example 8.12, and your inductive definition of the append function from Exercise 8.13.

9.9   Additional Exercises

1. Prove the following hold for all $n \geq 0$, by induction on n.

(a) $1^3 + 2^3 + 3^3 + \cdots + n^3 = (1+2+3+\cdots+n)^2$.
(This is known as *Nicomachus's Theorem*.)

(b) $1^2 + 3^2 + 5^2 + \cdots + (2n-1)^2 = \dfrac{n(2n-1)(2n+1)}{3}$.

(c) $1^3 + 3^3 + 5^3 + \cdots + (2n-1)^3 = n^2(2n^2 - 1)$.

(d) $1\cdot2\cdot3 + 2\cdot3\cdot4 + 3\cdot4\cdot5 + \cdots$
$$\cdots + n(n+1)(n+2) = \frac{n(n+1)(n+2)(n+3)}{4}.$$

(e) $1(1!) + 2(2!) + 3(3!) + \cdots + n(n!) = (n+1)! - 1$.

(f) $\dfrac{1}{4(1^2) - 1} + \dfrac{1}{4(2^2) - 1} + \dfrac{1}{4(3^2) - 1} + \cdots + \dfrac{1}{4(n^2) - 1} = \dfrac{n}{2n+1}$.

2. Show that every $n > 0$ can be expressed uniquely as

$$c_1(1!) + c_2(2!) + c_3(3!) + \cdots + c_n(n!)$$

where $0 \le c_j \le j$.

3. Prove, by induction on $(m+n)$, that for all $m, n \ge 0$:

$$1\cdot2\cdot3\cdots\cdots m \ + \ 2\cdot3\cdot4\cdots\cdots(m+1) \ + \ 3\cdot4\cdot5\cdots\cdots(m+2)$$
$$+ \ \cdots \ + \ n(n+1)(n+2)\cdots(n+m-1)$$
$$= \ \frac{n(n+1)(n+2)(n+3)\cdots(n+m)}{m+1}.$$

4. Prove, by induction on n, that a finite set with n elements has 2^n subsets.

5. Define the sequence $\langle g_0, g_1, g_2, \ldots \rangle$ as follows: $g_0 = 0$; $g_1 = 1$; $g_2 = 1$; and for all $n \ge 2$,

$$g_{2n-1} = g_{n-1}^2 + g_n^2 \qquad \text{and} \qquad g_{2n} = g_{n+1}^2 - g_{n-1}^2.$$

Thus for example:

$$n=2: \qquad g_3 = g_1^2 + g_2^2 \qquad g_4 = g_3^2 - g_1^2$$
$$n=3: \qquad g_5 = g_2^2 + g_3^2 \qquad g_6 = g_4^2 - g_2^2$$
$$n=4: \qquad g_7 = g_3^2 + g_4^2 \qquad g_8 = g_5^2 - g_3^2$$

Show, by induction on n, that $g_n = f_n$ for all $n \ge 0$.

6. Define the sequence $\langle x_0, x_1, x_2, \ldots \rangle$ as follows:

$$x_0 = 0; \qquad x_{n+1} = \frac{1}{1 + x_n} \qquad (n \ge 0).$$

Show, by induction on n, that $x_n = \dfrac{f_n}{f_{n+1}}$ for all $n \ge 0$.

7. Provide a correct proof for the claim made in Example 9.16.

8. Suppose that in a particular country, every road is one-way, and every pair of cities is connected by exactly one direct road. Show, by induction on the number n of cities, that there exists a city which can be reached from every other city either directly or via only one other city.

9. Imagine drawing n straight lines in the plane (extending to infinity in both directions). The resulting configuration is to be coloured like a map, with no two bordering "countries" having the same colour (but two countries which meet at a single point may have the same colour).

Show, by induction on n, that only two colours are needed.

(Hint: Suppose you have such a coloured plane with n lines, and you draw a new line; clearly the colouring condition fails nowhere except across this new 'border'. How can you restore the colouring condition without altering the colours on one side of this border?)

10. A collection of n circles drawn in the plane divide the plane into parts. Show that you can colour the parts with two colours so that no two parts with a common boundary line are coloured the same way.

(Hint: Similar to the previous exercise.)

11. You are given a $2^n \times 2^n$ checkerboard with one black square arbitrarily placed on the board and the remaining $4^n - 1$ squares white. You are also given a supply of tiles which look like 2×2 checkerboards with one corner square removed. You want to tile the checkerboard so that each white square is covered exactly once, while the black square remains uncovered.

Show, by induction on n, that the $2^n \times 2^n$ checkerboard can be so tiled, for all $n \geq 0$.

(Hint: For the inductive step, place the first tile in the centre of the board with the gap in the quadrant containing the black square, and look at the four $2^{n-1} \times 2^{n-1}$ quadrants.)

12. You are given a checkerboard in the shape of an equilateral triangle with sides of length 2^n made up of smaller equilateral triangles with sides of unit length. The topmost equilateral triangle is black but all others are white. You are also given a supply of tiles in the form of bucket-shaped trapeziums made from three small equilateral triangles.

You want to tile the large triangular-shaped checkerboard so that each white triangle is covered exactly once, while the black triangle remains uncovered.

Show, by induction on n, that the whole checkerboard can be so tiled, for all $n \geq 0$.

13. There are n identical cars on a circular track. Among all of them, they have just enough petrol for one car to complete a lap. Show that there is a car which can complete a lap by collecting petrol from the other cars on its way around.

   (Hint: For the induction step, first argue that there is a car A which can reach the next car B. Then consider removing B from the track, emptying its petrol into A.)

14. I put two cards on a table and tell two people that the cards have different positive integers written on their undersides. I tell them to take one card each at random and secretly look at the number written on their card. They are then put in a room with a clock which rings a bell every minute. They are not allowed to communicate in any way, but are instructed to wait in the room until one of them knows which card has the lower number and which has the higher number, and then to announce this fact the next time the clock rings.

   There seems to be no escape for these two people, as there seems to be no way for either of them to discover who has the larger number. Imagine being one of the two, sitting with a card with the number 26 written on it; how could you possibly determine whether the card held by the other person has a number which is smaller than this or greater than this? Paradoxically, it is doable.

   Prove, by induction on $n \geq 1$, that if n is the lower of the two numbers written on the two cards, then the person who has this card will announce that he has the card with the lower number after the bell rings n times.

15. What is wrong with the following "proof" that

$$1 + 2 + 3 + \cdots + n = \frac{(n-1)(n+2)}{2}.$$

   Proof: By induction on n.

   $1 + 2 + 3 + \cdots + n$

   $= (1 + 2 + 3 + \cdots + (n-1)) + n$

   $= \frac{(n-2)(n+1)}{2} + n \qquad$ *(by inductive hypothesis)*

   $= \frac{(n-1)(n+2)}{2}.$ $\qquad\qquad\qquad$ $\square$

16. What is wrong with the following "proof" that every natural number

is interesting[†].

Proof: By induction on n.

Base Case $(n = 0)$:   0 is interesting as it is the smallest natural
   number.

Induction Step: $(n > 0)$:

   Suppose every number less than n is interesting. If n itself is interesting for some reason, then we are done. On the other hand,
   if there is nothing interesting about n, then it is in fact the first
   natural number which is not interesting, which makes it an interesting number indeed!   □

17. What is wrong with the following "proof" that $x=2x$ for all real num-
    bers $x \geq 0$?

    Proof: By induction on x.

    Base Case $(x = 0)$: $x = 0 = 2 \cdot 0 = 2x$.

    Induction Step: $(x > 0)$:

       Suppose $y = 2y$ for every positive real number y less than x.
       In particular, since $\frac{x}{2} < x$, $\frac{x}{2} = 2(\frac{x}{4}) = x$.
       But then $x = 2(\frac{x}{2}) = 2x$ (by induction).   □

18. Despite seeming more powerful, the principle of *strong* induction fol-
    lows from *ordinary* induction, and hence provides added convenience
    but not added power. This can be demonstrated as follows.

    Suppose, for a property $P(n)$ of natural numbers, the premise of strong
    induction holds:

    $$\forall n \Big((\forall k < n\, P(k)) \Rightarrow P(n) \Big)$$

    That is, $P(n)$ holds of a particular value n whenever it holds for *all*
    smaller values. We will show, by *ordinary* induction, that $\forall n\, P(n)$ is
    true. Let $Q(n)$ be the property $\forall k < n\, P(k)$.

    (a) Show that $\forall n\, P(n) \Leftrightarrow \forall n\, Q(n)$ *without* using induction.
    (b) Show that $\forall n\, Q(n)$ by *ordinary* induction. Thus, by part (a),
        $\forall n\, P(n)$.

[†]This proof is clearly wrong, as *no* number is interesting. Proof: Suppose some numbers
are interesting; then there must be a *smallest* interesting number n. So what, who cares?

Chapter 10

Games and Strategies

You have to learn the rules of the game. And then you have to play better than anyone else.

- Albert Einstein.

Games-of-chance derive this title from the fact that *luck* plays a part in deciding the winner of a play of the game. Sometimes the game consists solely of luck, as with COIN-FLIPPING (*"heads wins"*) or CARD-CUTTING (*"highest card wins"*). Typically, though, this isn't the case, and a sensible *strategy* is needed to beat a good player who isn't burdened by a string of extraordinarily bad deals of the cards (in the case of, e.g., POKER or BRIDGE), or throws of the dice (in the case of, e.g., BACKGAMMON or MONOPOLY). However, casinos operate (very successfully!) on the premise that (most of) their clientele do not play with luck on their side.

What we might call *games-of-no-chance* are those games for which the winner is decided based solely on *ability*. Examples of such games are CHESS and GO. They involve no decisions taken on the results of random events such as the deal of cards or the throw of dice, and no information is hidden from the players (apart from what moves the other player will choose to make during the play of the game).

For example, in the children's paper-and-pencil game NOUGHTS AND CROSSES (also known as TIC-TAC-TOE), two players alternately place crosses (×) and noughts (o) in nine square spaces arranged in a 3 × 3 grid. The goal of the first player is to align three crosses in a line (row, column or diagonal), and the goal of the second player is to align three noughts in a line (row, column or diagonal). A player wins the game if they achieve their goal before the other player does so. A game that ends with a full grid without a line of crosses or noughts is a draw.

When children first learn to play this game, the outcomes will be variable; sometimes the first player wins, sometimes the second player wins, and sometimes the game ends in a draw. However, every child eventually becomes bored with this game, as they discover that they can only win if their opponent makes a silly error. This is regardless of whether they are

F. Moller, G. Struth, *Modelling Computing Systems*,
Undergraduate Topics in Computer Science,
DOI 10.1007/978-1-84800-322-4_11, © Springer-Verlag London 2013

playing as first player or second player, though it seems that the first player should have a distinct advantage.

10.1 Strategies for Games-of-No-Chance

In this chapter we shall be interested in such two player games-of-no-chance. We shall typically refer to the first player as A (for Alice) for whom we shall use female pronouns (she, her), and the second player as B (for Bob) for whom we shall use male pronouns (he, his). Furthermore, these will be games of *perfect information*, meaning that both players will be aware of all aspects of the game: at every point in the game, both players know what moves have been made up to that point in time, as well as what moves their opponent can make in response to any move that they themselves make. The game of PAPER-SCISSORS-ROCK, for example, is not a game of perfect information, as neither player has information regarding the move being made by the other player. While there is no element of chance in the players' decision making, as each player is free to choose whatever move they wish, the lack of information about the opponent's move makes luck a factor in this game.

Another typical feature of the games that we shall consider is finiteness. A *finite game* is one that is guaranteed to terminate within a finite number of steps. This isn't true of many games, for example CHESS (unless some rule is introduced which declares a game to be a draw if it continues indefinitely, the standard rule being that a draw is declared if 50 consecutive moves have been made without a piece being captured nor a pawn being moved). If a play of a particular game may continue indefinitely, we will rule infinite plays to be predetermined in some way; that is, either the first player wins every infinite play, or the second player wins every infinite play, or the game is declared to be a draw. For example, we may declare that every infinite play of the game of CHESS is ruled to be a draw.

We shall at times consider games in which the first player is in the role of an attacker; she makes attacking moves which the second player, in the role of a defender, must defend against with his responses. We may refer to such games as *attacker-defender games*. The first player's aim is to achieve some goal (which will end the game), while the second player's aim is to prevent her from doing this. The important aspect of these games is that a play which continues forever is a positive result for the second player; that is, every infinite play of an attacker-defender game is ruled to be a win for the second player.

A *strategy* for a player in a game is a rule which tells that player what move to make each time it is their turn to move. A strategy which guarantees that you will win the game regardless of what moves your opponent makes is referred to as a *winning strategy*. If a game may end in a draw, then a strategy which guarantees that your opponent will not win (without guaranteeing that you will win) is referred to as a *drawing strategy*.

A position in a game is a *winning position* if the player whose turn it is has a winning strategy from this position; it is a *losing position* if the other player (whose turn it is not) has a winning strategy from this position; and finally, it is a drawing position if neither player has a winning strategy from this position. Clearly, from a winning position there must be a move to a losing position, while every move from a losing position must lead to a winning position. From a drawing position there must be some move to a drawing position, perhaps some moves to winning positions, but no moves to losing positions.

For a given game it is not possible for both players to have a winning strategy, though it is possible that neither player has one. For example, we noted above that neither player has a winning strategy in NOUGHTS AND CROSSES; the game can be played out through the maximum nine moves filling in all nine squares in the grid without either player winning, regardless of how cleverly they play. The *first* player does not have a winning strategy because:

1. no matter what the first player does
2.   there is something that the second player can do such that
3.   no matter what the first player does
4.   there is something that the second player can do such that
5.   no matter what the first player does
6.   there is something that the second player can do such that
7.   no matter what the first player does
8.   there is something that the second player can do such that
9.   no matter what the first player does
                  she will not have formed a line of crosses.

Similarly, the *second* player does not have a winning strategy because:

1. there is something that the first player can do such that
2.   no matter what the second player does
3.   there is something that the first player can do such that
4.   no matter what the second player does
5.   there is something that the first player can do such that
6.   no matter what the second player does
7.   there is something that the first player can do such that

8.   no matter what the second player does
            he will not have formed a line of noughts.

The simplicity of this game makes it easy to analyse; a play consists of (at most) nine moves, and the game is very symmetric. Thus a drawing strategy for both players is easy to discover, which ultimately renders the game uninteresting to play.

All such games are boring in this sense. At most one of the two players has a winning strategy; and by following this strategy, they ensure that the other player cannot do anything to avoid losing. If draws are possible, then both players may have strategies which prevent the other from winning. This important fact is recorded by the following theorem.

Theorem 10.1

In any two-player game-of-no-chance of perfect information, either one of the two players has a winning strategy, or they both have drawing strategies.

Proof: Clearly, if one of the two players has a winning strategy, then the other player cannot have a winning strategy: Fixing the strategies of the two players, only one of the two players can win the game, so only one of these two strategies can be a winning strategy.

Assume, then, that neither player has a winning strategy. That the first player does not have a winning strategy means that the second player may respond to each move made by the first player in such a way that either:

- the game ends in a draw or as a win for the second player;   or
- The game continues forever, and infinite games are either ruled to be draws or ruled to be wins for the second player.

That is to say, the second player has a strategy for ensuring that the first player does not win. Equally, that the second player does not have a winning strategy means that the first player has a strategy for ensuring that the second player does not win. Each of these strategies, therefore, must be a drawing strategy for its associated player.   □

Corollary 10.1

If a game cannot end in a draw, then one of the two players has a winning strategy.

Example 10.1

Consider the following game: starting with a pile of 10 coins, two players take turns removing either 2 coins or 3 coins from the pile. The player who takes the last coin wins; if one coin remains, then the game is a draw.

We can systematically analyse this game from the end backwards as follows:

(1) If there is 1 coin left, then the game is a draw. This is thus a *drawing* position.

(2) If there are 2 coins left, then you can win the game by taking both coins. This is thus a *winning* position.

(3) If there are 3 coins left, then you can win the game by taking all three coins. This is thus a *winning* position.

(4) If there are 4 coins left, then you can either:

    – take 2 coins and leave 2, thus leaving the other player in what we know from (2) above is a winning position; or

    – take 3 coins and leave 1, thus leaving the other player in what we know from (1) above is a drawing position.

Clearly the latter option is the correct one to make, and this is thus a *drawing* position.

(5) If there are 5 coins left, then you can either:

    – take 2 coins and leave 3, thus leaving the other player in what know from (3) above is a winning position; or

    – take 3 coins and leave 2, again leaving the other player in what we know from (2) above is a winning position.

Whatever you do will leave the other player in a winning position. This is thus a *losing* position.

(6) If there are 6 coins left, then you can either:

    – take 2 coins and leave 4, thus leaving the other player in what we know from (4) above is a drawing position; or

    – take 3 coins and leave 3, thus leaving the other player in what we know from (3) above is a winning position.

Clearly the first option is the correct one to make, and this is thus a *drawing* position.

(7) If there are 7 coins left, then you can take 2 coins and leave 5, which we know from (5) above is a losing position. This is thus a *winning* position.

(8) If there are 8 coins left, then you can take 3 coins and leave 5, which we know from (5) above is a losing position. This is thus a *winning* position.

(9) If there are 9 coins left, then you can either:

- take 2 coins and leave 7, thus leaving the other player in what we know from (7) above is a winning position; or
- take 3 coins and leave 6, thus leaving the other player in what we know from (6) above is a drawing position.

Clearly the latter option is the correct one to make, and this is thus a *drawing* position.

(10) If there are 10 coins left (that is, if you are at the start of the game), then you can either:

- take 2 coins and leave 8, thus leaving the other player in what we know from (8) above is a winning position; or
- take 3 coins and leave 7, again leaving the other player in what we know from (7) above is a winning position.

Whatever you do will leave the other player in a winning position. This is thus a *losing* position.

We can summarise this analysis concisely in the following table:

0	1	2	3	4	5	6	7	8	9	10
L	*D*	*W*	*W*	*D*	*L*	*D*	*W*	*W*	*D*	*L*
−	−	2	3	3	−	2	2	3	3	−

The top row indicates the running total; the middle row indicates whether the current player is in a winning position (*W*), or a losing position (*L*), or in a drawing position (*D*); and the bottom row indicates how many coins (2 or 3) the player should remove from the pile in that turn (there being no entry in the cases where the player is in a losing position).

Figure 10.1 depicts this game as a so-called *game tree* for this game. The nodes of this tree represent positions in the game (labelled by the number of coins remaining in the pile), and the arrows represent the possible moves which a player can make in the given position (labelled by the number of coins removed by that move). The winning positions are depicted by circled nodes, while the losing positions are depicted by boxed nodes; the nodes which are neither circled nor boxed depict drawing positions. The important observations to make are:

1. every winning position has at least one move leading to a losing position, (that is, every circled node has an arrow leading to a boxed node, emphasised in the figure by a double arrow);

2. every move from a losing position leads to a winning position (that is, every arrow from a boxed node leads to a circled node); and

3. every drawing position has a move to another drawing position; possibly a move to a winning position; but no move to a losing position

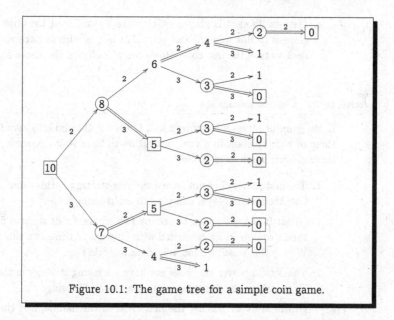

Figure 10.1: The game tree for a simple coin game.

(that is, every undecorated node has an arrow to another undecorated node, emphasised in the figure by a double arrow; possibly an arrow to a circled node; but no arrow to a boxed node).

These three observations respectively define what it means for a position to be a winning position, a losing position, or a drawing position.

Exercise 10.1   (Solution on page 453)

In the game of TAKE-3, there is a single pile of coins, and two players alternately remove either 1, 2, or 3 coins from the pile. The player who takes the last coin wins.

1. For each number n from 1 to 10, explain who has the winning strategy in TAKE-3 starting from a pile of n coins. In the cases in which the first player has the winning strategy, state how many coins (1, 2 or 3) the first player should take.

2. Generalise the above by explaining who has the winning strategy in TAKE-3 starting from a pile of n coins for an arbitrary n.

3. Generalise the above further by explaining who has the winning strategy in TAKE-k starting from a pile of n coins for an arbitrary n, but where players may alternately remove between 1 and k coins (above, we had $k=3$).

4. Misère Take-3 is played exactly like Take-3, but the object of the game is to *not* take the last coin; that is, you wish to force your opponent to take the last coin. How does this change the above analysis?

Exercise 10.2   (Solution on page 454)

In the game of Misère Noughts and Crosses, the aim is to *avoid* placing three of your symbols in a row, but rather to force your opponent to place three of their symbols in a row.

1. The first player does not have a winning strategy in this game. Explain how the second player can play to avoid losing.

   (Hint: It is a good idea to occupy two adjacent side squares first, and then a square which is aligned with only one of these two side squares. Why is this possible, and why does it work?)

2. The second player also does not have a winning strategy in this game. Explain how the first player can play to avoid losing.

   (Hint: Start by placing the first cross in the middle, and then "mirroring" every move of the second player by placing each subsequent cross directly opposite to where the second player places his noughts. Why is this a good idea?)

Exercise 10.3   (Solution on page 454)

The game of Clock-2-3 is played on a board which looks like the face of a 12-hour clock such as depicted as follows:

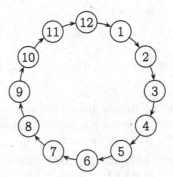

A token is placed on one of the hours (1 through 12) and the players take turns moving the token either 2 or 3 hours forward (i.e., in a clockwise fashion). The player who moves the token onto the 12 o'clock slot wins the game.

Explain who has the winning strategy in CLOCK-2-3 starting from each of the 12 hours. In the cases in which the first player has the winning strategy, state how many hours forward (2 or 3) the first player should move the token. (As a start, the first player clearly has a winning strategy starting from either 9 o'clock or 10 o'clock, by moving 3 and 2 hours, respectively, to land on 12 o'clock. Thus the second player has the winning strategy starting from 7 o'clock, as the first player will be forced to move the token to either 9 o-clock or 10 o'clock.)

Exercise 10.4   (Solution on page 456)

The following depicts a simple variant of the children's board game SNAKES AND LADDERS.

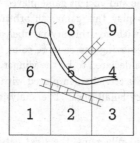

In this game, a single shared counter is started on square 1, and two players take turns moving the counter either *one* or *two* spaces forward (with the player moving deciding whether to move one or two spaces). If the counter lands at the foot of a ladder, it climbs to the top of the ladder; and if the counter lands on the head of a snake, it slides down to the tail of the snake. The object of this game is to be the one to move the counter to the final square number 9.

Identify which of the positions are winning positions; which are losing positions; and which are drawing positions. (Recall that a winning position is one from which there is a move to a losing position, whereas a losing position is one from which every move leads to a winning position; all other positions are drawing positions, as from these you cannot force a win nor be forced to lose.) As a start, 9 is a losing position in both games, while 8 is a winning position in both games, as you can win by moving one space forward. For the non-losing positions, indicate the optimal move(s).

The CHESS-playing computer Deep Blue attributes a large part of its success in its ability to search for a winning strategy in a manner similar to the above analysis. The salvation for such games comes from the fact that there are astronomically-many configurations to consider, far too many for a modern (and indeed any conceivable) computer to analyse. Today's Kasparovs are safe in the fact that CHESS-playing computers such as Deep Blue must still invoke questionable decision-making procedures, but perhaps one day a Very Deep Blue will render CHESS-playing, like playing NOUGHTS AND CROSSES, a pointless activity.

In the rest of this chapter we consider several moderately-simple two-player games-of-no-chance, and try to understand the strategies which a player should use in order to win them.

(10.2) Nim

NIM is a simple and ancient game played with coins, thought to be Chinese in origin. To play this game, an arbitrary number of piles of coins are formed, each with an arbitrary number of coins in them, and two players alternate in removing one or more coins from any one pile. Whoever takes the last coin is declared to be the winner.

This game is trivial when played with only one or two piles of coins, or with three very small piles. The analysis of the game in these cases is as follows.

1. In the one-pile game, the *first* person has a trivial winning strategy: take all the coins in the first move.

2. In the two-pile game,

    (a) if the piles contain an equal number of coins then the *second* player has a winning strategy: always take the same number of coins as the first player, repeatedly leaving the first player with equal-sized piles.

    (b) if the piles contain an unequal number of coins, then the *first* player has a winning strategy: start by taking coins from the larger pile to leave equal-sized piles, and then use the strategy described in 2(a) for the second player.

3. In the three-pile game,

    (a) if two of the piles are equal, then the *first* player has a winning strategy: take all of the coins in the third pile, leaving just the two equal-sized piles (and one empty pile), and then use the strategy described in 2(a) for the second player.

    (b) if the piles contain one, two, and three coins, respectively, then the *second* player has a winning strategy:

    i. if the first person takes the whole of one of the piles, then there will be just two unequal piles left (and one empty pile), and the second player can win using the strategy described in 2(b) for the first player.

    ii. if the first player takes just part of one of the piles, then there will be three non-empty piles remaining, two of which must be equal, and the second player can win using the strategy described in 3(a) for the first player.

The game is traditionally played with three piles, containing three, four, and five coins, respectively, and even here its complexity starts to become convincing; after playing many times, it remains difficult to glean any good long-term strategy. The only approach to the game which comes immediately to mind, reminiscent of games like CHESS, is to look ahead several moves, anticipating the moves of the other player, in order to avoid bad positions. With time, it is possible to recognise more and more bad positions, and become better at avoiding these. However, the character of the game changes with four, five, or more piles.

In fact there is a straightforward winning strategy for this game, either for the first player or the second player, depending on the number of piles and the number of coins in each pile. To see this, write out the numbers of coins in the piles in binary notation, one above the other, and add up the columns modulo 2; that is, compute the sum of a column to be 0 if it has even parity (i.e., there are an even number of 1's in the column), and 1 if it has odd parity (i.e., there are an odd number of 1's in the column). If all columns have an even parity, we shall say that the position is *balanced*; otherwise we say that the position is *unbalanced*. The following observations can be made:

1. If every column has even parity (i.e., we are in a balanced position), then *every* move will result in some column having odd parity (i.e., every move leads to an unbalanced position).

2. If one or more columns have odd parity (i.e., we are in an unbalanced position), then *some* move will result in every column having even parity (i.e., some move leads to a balanced position).

For example, in the 3-4-5-7 game, the first and third columns have odd parity (while the second column has even parity). By taking 3 coins from the second pile, we give the first and third columns even parity (while leaving the parity of the second column even).

$$
\begin{array}{ll}
3: 0\,1\,1 & 3: 0\,1\,1 \\
4: 1\,0\,0 \Longrightarrow & 1: 0\,0\,1 \\
5: 1\,0\,1 & 5: 1\,0\,1 \\
7: 1\,1\,1 & 7: 1\,1\,1 \\
\hline
1\,0\,1 & 0\,0\,0 \\
\uparrow\ \uparrow &
\end{array}
$$

From this new position, whatever coins are removed, there will result at least one column with odd parity.

With the above two observations, along with the insight that the ultimate goal of the game is to make all columns add up to zero, and hence an even number, it is clear that:

1. the *first* player has a winning strategy if one or more of the columns
   has odd parity: the correct move is to remove coins so as to leave all
   columns with even parity;

2. the *second* player has a winning strategy if all of the columns have
   even parity: regardless of what move is made, the resulting parity of
   at least one column will be odd.

Exercise 10.5   (Solution on page 456)

If a player is in a winning position in NIM, then there will in general be more
than one winning move. (Two moves are different if they involve different
piles, or if they involve the same pile but removing different numbers of
coins.) What is the maximum number of different winning moves possible
from a NIM position with n piles? Justify your answer.

Exercise 10.6   (Solution on page 456)

Suppose we change the rules of NIM slightly so that the first player, instead
of removing some coins from a pile, has the additional option of *creating a
new pile* of any size (with at least one coin in it); the first player may do
this at most once during a play of the game. Under which circumstances can
the first player force a win with the help of this extra move? (Consider, in
particular, the two situations in which the game starts from an unbalanced,
respectively a balanced, position.) Justify your answer.

★ **10.3**   **Fibonacci Nim**

The next game we consider is a variation on NIM called FIBONACCI NIM.
In this game we have a single pile containing $n \geq 2$ coins. The first player
removes one or more coins but not the whole pile. From then on, the players
alternate moves, each person removing one or more coins, but not more than
twice as many coins as the other player has taken in the preceding move.
The player who removes the last coin wins.

The analysis of this game is complicated by the fact that a player's
available moves depend on the opponent's last move. However, we can
nonetheless easily analyse small instances of this game:

2 coins: the first player must take 1 coin, leaving the second player to take
the last coin. Hence in this case, the *second* player has a (trivial)
winning strategy.

3 coins: The first player must take 1 or 2 coins; in either case the second player can take all remaining coins. Hence in this case, the *second* player again has a (trivial) winning strategy.

4 coins: The first player can take 1 coin, leaving the second player to take either 1 or 2 of the remaining 3 coins; the second player can thus not avoid losing as described in the 3-coin game for the first player. Hence in this case, the *first* player has a winning strategy.

5 coins: If the first player takes more than 1 coin, then the second player will be able to take all remaining coins; thus, in order to win the first player must take only 1 coin, leaving 4 coins. But then the second player can win by using the strategy described in the 4-coin case for the first player. Hence in this case, the *second* player has a winning strategy.

6 coins: The first player can take 1 coin, leaving 5 coins to the second player. The second player can then not avoid losing as described in the 5-coin case for the first player. Hence in this case, the *first* player has a winning strategy.

7 coins: The first player can take 2 coins, leaving 5 coins to the second player. The second player can then not avoid losing as described in the 5-coin case for the first player. Hence in this case, the *first* player has a winning strategy.

8 coins: If the first player takes more than 2 coins, then the second player will be able to take all remaining coins; thus, in order to win the first player must take only 1 or 2 coins, leaving either 7 or 6 coins. The second player can then win by using the strategy described in the 7-coin or 6-coin case for the first player. Hence in this case, the *second* player has a winning strategy.

9 coins: The first player can take 1 coin, leaving 8 coins to the second player. The second player can then not avoid losing as described in the 8-coin case for the first player. Hence in this case, the *first* player has a winning strategy.

We can exhaustively work out winning strategies this way, but the reasoning is indeed exhausting. It would be a major effort, for example, to work out if we have a winning strategy as the first player starting with 100 coins, and if so how many coins we should take. There is, however, a straightforward way to work out who has the winning strategy, and what the winning move is if one exists. To determine this, we first recall the following.

Theorem 10.6   The Fibonacci Number System

Every integer $N \geq 0$ can be expressed uniquely as a sum of Fibonacci numbers

$$N = f_{k_1} + f_{k_2} + f_{k_3} + \cdots + f_{k_n}$$

where $0 \ll k_1 \ll k_2 \ll k_3 \ll \cdots \ll k_n$.   (Here, $i \ll j$ means that $i \leq j-2$.)

For example, $100 = 3 + 8 + 89 = f_4 + f_6 + f_{11}$.

Proof: This is Zeckendorf's Theorem which we proved in Example 9.15.

$\square$

Theorem 10.7

The first player has a winning strategy in FIBONACCI NIM starting with n coins if, and only if, n is not a Fibonacci number. In this case, the winning strategy, when n coins remain, is always to take f_{k_1} coins, where $n = f_{k_1} + f_{k_2} + \cdots + f_{k_r}$ (with $0 \ll k_1 \ll k_2 \ll \cdots \ll k_r$) is the representation of n in the Fibonacci number system.

For example, in the game starting with $100 = f_4 + f_6 + f_{11}$ coins, the winning opening move is to take $f_4 = 3$ coins.

Exercise 10.7   (Solution on page 457)

Prove Theorem 10.7.

10.4 Chomp

In the game of CHOMP, we have an $m \times n$ chocolate bar, in which the leftmost-topmost square $(1, 1)$ is poisonous. Two players take turns taking bites out of the chocolate bar, with each player having to choose a remaining square and eat it along with all remaining squares below and to the right. The goal is to force the other player to eat the poisonous square.

As before, we can easily analyse small instances of this game.

1. In the 1×1 case, the first player loses right away; hence the *second* player has a trivial winning strategy.

2. In the $1 \times n$ case with $n > 1$ (or, similarly, the $m \times 1$ case with $m > 1$), the *first* player has a trivial winning strategy: bite off all but the poisonous square, leaving just the poisonous square for the second player to take.

3. In the $2 \times n$ (or $m \times 2$) case, the *first* player has a simple winning strategy: bite off one square, leaving a $2 \times n$ rectangle with the bottom-right square missing.

   (a) if the remaining chocolate is a $2 \times k$ rectangle with the bottom-right square missing, then *every* move will result in a shape *different* from this.

   (b) if the remaining chocolate is *not* in the shape of a $2 \times k$ rectangle with only the bottom-right square missing, then *some* move will result in a shape of this form.

   With this observation, along with the insight that the ultimate goal of the game is to leave just the poisonous square which has the shape of a 2×1 rectangle with the bottom-right square missing, it is clear that the first person has a winning strategy.

4. In the $n \times n$ (square) case, the *first* player again has a simple winning strategy: bite off the $(n-1) \times (n-1)$ sub-square, leaving just the top row and left column. From here, just mimic every move of the second player, biting off as many squares from the row (respectively, column) as the second player bites off the column (respectively, row).

Apart from these special cases, very little is known about winning strategies in this game. The only way to find the winning strategy is to explore moves, and responses to moves, and responses to responses to moves, etc. For example consider the 3×4 game. In this case, it is not a good idea to take just one square (as was the strategy in case 3 above), nor to take all but the first row and first column (as was the strategy in case 4 above). However, the first player does have a winning strategy, which starts by biting off a 2×2 square. This leaves the second player with 7 moves to choose from; whatever move the second person takes, though, will be bad, as can be seen in Figure 10.2.

Despite the difficulty of this game, we can easily prove the following remarkable fact.

Theorem 10.8

Except for the degenerate 1×1 case, the first player always has a winning strategy.

Proof: Suppose, for the sake of argument, that the *second* player has a winning strategy. This means, in particular, that whatever move the first person opens the game with, the second person has a response which will leave the chocolate in a configuration from which the first person cannot win.

Consider the response that the second person makes using this winning strategy if the first person opens by biting off just a single square. Whatever

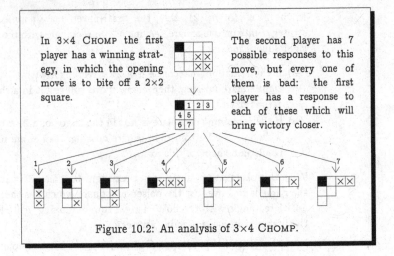

In 3×4 CHOMP the first player has a winning strategy, in which the opening move is to bite off a 2×2 square.

The second player has 7 possible responses to this move, but every one of them is bad: the first player has a response to each of these which will bring victory closer.

Figure 10.2: An analysis of 3×4 CHOMP.

this response, it is a move which the first player could equally have opened the game with, thus leaving the second player to play from the losing configuration.

This contradicts the assumption that the second player has the winning strategy.   □

This is indeed an interesting state of affairs. In this game, we know that the first player has a winning strategy, but apart from exhaustively analysing all possible plays of the game, there is no way of knowing how to win as the first player.

10.5 Hex

The game of HEX is played on a board consisting of an $n \times n$ grid of hexagons, as shown in Figure 10.3. At the beginning of the game, the first player is considered to own the territories to the North-East and South-West of the board (the two sides labelled with crosses × in the figure), while the second player is considered to own the territories to the North-West and South-East of the board (the two sides labelled with noughts ○ in the figure). The object of the game for each player is to create a path through the board joining their disconnected territories. The players alternate moves; the first player places a cross × in a vacant hexagon, and the second player follows on by placing a nought ○ in a vacant hexagon. The winner of the game is the first player to connect their two sides of the board with a contiguous chain of hexagons labelled with their symbol.

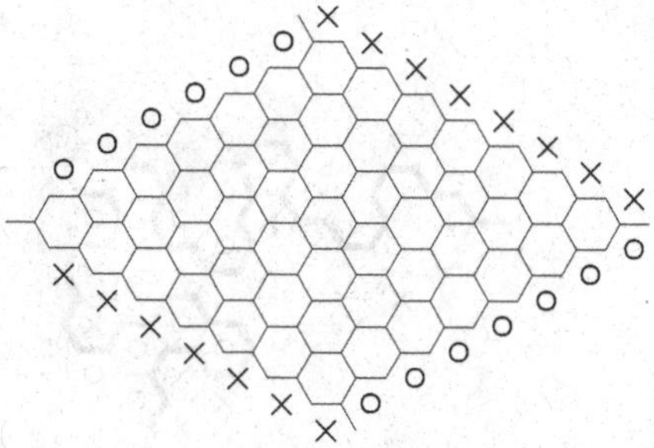

Figure 10.3: The Hex board.

Theorem 10.9

The game of Hex can never end in a draw.

Proof: An informal argument runs as follows. Think of the crosses as land and the circles as water; when all the hexagons are labelled, either there is an isthmus connecting the two continents. or else water flows between the two oceans, In the first case, the first player has a winning chain of ×-labelled hexagons, while in the latter case the second player has a winning chain of o-labelled hexagons.

A formal proof would require a fair amount of explanation; here we provide only an outline. Assuming that every hexagon is labelled with a cross × or a nought o, we show that one of the opposite pairs of sides is connected in a winning fashion. To see this, we imagine tracing a path along the boundaries of the hexagons, entering the grid at the left-most corner. At each junction we look at the territory we are facing; if it is labelled × then we turn left, and if it is labelled o then we turn right. If we do this, then we shall trace a path which always has × on its right and o on its left, and the path will exit the grid either at the top or at the bottom. In the first case, the ×-hexagons to the right of the path include a winning path for the first player, and in the second case, the o-hexagons to the left of the path include a winning path for the second player. Figure 10.4 gives an example in which the second player has a winning chain.   □

Knowing that this game can never ends in a draw, we can then prove the following.

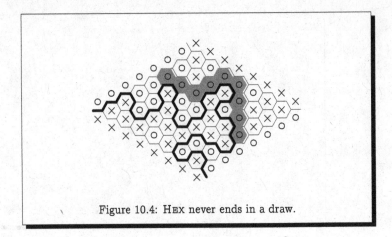

Figure 10.4: HEX never ends in a draw.

Theorem 10.10

The first player always has a winning strategy for HEX.

Proof: Suppose for the purpose of argument that the second player has a winning strategy. Then the first player may play as follows.

1. She may label *any* hexagon chosen at random, and then forget that she has done this.

2. She may then pretend from this point on that she is playing the game as the *second* player, using the (supposed) winning strategy for the second player.

3. If at any time this strategy dictates that she should label the pre-labelled hexagon, then she should simply label any other unlabelled hexagon at random, pretend that *it* isn't labelled, and pretend that her response was to label the pre-labelled hexagon.

In this fashion, the first player is *stealing* the winning strategy from the second player, and using it to win the game. This proves that if the second player has a winning strategy, then the first player has a winning strategy, which of course is a contradiction.   □

Again, as with CHOMP, we are able to prove that the first player has the winning strategy in HEX, but our proof gives no indication as to what that strategy might be!

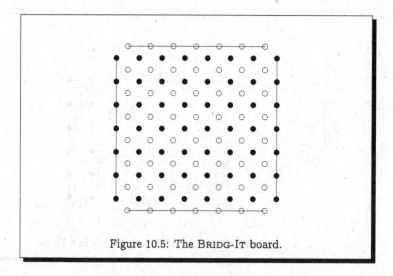

Figure 10.5: The BRIDG-IT board.

10.6 Bridg-It

The game of BRIDG-IT is similar to HEX, but is played on a staggered $n \times n$ board as depicted in Figure 10.5. The goal of the first player is to link the left- and right-hand borders, while the goal of the second player is to link the top and bottom borders. The two players alternate moves; the first player joins two neighbouring spots • either horizontally or vertically, and the second player joins two neighbouring circles ○ either horizontally or vertically. Neither player can cross a link previously made by the other player.

Theorem 10.11

The game of BRIDG-IT *can never end in a draw.*

Proof: Assuming that no further moves can be made, the board will depict a simple maze pattern, such as that given in Figure 10.6. Entering the maze from the bottom left, there is a unique path through the maze, which always has the first player's •-links on its left and the second player's ○-links on its right. This path must exit the maze at either the bottom right or the top left. (The path cannot exit the maze at the top right, as then it would end with ○-links on its left and •-links on its right.) In the first case, the •-links to the left of the path contain a winning path for the first player, and in the second case, the ○-links to the right of the path contain a winning path for the second player. □

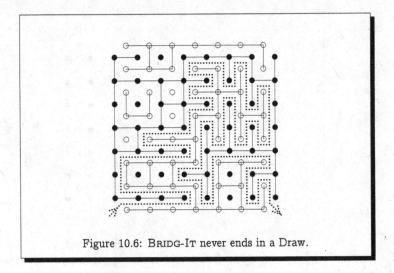

Figure 10.6: BRIDG-IT never ends in a Draw.

Knowing that this game can never ends in a draw, we can then prove the following.

Theorem 10.12

The first player always has a winning strategy for BRIDG-IT.

Proof: The reasoning is identical to that used in the proof of the analogous result for HEX.   □

Yet again this proves that the first player has a winning strategy without giving any indication as to what that strategy might be. However, in this case, we can explicitly describe a winning strategy for the first player. Referring to Figure 10.7, the first player should open with the link indicated. From that point onwards, each link that the second player makes will touch the end of one of the dotted lines depicted in Figure 10.7; in response, the first player should add the link which touches the other end of this dotted line. In this way, the first player will successfully block any attempt by the second player to create a path linking the top and bottom borders, and hence she will herself eventually win.

Exercise 10.12   (Solution on page 459)

Argue that the above does indeed describe a winning strategy for the first player.

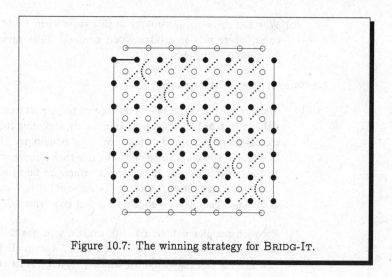

Figure 10.7: The winning strategy for BRIDG-IT.

10.7 Additional Exercises

1. Consider the following game:

   > Starting with 5 coins, two players take turns taking 1 or 2 coins; and whoever ends up with an *odd* number of coins wins.

   (a) Draw the complete game tree for this game, and determine who has the winning strategy.

   (b) Who has the winning strategy in this game when started with n coins, where n is an arbitrary odd number? (Hint: for each $n = 1, 2, 3, \ldots$, determine whether or not you have a winning move if it is your turn and there are n coins left. and you are currently holding an even number of coins; in parallel to this, determine whether or not there is a winning move if it is your turn and there are n coins left. and you are currently holding an odd number of coins. Look for a pattern.)

2. Consider the following game:

   > Starting with 5 coins, each player takes turns taking 1, 2 or 3 coins; and whoever ends up with an *odd* number of coins wins.

   (a) Draw the complete game tree for this game, and determine who has the winning strategy.

    (b) Who has the winning strategy in this game when started with n coins, where n is an arbitrary odd number? (The same hint as for question 1(b) applies.)

3. Consider the following game:

    Starting with a single pile of coins, two players alternate taking either 1 coin or half of the remaining coins, including the leftover coin if there is an odd number of coins remaining. Thus, for example, if there are 25 coins in the pile then a move consists of taking either 1 coin or 13 coins; if 13 coins are taken leaving 12 in the pile, then the next move will consist of taking either 1 coin or 6 coins. The player who takes the last coin wins.

    (a) For each number n from 1 to 10, explain who has the winning strategy in this game starting from a pile of n coins. In the cases in which the first player has the winning strategy, state how many coins the first player should take.

    (b) Argue that the first player has a winning strategy in the game starting with n coins if, and only if, the binary representation of n ends in an even number of 0's. Specifically,

        • if the binary representation of n ends in an even number of 0's, then either $n=1$ and you can win by taking the single coin, or there is a move which leaves a number of coins whose binary representation ends in an odd number of 0's; and

        • if the binary representation of n ends in an odd number of 0's, then every move leaves a number of coins whose binary representation ends in an even number of 0's.

4. Consider the following game:

    Starting with a pile of n coins, two players alternately remove a number of coins which is a power of 2. That is, a player may take 1 coin, or 2 coins, or 4 coins, or 8 coins, or 16 coins, or 2^k coins for any k. The player who takes the last coin wins.

Argue that the second player has a winning strategy if, and only if, n is a multiple of 3.

5. Consider the following game:

    Starting with a pile of n coins, two players alternately remove either 1 or 3 or 8 coins. The player who takes the last coin wins.

Argue that the second player has a winning strategy if, and only if, n is of the form $11k$ or $11k+2$ or $11k+4$ or $11k+6$.

6. (a) The game of CLOCK-1-3 is identical to the game of CLOCK-2-3 from Exercise 10.3 (page 258) except in this game the token moves either 1 or 3 hours forward.

   Work out who has the winning strategy in the game of CLOCK-1-3 starting from each of the 12 hours. In the cases in which the first player has the winning strategy, state how many hours forward (1 or 3) the first player should move the token.

   (b) The game of CLOCK-1-4 is identical to the game of CLOCK-2-3 from Exercise 10.3 (page 258) except in this game the token moves either 1 or 4 hours forward.

   Work out who has the winning strategy in the game of CLOCK-1-4 starting from each of the 12 hours. In the cases in which the first player has the winning strategy, state how many hours forward (1 or 4) the first player should move the token.

7. (a) The game of DAYS-OF-THE-YEAR is played by two players who take turns naming a date of the year starting from January 1st. On any move a player may increase the month or the day but not both. Thus, for example, the first player can start the game by naming any day in January (apart from the 1st), or the 1st of any month of the year (apart from January). The player who names December 31st wins.

   Work out who has the winning strategy for this game. For a start, you can note that there is a winning move from any date in December (apart from the 31st), as well as from the 31st of any month (apart from December).

   (b) The game of DAYS-OF-THE-CENTURY is played by two players who take turns naming a date in the 20th century starting from 1st January 1900. On any move a player may increase the month or the day or the year, but only one of these three. The player who names 31st December 1999 wins.

   Work out who has the winning strategy for this game.

8. In MISÈRE NOUGHTS AND CROSSES, it would seem sensible to avoid placing the first cross in the centre, as the centre is involved in the most winning lines. However, as the Hint in Exercise 10.2 suggests, a sensible opening move in MISÈRE NOUGHTS AND CROSSES is to place a cross in the centre.

   This is in fact the only sensible opening move. Suppose that the first player starts by placing a cross somewhere other than the centre, that is, in a corner or side square. Show that the second player has a winning strategy from this position.

9. The following depicts a simple variant of the children's board game SNAKES AND LADDERS.

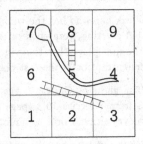

The rules of the game are as described in Exercise 10.4 (page 259).

Identify which of the positions are winning positions; which are losing positions; and which are drawing positions. (The game is played as described in Exercise 10.4 on page 259.) For the non-losing positions, indicate the optimal move(s).

10. Who has the winning strategy in NIM when you start with n piles each containing an equal number of coins? Justify your answer without referring to the general theory of NIM, that is, without referring to balanced versus unbalanced positions.

11. In MISÈRE NIM, the objective is to *not* take the last coin. What is the winning strategy for this variation?

12. The game of NIM-k is played just like NIM except that in a single move a player can remove (a different number of) coins from up to k different piles. Thus NIM-1 is the usual game of NIM.

Prove that there is a winning strategy in NIM-k from a given collection of piles if, and only if, when writing out the numbers of coins in the piles in binary notation, one above the other, and adding up the columns, the sum of at least one of the columns is not divisible by $k+1$.

13. Does the first player have any other safe opening moves in 3×4 CHOMP apart from the one outlined in Figure 10.2?

14. What are the possible safe opening moves in 3×5 CHOMP?

15. In this exercise we use a simple game to prove the result from Example 6.16 that the set $[0,1] = \{ x \ : \ 0 \le x \le 1 \}$ is uncountable.

In this game, a subset $S \subseteq [0,1]$ of real numbers between 0 and 1 is fixed, and the two players A and B take turns choosing real numbers $a_0, b_0, a_1, b_1, a_2, b_2, \ldots$ – with A choosing the a_is and B choosing the b_is

– starting from $a_0 = 0$ and $b_0 = 1$. When choosing a_i, A must choose a value satisfying $a_{i-1} < a_i < b_{i-1}$; and when choosing b_i, B must choose a value satisfying $a_i < b_i < b_{i-1}$. That is, A starts at 0 and B starts at 1, and they take turns moving towards – but never reaching – the other.

The increasing sequence $a_0, a_1, a_2, \ldots$ which A is choosing must converge towards a limit value a; that is, a is the smallest real value which is bigger than every a_i. If $a \in S$ then A wins the game; otherwise B wins the game.

(a) Prove that if S is countable then B has a winning strategy in this game. (Hint: Given an enumeration $s_1, s_2, s_3, \ldots$ of S, B's winning strategy is to choose $b_i = s_i$ whenever possible.)

(b) Deduce from the above that $[0, 1]$ is uncountable. (Hint: A clearly has the winning strategy when $S = [0, 1]$.)

16. This exercise exposes a paradox devised by the mathematician William Zwicker.

Professor Bertrand likes *every* game which can *never* be played forever; and he hates any game which may potentially go on forever. For example, he likes NIM, but he hates lawn tennis, as it could potentially get into an infinite "advantage-deuce" cycle.

Consider the game of RUSSELL whose rules are as follows:

- The first player chooses *any* game that must terminate.
- The two players play the chosen game, with the second player making the first move in the chosen game.

Does Professor Bertrand like the game of RUSSELL? Explain.

Part II

Modelling
Computing Systems

Chapter 11

Modelling Processes

If you can't describe what you are doing as a process, then you don't know what you're doing.

- W. Edwards Deming.

Having mastered the basic mathematical machinery presented in the first part of this book for modelling computing systems, the first question we must then address is: What exactly do we mean by a computing system? We are not speaking here of the various hardware components of a digital computer – as Edsger W. Dijkstra noted, computer science is no more about computers than astronomy is about telescopes. Rather what we have in mind is any computational process. Roughly speaking, a process describes the behaviour of a system as performing various actions that change the system's state. These changes are controlled by a set of rules which depend only on the state of the system and the state of its environment.

For example, if it is raining and we are outside holding a closed umbrella, then we should perform the action of opening the umbrella. This doesn't change the state of the environment – it continues to rain – but it changes our state; we are now under the protection of the umbrella. The rules that we abide by might then stipulate that we should close our umbrella once again if and when the rain stops, or when we enter a building.

As another example which has a more computational flavour, consider the simple calculator in Figure 11.1. The actions which may be performed are button-presses, which may change the state of the calculator – most obviously by changing the display, but also by changing the internal state of the calculator. Of course, if the calculator is off, then the only action which has any effect is pressing the $\boxed{\frac{ON}{OFF}}$ button which starts the calculator in its initial state; and at any time when the calculator is on, this button can be pressed to turn it off, or the "clear" button $\boxed{C}$ can be pressed to put the calculator into its initial state. Thus, pressing the $\boxed{C}$ button when the calculator is on has the same effect as pressing the $\boxed{\frac{ON}{OFF}}$ button twice.

Consider carrying out a simple calculation such as $123 \div 45$ using the following sequence of button presses (starting from the initial state of the

F. Moller, G. Struth, *Modelling Computing Systems*,
Undergraduate Topics in Computer Science,
DOI 10.1007/978-1-84800-322-4_12, © Springer-Verlag London 2013

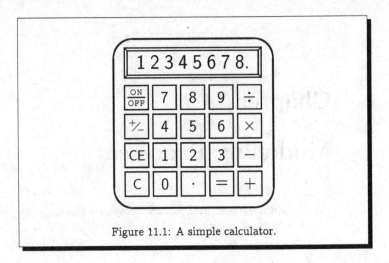

Figure 11.1: A simple calculator.

calculator):

$$\boxed{1} \quad \boxed{2} \quad \boxed{3} \quad \boxed{\div} \quad \boxed{4} \quad \boxed{5} \quad \boxed{=}$$

As you press the first three numeric buttons, the calculator simply accumulates these digits, displaying the sequence of digits as it increases in length. When you then press the $\boxed{\div}$ button, the calculator stores the number 123 in its memory, along with the operation of division, and awaits the entry of a second number made up from a further sequence of digits, in this instance the number 45. Pressing the $\boxed{=}$ button tells the calculator that the second number has been completely entered, and that the operation $\div$ in its memory should be applied between the first number 123 that it has stored in its memory and this second number 45. The calculator will respond by displaying the value 2.7333333.

There are many design decisions which must be made when describing the behaviour of a calculator, though for such a simple calculator as above most decisions are widely-accepted. For example, the sequence of button presses

$$\boxed{1} \quad \boxed{2} \quad \boxed{3} \quad \boxed{\times} \quad \boxed{\div} \quad \boxed{4} \quad \boxed{5} \quad \boxed{=}$$

is virtually universally accepted to mean $123 \div 45$, recognising that the user inadvertently pressed the $\boxed{\times}$ button and corrected this by subsequently pressing the $\boxed{\div}$ button to "overwrite" the operation[†]. However, we have

---

[†]This interpretation is generally, though less universally, accepted in the instance when the user presses $\boxed{-}$ after another operator button; the correct sequence of button-presses for calculating $123 \times (-45)$ is of course $\boxed{1} \ \boxed{2} \ \boxed{3} \ \boxed{\times} \ \boxed{4} \ \boxed{5} \ \boxed{+\!/\!-} \ \boxed{=}$.

barely started describing the behaviour of this calculator. To specify its complete behaviour as above would require many pages. It is not uncommon for realistic – yet nonetheless modest – systems to have specification documents running into several hundred pages.

Expressing the whole behaviour of a system in English prose as above quickly becomes a tedious, lengthy, and extremely error-prone activity. We clearly need a formal framework with which to describe the behaviour of such processes, as well as a language for expressing them. Also, we need an understanding as to when such a process is *correct*, that is, exhibits the behaviour that we expect (i.e., specify) that it should. These are all the concerns of the present chapter.

11.1 Labelled Transition Systems

In considering how we might wish to view a computational process, we can identify various of its underlying properties. Firstly, at any given moment in time, the process will be in a specific *state*. Secondly, in a given state, certain *events* or *actions* may happen which will cause the process to evolve into a new state. In fact, a state of a process may be completely determined by what actions may occur in that state, and to what new states each action might lead.

As a very simple example, consider a light switch that is either in the "off" position and may be switched on, or is in the "on" position and may be switched off. At any given moment in time the system (ie, the light) will be in one of two states, which we might refer to as OFF and ON. In the OFF state you can turn the light on (ie, do an *on* action) to take the system to the ON state, whereas in the ON state you can turn the light off (ie, do an *off* action) to take the system to the OFF state. We can picture this simple system as follows:

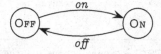

Here, the two states of the system are represented by circles, and there are arrows leading from one state to another; each arrow is labelled by the action which causes the process to make a transition from one state to the next. (For convenience we've also labelled the two states – by OFF and ON, respectively – but these labels are inessential: they do not add any information about what the process can do in any given state.)

As a slightly more complicated example, consider a simple drinks vending machine which accepts a 50p coin and allows the user to decide whether to press a coffee button or a tea button, before returning to its initial state. Its behaviour can be pictured as follows:

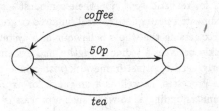

Again we have two states represented by circles, and arrows leading from one state to another, each labelled by the action which causes the change to the state. (In this case, we've not bothered labelling the states.)

This way of depicting processes is captured by the following definition of a *labelled transition system*.

Definition 11.1

A *labelled transition system (LTS)* is a triple $\mathcal{T} = (\text{States}, \text{Actions}, \rightarrow)$ consisting of:

- a set States of *states*;

- a set Actions of *actions or events*; and

- a set $\rightarrow\ \subseteq$ States $\times$ Actions $\times$ States of *transitions between states labelled by actions (a transition relation).*

We will generally write $s \xrightarrow{a} t$, or the more pictorial $\text{\textcircled{s}} \xrightarrow{a} \text{\textcircled{t}}$, instead of $(s, a, t) \in\ \rightarrow$, meaning that in state s, we may do the action a and thereby evolve into the state t. We will also write $s \xnrightarrow{a}$ to signify that there are no a-labelled transitions leading out of state s, and $s \nrightarrow$ to signify that there are no transitions leading out of s.

Definition 11.2

Given an LTS $\mathcal{T} = (\text{States}, \text{Actions}, \rightarrow)$, the *extended transition relation* $\rightarrow\ \subseteq$ States $\times$ Actions* $\times$ States is defined inductively as follows. (We use the notation introduced above in writing $s \xrightarrow{w} t$ instead of $(s, t) \in\rightarrow$; and give two clauses in our inductive definition: one base case for the empty string ϵ and one inductive case for aw where $a \in$ Actions and $w \in$ Actions*.)

- $s \xrightarrow{\epsilon} s$;   and
- $s \xrightarrow{aw} t$ if, and only if, $s \xrightarrow{a} s' \xrightarrow{w} t$ for some s'.

That is, for $w = a_1 a_2 \cdots a_k$, we have $s \xrightarrow{w} t$ if, and only if,

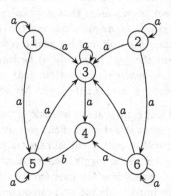

Figure 11.2: An example labelled transition system.

$$\bullet \ s \xrightarrow{a_1} \cdot \xrightarrow{a_2} \cdot \cdots \cdot \xrightarrow{a_k} t.$$

As depicted in the examples above, labelled transition systems are typically presented pictorially with states represented by circles, and transitions represented by arrows between states labelled by actions. States, actions and transitions are exactly the properties that define computational processes.

Example 11.2

Figure 11.2 depicts a labelled transition system with:

- state set States $= \{A, B, C, D, E, F\}$;
- action set Actions $= \{a, b\}$; and
- transition relation $\rightarrow = \{\ (A, a, A),\ (A, a, C),\ (A, a, E),$
  $(B, a, B),\ (B, a, C),\ (B, a, F),$
  $(C, a, C),\ (C, a, D),\ (D, b, E),$
  $(E, a, C),\ (E, a, E),$
  $(F, a, C),\ (F, a, D),\ (F, b, F)\ \}.$

Labelled transition systems provide an ideal tool for modelling situations which evolve over time, as in the following example.

Example 11.3   The Man-Wolf-Goat-Cabbage Riddle

The following is a very old riddle – in fact it was posed by Alcuin of York in the 8th century (and solved in 2009 by Homer Simpson in *The Simpsons* episode titled *Gone Maggie Gone*). It reads as follows.

*A man needs to cross a river with a wolf, a goat, and a cabbage.
His boat is only large enough to carry himself and one of his three
possessions, so he must transport them one at a time. However, if
he leaves the wolf and the goat together unattended, then the wolf
will eat the goat; similarly, if he leaves the goat and the cabbage
together unattended, then the goat will eat the cabbage. How can
the man get across safely with his possessions?*

Initially all four entities are on one side of the river (the left-hand side, say).
We can represent this state of affairs by $\langle$MWGC $\wr\wr\rangle$. By this we mean that
the man M, wolf W, goat G and cabbage C are all on the left-hand side of the
river (to the left of the wiggly lines representing the river), while nothing is
on the right-hand side of the river (to the right of the wiggly lines).

From the initial state the man can do one of four things.

1. He may cross the river with the goat, leaving the wolf and cabbage
   together on the left-hand side of the river. We represent the resulting
   state by $\langle$WC $\wr\wr$ MG$\rangle$, denoting that the wolf and cabbage are on the
   left-hand side while the man and goat are on the right-hand side. Note
   that this labelling of the state is purely for our benefit and is not itself
   a part of the definition of the process.

   Using g to represent the action of the man crossing the river with the
   goat, this gives us the following transition:

   $$\langle\text{MWGC } \wr\wr\rangle \xrightarrow{g} \langle\text{WC } \wr\wr \text{ MG}\rangle$$

2. He may cross the river with the wolf, leaving the goat and cabbage
   together on the left-hand side of the river. We represent the resulting
   state by $\langle$GC $\wr\wr$ MW$\rangle$. We shade this state to indicate that this is an
   unacceptable state of affairs, as the goat will in this instance eat the
   cabbage. Note however that this shading – just like the labelling of
   the state – is not in itself a part of the definition of the process.

   Using w to represent the action of the man crossing the river with the
   wolf, this gives us the following transition:

   $$\langle\text{MWGC } \wr\wr\rangle \xrightarrow{w} \langle\text{GC } \wr\wr \text{ MW}\rangle$$

3. He may cross the river with the cabbage, leaving the wolf and goat
   together on the left-hand side of the river. The resulting state will
   then be represented by $\langle$WG $\wr\wr$ MC$\rangle$. Again the shading indicates that
   this is an unacceptable state of affairs, as the wolf will eat the goat.

   Using c to represent the action of the man crossing the river with the
   cabbage, this gives us the following transition:

   $$\langle\text{MWGC } \wr\wr\rangle \xrightarrow{c} \langle\text{WG } \wr\wr \text{ MC}\rangle$$

4. He may cross the river on his own, leaving the wolf, goat and cabbage together on the left-hand side of the river. The resulting state will then be represented by ⬤ WGC ⫽ M. Once again the shading indicates that this is an unacceptable state of affairs, as the wolf will eat the goat, which may itself have had the time and opportunity to first eat the cabbage.

Using m to represent the action of the man crossing the river alone, this gives us the following transition:

It is clear from the above considerations that the man initially has just one viable option, which is to cross the river with the goat.

In this fashion, we can model the problem using a labelled transition system, using states to represent the possible states of affairs, and transitions to represent the actions available to the man. There will be a total of 16 states in this LTS:

- The initial state is represented by the state ⟨ MWGC ⫽ ⟩, and the desired final state is represented by the state ⟨ ⫽ MWGC ⟩.

- There are 8 further safe states, namely

| MWG ⫽ C | MWC ⫽ G | MGC ⫽ W | MG ⫽ WC |
| C ⫽ MWG | G ⫽ MWC | W ⫽ MGC | WC ⫽ MG |

- Finally there are 6 dangerous states, namely

The transitions will be labelled according to which of the four actions the man takes:

m: the man crosses the river on his own.

w: the man crosses with the wolf.

g: the man crosses with the goat.

c: the man crosses with the cabbage.

The resulting LTS is presented in Figure 11.3. From this LTS we can readily read off a solution to the riddle (which is not unique) by following a path from the initial state to the final state which passes only through safe states.

Again note that the labelling of the states, including the shading of what we recognise to be dangerous states, is not part of the definition of an LTS. This is included solely for our own convenience.

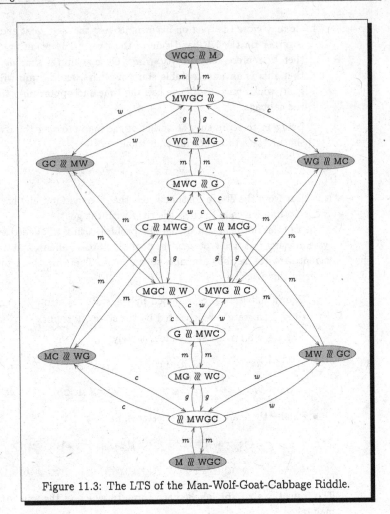

Figure 11.3: The LTS of the Man-Wolf-Goat-Cabbage Riddle.

Exercise 11.3 (Solution on page 459)

Three missionaries are travelling with three cannibals when they come upon a river. They have a boat, but it can only hold two people. The river is filled with piranha, so they all must eventually cross in the boat; no one can cross the river by swimming. The problem is: should the cannibals ever outnumber the missionaries on either side of the river, the outnumbered missionaries would be in deep trouble. Each missionary and each cannibal can row the boat.

How can all six get across the river safely?

Exercise 11.4   (Solution on page 461)

In the 1995 film *Die Hard: With a Vengence*, New York Detective John Mc-Clane (played by Bruce Willis) and Harlem dry cleaner Zeus Carver (played by Samuel L. Jackson) had to solve the following problem in order to prevent a bomb from exploding at a public fountain. Given only a five-gallon jug and a three-gallon jug, neither with any markings on them, they had to fill the larger jug with *exactly* four gallons of water from the fountain, and place it onto a scale in order to stop the bomb's timer and prevent disaster. How did they manage this feat?

Exercise 11.5   (Solution on page 461)

You are sitting in a pub wearing a blindfold, and I put in front of you a square tray with a beer mat in each of the four corners, each of which is either face-up or face-down, but not all the same.

You reach out and – blindly feeling your way around the tray – you turn over as many beer mats as you wish. When you are through, if the beer mats are all oriented the same way (either all face-up or all face-down) then you win. Otherwise, I will rotate the tray by an arbitrary amount, and let you try again.

What strategy will *guarantee* that you win the game?

11.2 Computations and Processes

Consider the following algorithm, attributed to Euclid (c. 300 BC), for computing the greatest common divisor (GCD) of two numbers x and y, that is, the largest integer which evenly divides both x and y. (In the code below, the modulus function x mod y simply returns the remainder when dividing x by y, and := represents the assignment operation.)

```
loop begin
    x := x mod y;
    if x=0 then return y;
    y := y mod x;
    if y=0 then return x
loop end
```

This algorithm repeatedly "executes" the four lines of code between "loop begin" and "loop end" until a value is returned. For example, if we apply this algorithm to the values $x=246$ and $y=174$, we get the value 6 returned, which is indeed the GCD of 246 and 174.

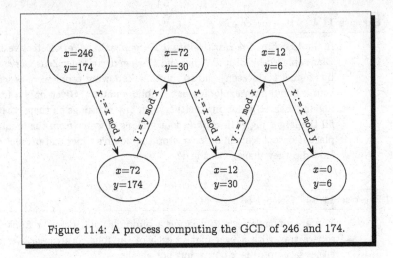

Figure 11.4: A process computing the GCD of 246 and 174.

To understand how this program works, we might try hand-turning it, keeping track of the state (i.e., values) of the variables. For example, starting in the state in which the variables have the values $x=246$ and $y=174$, the first action which takes place is the assignment x := x mod y; this action has the effect of changing the state of the system by updating the value of x.

This computation is captured by the labelled transition system depicted in Figure 11.4.

Exercise 11.6 (Solution on page 462)

Consider the transition system depicted in Figure 11.4.

1. How many states are there? List them.

2. How many distinct actions are there? List them.

3. How many transitions are there? List them.

As an example of a more abstract process, consider the workings of the simple table lamp represented in Figure 11.5 which has a string to pull for turning the light on and off, and a reset button which resets the circuit if a built-in circuit breaker breaks when the light is on. At any moment in time the lamp can be in one of three states:

- OFF – in which the light is off (and the circuit breaker is set);

- ON – in which the light is on (and the circuit breaker is set); and

- BROKEN – in which the circuit breaker is broken (and the light is off).

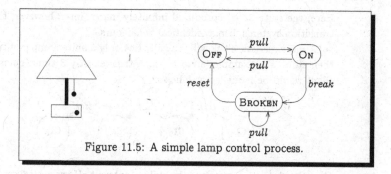

Figure 11.5: A simple lamp control process.

In any state the string can be pulled, causing a transition into the appropriate new state (from the state BROKEN, the new state is the same state BROKEN). In the state ON, the circuit breaker may break, causing a transition into the state BROKEN in which the reset button has popped out; from this state, the reset button may be pushed, causing a transition into the state OFF.

Exercise 11.7   (Solution on page 463)

Extend the lamp process by adding actions "*blow*" and "*replace*", which model the blowing and replacing of the light bulb. Assume that the bulb can only blow when the light is on, and that only a blown bulb can be replaced. Keep in mind that the string can still be pulled even if the bulb is blown, and that when a bulb is replaced, the lamp may be on or off depending on the pulls of the string.

Example 11.7

We can model a very simple clock that does nothing but tick repeatedly forever as follows:

$$tick \circlearrowright C1$$

This simple transition system has only one state C1 and one transaction C1 $\xrightarrow{tick}$ C1, but it can be "unrolled" into the following infinite-state transition system:

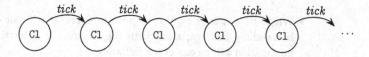

Here, the state Cl is reproduced infinitely-many times; however, these two transition systems display identical behaviours.

A more realistic clock will typically tick only a finite number of times and then stop. For example, a clock which ticks exactly 3 times before falling silent would be modelled as follows:

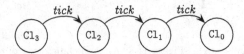

Here the state Cl_3 represents the state of interest. However, we can see that the additional states Cl_2, Cl_1 and Cl_0 represent clocks which tick exactly 2, 1, and 0 times, respectively. We can immediately generalise this example to model an infinite number of states Cl_n (for $n \in \mathbb{N}$) where state Cl_n represents a clock which ticks exactly n times before falling silent.

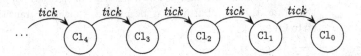

This is very similar to the unrolled version of the infinitely-ticking clock above; they both have an infinite number of states, with each state making a single *tick* action to get to the next state. However, in this system with the different states Cl_n, each *tick* action takes us one step closer to the state Cl_0 in which the clock stops. In the unrolled system above each state is the same as any other state in that an infinite number of *tick* actions can be performed from it. In particular, there is no state like Cl_0 from which no *tick* action can occur.

Exercise 11.8   (Solution on page 463)

Draw a model of a Clock Cl_* which can tick any number of times, but may stop ticking after any tick. How many states, actions, and transitions does your model have?

Example 11.8

In this example we consider a simple elevator which moves between three floors. The state of the elevator reflects three entities:

- Which floor it is at or, if it is between floors, which pair of floors it is travelling between; along with the direction it is travelling. This information will be one of the following:

$$1: \text{at 1 (heading up)}; \qquad 1^+: \text{moving from 1 to 2};$$
$$2^\uparrow: \text{at 2 (heading up)}; \qquad 2^+: \text{moving from 2 to 3};$$
$$2^\downarrow: \text{at 2 (heading down)}; \qquad 2^-: \text{moving from 2 to 1};$$
$$3: \text{at 3 (heading down)}; \qquad 3^-: \text{moving from 3 to 2}.$$

- Whether the door is open or closed.

- Which set of floors it has to travel to – and in which direction – due either to a call button being pressed on a floor, or a floor button being pressed in the elevator. This information will be a subset of the following:

$$1: \text{collect from or drop off at floor 1};$$
$$2^\uparrow: \text{collect from or drop off at floor 2 while heading up};$$
$$2^\downarrow: \text{collect from or drop off at floor 2 while heading down};$$
$$3: \text{collect from or drop off at floor 3}.$$

A state, therefore, will be of the form (*floor*, *door*, *stops*) where

$$floor \in \{1, 2^\uparrow, 2^\downarrow, 3\} \cup \{1^+, 2^+, 2^-, 3^\downarrow\};$$
$$door \in \{open,\ closed\};\ \text{and}$$
$$stops \subseteq \{1, 2^\uparrow, 2^\downarrow, 3\}.$$

There will therefore be as many as $8 \times 2 \times 2^4 = 256$ states.

Exercise 11.9   (Solution on page 463)

Augment the above description of the states of the elevator system by describing: the set of possible actions; when (in which states) each possible action can occur; and how the state changes when that action occurs. (Of course, with 256 states, it is unreasonable to draw out this labelled transition system, so don't even try!)

Exercise 11.10   (Solution on page 466)

Consider the process of flipping a coin, in which the following three actions are possible at different times:

- a *toss* action in which the coin is tossed into the air;

- a *heads* action in which the coin lands with heads showing; and

- a *tails* action in which the coin lands with tails showing.

Upon doing a *toss* action, either a *heads* action or a *tails* action will occur, and the process will be back in its initial state in which the coin may once again be flipped.

Draw two different models of this process: the first in which the outcome of the flip is determined already when the coin is tossed, and the second in which the outcome of the flip is determined only when it is observed. What are the implications of these different interpretations of determinism? Which do you consider to be the most realistic model?

11.3 A Language for Describing Processes

Drawing processes graphically is fine for small examples. However, you would never draw the labelled transition system for even moderately-complex processes. For example, drawing the labelled transition system for the elevator in Example 11.8 above would not only be tedious and error-prone, it would not be very insightful.

We will need a proper language for describing bigger and more complicated systems. A formal description language can also be programmed and analysed (verified) on a machine. In this section we shall present such a language. The language, which we refer to as PROC, will have

- *(process) variables*, such as OFF, ON, and BROKEN; *and*
- *events* or *actions*, such as *pull*, *break*, and *reset*.

Every expression in the language will represent a state in a labelled transition system. Each process variable is itself an expression in the language, and all of the expressions in the language will be built up from actions and process variables using simple *operations* for combining them. In the remainder of this section we shall explore the two basic operations in the language: *action prefix* and *choice*, as well as the means by which processes are defined.

11.3.1 The Nil Process 0

The most basic process expression in the language is 0, which is referred to as the *nil* process and represents a state which has no transitions leading out of it:

For example, the state Cl_0 from Example 11.7 is an example of the nil process. A process which evolves into such a state 0 is said to have *deadlocked*.

11.3.2 Action Prefix

If a is an action and E is a process expression, then the *action prefix* expression $a.E$ represents a state in a process which has one transition: an a-transition leading to the state represented by E:

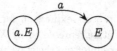

As an example, the clock Cl_1 which ticks once and then falls silent is represented by the expression

$tick.0$

which depicts the following process:

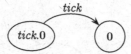

By repeatedly applying the action prefix operation, we can express the clock Cl_3 which ticks three times before falling silent as follows:

$tick.tick.tick.0$

which depicts the following process:

For an example based on the lamp process of Figure 11.5, if *pull* is an action and On is a process variable (and hence a valid expression), the action prefix expression

$pull.\text{On}$

represents the following state with a single *pull* transition:

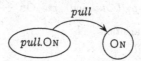

11.3.3   Process Definitions

Each process variable, being a process expression, represents a state in a labelled transition system, and as such stands for some process. Every process variable X must therefore have a *defining equation*

$$X \stackrel{\text{def}}{=} E$$

where E is the process expression for which X stands. The transitions leading from the state represented by X are determined by (that is, are the same as) those leading from the state represented by E.

For example, in the simple table lamp process of Figure 11.5, the state represented by the process variable OFF has a single transition leading out of it labelled *pull* and leading into the state represented by the variable ON. The process variable OFF can thus be defined as follows:

$$\text{OFF} \stackrel{\text{def}}{=} pull.\text{ON}$$

The one and only transition from the state represented by the expression *pull*.ON is

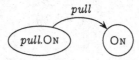

and since OFF – by definition – has the same transitions leading out of it as *pull*.ON, the one and only transition from the state represented by the expression OFF is

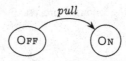

The *process definition* $X \stackrel{\text{def}}{=} E$ thus defines which transitions are possible from the state represented by the variable X, namely precisely those possible from the expression E. In other words, X is defined to be identical in behaviour to E. Formally,

if $X \stackrel{\text{def}}{=} E$ and $E \stackrel{a}{\rightarrow} E'$

then $X \stackrel{a}{\rightarrow} E'$.

In the lamp process, since

$$\text{OFF} \stackrel{\text{def}}{=} pull.\text{ON} \quad \text{and} \quad pull.\text{ON} \stackrel{pull}{\longrightarrow} \text{ON},$$

we have the transition

$$\text{OFF} \stackrel{pull}{\longrightarrow} \text{ON}.$$

11.3.4 Choice

The state OFF in the lamp process is particularly simple as there is only one transition leading out of it. However, the other two states offer a *choice* of transitions:

- from the state ON there is a transition *pull* leading to the state OFF and a transition *break* leading to the state BROKEN; and

- from the state BROKEN there is a transition *pull* leading back to the state BROKEN and a transition *reset* leading to the state OFF.

Thus, the state ON may behave either like the state *pull*.OFF or like the state *break*.BROKEN, and the state OFF may behave either like the state *pull*.BROKEN or like the state *reset*.OFF.

Such choices between behaviours are catered for in the language PROC by the *choice* operation: given expressions E and F, the expression $E + F$ represents the process state which has all of the transitions of E and of F; in essence, it can behave as either E or as F, with the choice being taken at the moment the first transition occurs. More formally,

if $E \xrightarrow{a} E'$ then $E + F \xrightarrow{a} E'$, and

if $F \xrightarrow{a} F'$ then $E + F \xrightarrow{a} F'$.

Referring to the lamp example, from the state ON we can either perform a *pull* action to go to state OFF or a *break* action to go to state BROKEN; and from the state BROKEN we can either perform a *pull* action to go to state BROKEN or a *reset* action to go to state OFF. The two process variables ON and BROKEN thus have the following definitions:

$$\text{ON} \overset{\text{def}}{=} pull.\text{OFF} \quad + \quad break.\text{BROKEN}$$
$$\text{BROKEN} \overset{\text{def}}{=} pull.\text{BROKEN} + reset.\text{OFF}$$

By the above rules for choice, we thus have the transitions

$$pull.\text{OFF} + break.\text{BROKEN} \xrightarrow{pull} \text{OFF} \quad \text{and}$$
$$pull.\text{OFF} + break.\text{BROKEN} \xrightarrow{break} \text{BROKEN}$$

and hence (by the process definition operation) the transitions

$$\text{ON} \xrightarrow{pull} \text{OFF} \quad \text{and} \quad \text{ON} \xrightarrow{break} \text{BROKEN}.$$

From state ON we can chose to perform action pull to go into state OFF or to perform action break to go into state BROKEN.

By analogous reasoning we can infer the following two transitions for the state represented by the variable BROKEN:

$$\text{BROKEN} \xrightarrow{pull} \text{BROKEN} \quad \text{and} \quad \text{BROKEN} \xrightarrow{reset} \text{OFF}.$$

Exercise 11.11 (Solution on page 467)

Explain how these last two transitions for BROKEN can be inferred.

We can extend the choice operation in the natural way to any finite sum

$$E_1 + E_2 + \cdots + E_n$$

to describe a choice of doing the actions of any of the summand process terms E_i (where $1 \leq i \leq n$). We shall also sometimes write the choice operation as an indexed sum; that is, instead of writing $E_1 + E_2 + \cdots + E_n$ we may write, for example, any of the following:

$$\sum_{i=1}^{n} E_i \quad \text{or} \quad \sum_{1 \leq i \leq n} E_i \quad \text{or} \quad \sum \{E_1, E_2, \ldots, E_n\} \quad \text{or} \quad \sum \{E_i \ : \ 1 \leq i \leq n\}.$$

More generally, given any (possibly infinite) set S of process expressions, we can write $\sum S$ to represent a choice over all of these. For example, instead of writing

$$\text{ON} \stackrel{\text{def}}{=} pull.\text{OFF} + break.\text{BROKEN}$$

we could write

$$\text{ON} \stackrel{\text{def}}{=} \sum \{ pull.\text{OFF}, \ break.\text{BROKEN} \}.$$

The transitions that the process expression $\sum S$ can perform are determined by the process expressions in S. Formally,

$$\text{if} \ E \in S \ \text{and} \ E \stackrel{a}{\rightarrow} E' \ \text{then} \ \sum S \stackrel{a}{\rightarrow} E'.$$

Furthermore, as S may be an infinite set of process expressions, we may use this notation to express *infinite* choices. For example, the infinite choice

$$E_1 + E_2 + E_3 + E_4 + \cdots$$

can be written as $\sum_{i \geq 1} E_i$.

An interesting case of this generalised choice operation occurs when S is the empty set $\emptyset$. The process expression $\sum \emptyset$, by definition, can make no transitions, and thus provides a definition of the *nil* process: $0 \stackrel{\text{def}}{=} \sum \emptyset$.

The syntax (notation) and semantics (meaning) of the language PROC is summarised in Figure 11.6.

Example 11.11

Continuing with Example 11.7, we can give the following process definitions to the simple clock Cl that ticks forever, and the clocks Cl_n (for $n \in \mathbb{N}$) which tick exactly n times:

$$\text{Cl} \stackrel{\text{def}}{=} tick.\text{Cl} \qquad \text{Cl}_0 \stackrel{\text{def}}{=} 0 \qquad \text{Cl}_{n+1} \stackrel{\text{def}}{=} tick.\text{Cl}_n$$

Name	Syntax	Semantics
PROCESS VARIABLE	X	If $E \xrightarrow{a} E'$ and $X \stackrel{\text{def}}{=} E$ then $X \xrightarrow{a} E'$
ACTION PREFIX	$a.E$	$a.E \xrightarrow{a} E$
CHOICE (1)	$E + F$	If $E \xrightarrow{a} E'$ then $E + F \xrightarrow{a} E'$ If $F \xrightarrow{a} F'$ then $E + F \xrightarrow{a} F'$
CHOICE (2)	$\sum\{E_i : i \in I\}$	If $E_j \xrightarrow{a} E'$ with $j \in I$ then $\sum\{E_i : i \in I\} \xrightarrow{a} E'$
NIL	0	no transitions $(X \stackrel{\text{def}}{=} \sum \emptyset)$

Figure 11.6: Syntax and semantics of the process language PROC.

Consider now a clock, which we will call Clock, which may tick some finite but indeterminate number of times depending on the amount of energy powering it and then stop. There is no way of knowing a priori how many times it will tick; it will tick once when it is started, and then continue to tick until its energy source is depleted. Thus, after the first tick, the clock will be in a state in which it will tick again some precise finite number of times. That is, it will be in the state Cl_n for some $n \in \mathbb{N}$. Its definition as a process is given as follows:

$$\text{Clock} \stackrel{\text{def}}{=} \sum_{i \geq 0} tick.Cl_i.$$

Finally, consider yet another clock, which we will call Clock$_*$, which may behave like Clock, but might decide – upon performing the first tick – to continue ticking forever. That is, it has the possibility to evolve into the state Cl after the first tick. Its definition as a process is given as follows:

$$\text{Clock}_* \stackrel{\text{def}}{=} \sum_{i \geq 0} tick.Cl_i + tick.Cl.$$

These processes all appear in the transition system depicted in Figure 11.7.

Exercise 11.12   (Solution on page 467)

Give a process definition for the Clock process Cl_* which you defined in Exercise 11.8.

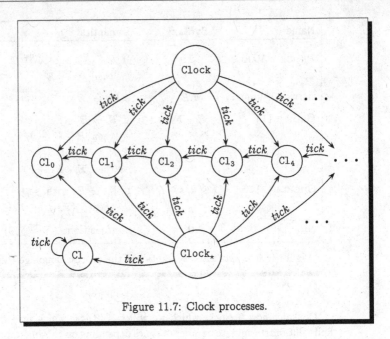

Figure 11.7: Clock processes.

Exercise 11.13   (Solution on page 467)

Design a simple change-making process which will initially accept a 5p, 10p or 20p coin, and dispense any sequence of 1p, 2p and 5p coins which sum up to the value of the coin inserted, before returning to its initial state.

To do this, introduce the process variables C_n for $n \in \{0, 1, 2, \ldots, 20\}$, and the following actions:

i_5: insert a 5p coin	d_1: dispense a 1p coin
i_{10}: insert a 10p coin	d_2: dispense a 2p coin
i_{20}: insert a 20p coin	d_5: dispense a 5p coin

Each variable C_n is to represent the process in the state in which n pence remains to be dispensed. In particular, the process variable C_0 is to represent the initial state of the process, and has the following definition:

$$C_0 \stackrel{\text{def}}{=} i_5.C_5 \ + \ i_{10}.C_{10} \ + \ i_{20}.C_{20}$$

1. Give the definitions for the remaining process variables $C_1, C_2, \ldots, C_{20}$.

2. Draw the labelled transition system representing this process.

$$V_1 \stackrel{\text{def}}{=} 10p.10p.(\ \texttt{coffee.collect}.V_1$$
$$+\ \texttt{tea.collect}.V_1\ )$$

$$V_2 \stackrel{\text{def}}{=} 10p.(\ 10p.\texttt{coffee.collect}.V_2$$
$$+\ 10p.\texttt{tea.collect}.V_2\ )$$

$$V_3 \stackrel{\text{def}}{=} 10p.10p.\texttt{coffee.collect}.V_3$$
$$+\ 10p.10p.\texttt{tea.collect}.V_3$$

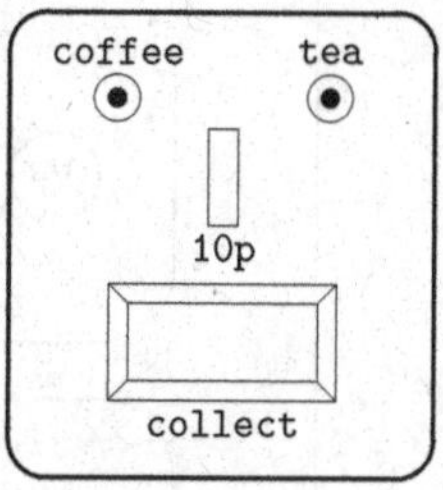

Figure 11.8: Three implementations of a vending machine.

11.4 Distinguishing Between Behaviours

Consider the problem of designing a simple vending machine which will allow a user to insert two 10p coins in succession, and then push a coffee *or* a tea button; the user will then be allowed to collect the relevant beverage, after which the machine will return to its original state, permitting the next person to use it.

This informal description is typical of how an actual "specification" may appear, but demonstrates the sort of ambiguity which arises in such loose specifications. The problem in this case stems from the ambiguity of the word "*or*." The vending machine might be implemented by any one of the three programs in Figure 11.8. We can draw these three processes as in Figure 11.9. Note that the states of these transition systems, though not spelt out, all represent expressions of the language PROC. For example, the four states of the first process, from left to right, represent the following expressions:

V_1
10p.(coffee.collect.V_1 + tea.collect.V_1)
coffee.collect.V_1 + tea.collect.V_1
collect.V_1

The transitions are easily derived using the semantic rules for inferring transitions. For example,

$$V_1 \xrightarrow{\ 10p\ } 10p.(\texttt{coffee.collect}.V_1 + \texttt{tea.collect}.V_1)$$

since

$$V_1 \stackrel{\text{def}}{=} 10p.10p.(\texttt{coffee.collect}.V_1 + \texttt{tea.collect}.V_1)$$

and by the action prefix rule,

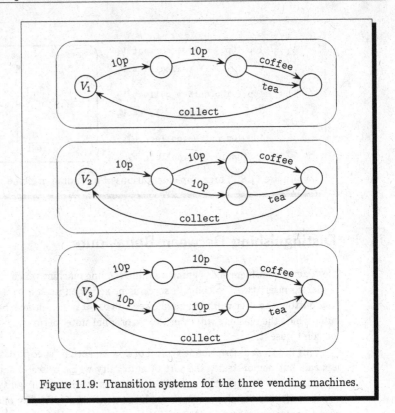

Figure 11.9: Transition systems for the three vending machines.

$$10p.10p.(\texttt{coffee.collect}.V_1 + \texttt{tea.collect}.V_1)$$

$$\xrightarrow{10p} 10p.(\texttt{coffee.collect}.V_1 + \texttt{tea.collect}.V_1).$$

Exercise 11.14   (Solution on page 467)

List the states of the other two vending machine processes.

Clearly the behaviour of V_1 is different from the behaviour of V_2 and V_3. Specifically, the following property is true of V_1 but not true of V_2 nor of V_3.

> *After inserting two* 10p *coins,*
>   *we are* guaranteed *to be able to press the* coffee *button.*

In other words,

> *No matter how we do a* 10p *action,*
> *we must end up in a state in which,*

> no matter how we do a 10p action,
> we must end up in a state in which
> we may do a coffee action.

In contrast, the following property is the *negation* of the above property, and as such is true of V_2 and V_3 but not V_1.

> We may do a 10p action and end up in a state in which
> we may do a 10p action and end up in a state in which
> we cannot do a coffee action.

Notice that the negation of a *must* property (i.e., a necessity) is a *may* property (a possibility), and vice versa.

It is less clear, but still true, that the behaviour of V_2 is different from the behaviour of V_3. In particular, the following property is true of V_3 but not true of V_1 nor of V_2.

> Already after inserting the first 10p coin,
> we have lost *either the* possibility *of selecting a* coffee,
> *or the* possibility *of selecting a* tea.

Even simpler,

> We may do a 10p action and end up in a state in which,
> no matter how we do a 10p action,
> we must end up in a state in which
> we cannot do a tea action.

Exercise 11.15 (Solution on page 468)

Negate this property to get a property which is true of V_1 and V_2 but not true of V_3.

The question then is: *How do we formally distinguish between processes?* Clearly, the answer to this question lies at the heart of the problem of verifying the correctness of systems. Answering this question is the goal of the next chapter.

Exercise 11.16 (Solution on page 469)

Let

$$A \stackrel{\text{def}}{=} b.c.0 + b.d.0 \qquad C \stackrel{\text{def}}{=} a.B + a.A$$
$$B \stackrel{\text{def}}{=} A + b.(c.0 + d.0) \qquad D \stackrel{\text{def}}{=} a.B$$

1. Draw a transition system which includes the above states A, B, C and D.

2. Explain clearly how the states C and D behave differently.

11.5 Equality Between Processes

At the start of Section 11.1 we posited that a state of a process is completely determined by what actions may occur in that state, and what new states each action might lead to. With this in mind, we would naturally consider two states E and F to be equal, $E = F$, whenever it is the case that for all actions a and states G: $E \xrightarrow{a} G$ if, and only if, $F \xrightarrow{a} G$. That is, $E = F$ whenever the following is true:

- if $E \xrightarrow{a} G$ then $F \xrightarrow{a} G$; and
- if $F \xrightarrow{a} G$ then $E \xrightarrow{a} G$.

Exercise 11.17 (Solution on page 469)

Show that this notion of equality is an equivalence relation over process expressions; namely that it is reflexive, symmetric and transitive.

By equating states of a process in this way, we can show that the following equations are true of process terms defined in the language PROC.

(S_1) $E + 0 = E$.

(S_2) $E + E = E$.

(S_3) $E + F = F + E$.

(S_4) $(E + F) + G = E + (F + G)$.

(S_5) If $X \stackrel{\text{def}}{=} E$ then $X = E$.

Each of these equations is easily justified by considering the rules by which transitions can be inferred.

Example 11.17

To show that $E + 0 = E$, we need to confirm the validity of the following two propositions:

- if $E + 0 \xrightarrow{a} G$ then $E \xrightarrow{a} G$; and
- if $E \xrightarrow{a} G$ then $E + 0 \xrightarrow{a} G$.

The second proposition follows immediately from the rule for choice.

For the first proposition, if $E + 0 \xrightarrow{a} G$ then the rule for choice says that either $E \xrightarrow{a} G$ or $0 \xrightarrow{a} G$; but since there are no transitions leading out of 0, we must have that $E \xrightarrow{a} G$ as required.

Exactly when two states should be deemed equal is explored in detail in the next chapter; however, the above equations will certainly be true. Even further, we can recursively extend the notion of equality between states by declaring two states E and F to be equal, $E = F$, not only if they possess exactly the same transitions, but whenever each transition of one can be matched, up to equality, by the other; that is, $E = F$ whenever the following is true:

- if $E \xrightarrow{a} E'$ then $F \xrightarrow{a} F'$ for some F' such that $E' = F'$; and
- if $F \xrightarrow{a} F'$ then $E \xrightarrow{a} E'$ for some F' such that $E' = F'$.

We can represent this situation pictorially as follows:

$$
\begin{array}{ccc}
E & \cdots\cdots & F \\
\Big\downarrow a & = & \Big\downarrow a \\
E' & \cdots\cdots & F'
\end{array}
$$

With this refinement to the notion of equality between states, we can show the following equations to be true.

(C_1) If $E = F$ then $E + G = F + G$.

(C_2) If $E = F$ then $a.E = a.F$.

These are important properties for any notion of equality between process terms, as they ensure that a term does not change when we replace a subterm within the term by an equal subterm.

Exercise 11.18 (Solution on page 469)

Consider the following processes, all of which perform a-transitions over and over, ad infinitum.

$A \stackrel{\text{def}}{=} a.A$, and

$A_i \stackrel{\text{def}}{=} a.A_{i+1}$ for each $i \in \mathbb{N}$.

Clearly, A and A_0 exhibit the same behaviour. However, explain why we cannot infer that $A = A_0$.

11.6 Additional Exercises

1. As we saw from Example 11.3, modelling puzzles have a long history. Water jug puzzles of the type presented in Exercise 11.4 are referred to as *Tartaglian water measuring problems* as they were favourites of the 16th-century Italian mathematician Niccolò Tartaglia (though these days you'd no doubt be more successful searching online for "Diehard water puzzle" than "Tartaglian water puzzle"). In fact, the

problem faced by John and Zeus in Exercise 11.4 was adapted from the following puzzle posed by Abbot Albert in the 13th century.

Given an eight-unit jug filled with water, an empty five-unit jug and an empty three-unit jug, how can we divide the water into two parts, each exactly four-unit? (None of the jugs have any markings on them, and we cannot estimate quantities by eye; we can only measure exact quantities by pouring water from one jug to another until one of the two jugs becomes either full or empty.)

2. (a) Three married couples wish to cross a river. Their boat, however, can only carry two people at a time. Also, the husbands are very jealous: each one of them refuses to let his wife be in the presence of another man unless he himself is present as well. How can they cross the river using the fewest number of trips?

   (b) Argue that the above problem cannot be solved if you have four couples wanting to cross the river.

   (c) Show that five couples can cross the river in a boat that can carry three, but that six couples cannot.

3. Alice, Bob, Carol and Dave want to cross a river in a boat. However, their boat can only hold 100 kg. Alice is 46 kg, Bob is 49 kg, Carol is 52 kg and Dave is 60 kg. Also, Bob has a broken arm and can't row.

   How can they all get across the river?

4. Alice, Bob, Carol and Dave want to cross a bridge in the dark of night. However, the bridge is rigged with a bomb which is due to explode, destroying the bridge, in 17 minutes. They have one flashlight which must be used when crossing the bridge, but the bridge can hold only two people at once. Their walking speeds allow them to cross in 1, 2, 5 and 10 minutes, respectively; when two of them cross together, they must walk together with the flashlight.

   How can they all get safely across the bridge?

5. This question considers further bridge-crossing problems based on that in question 4.

   (a) How quickly can six people cross a bridge two-at-a-time aided by a single flashlight, if their crossing times are 1, 3, 4, 6, 8 and 9 minutes, respectively?

   (b) How quickly can seven people cross a bridge three-at-a-time aided by a single flashlight, if their crossing times are 1, 2, 6, 7, 8, 9 and 10 minutes, respectively?

6. A queen, her son, and her daughter are being held captive in the tower of a castle. Outside the tower window is a rope running over a pulley with baskets of equal weight attached to the ends of the rope. One

basket is empty and is outside the window, while the other basket is on the ground with a 30 kg rock in it. One basket can be safely lowered to the ground using the other basket as a counterbalance as long as the difference in weight between the two baskets does not exceed 6 kg; if one basket is more than 6 kg heavier than the other, the heavier basket will crash to the ground. The queen weighs 78 kg, her daughter 42 kg and her son 36 kg. Each basket can hold two people, or one person and the rock.

How can the queen and her children escape to the ground using the smallest number of steps?

7. We can compute the value of x mod y using the following simple algorithm:

```
while x ≥ y do x := x-y
return x
```

   (a) Draw the transition system associated with the computation of the value 72 mod 30.

   (b) List the states, actions and transitions of this transition system.

8. Consider the following process definition.

$$X \stackrel{\text{def}}{=} a.0 + a.Z \qquad Y \stackrel{\text{def}}{=} a.Z \qquad Z \stackrel{\text{def}}{=} a.Z$$

   (a) Draw the labelled transition system for the above process.

   (b) Explain in words how states X and Y differ, behaviourally.

9. Argue that the process Clock2 given by the process definition

$$\text{Clock2} \stackrel{\text{def}}{=} \sum_{i \geq 0} \text{Cl}_i$$

defines the same process as the process Clock from Example 11.11.

10. Design a keypad lock which has three buttons labelled A, B and C. Any of the keys can be pressed at any time, and if the correct sequence of 5 key presses, namely $BBCBA$, is keyed in, then the lock will open.

11. In this question, we study the specification of a car safety system, in which a bell rings (repeatedly) whenever the ignition is on while the door is open or the seat belt is unbuckled.

   The labelled transition system for this specification is pictured in Figure 11.10 Here we have a system with

   - eight states $S = \{ X_1, X_2, X_3, X_4, X_5, X_6, X_7, X_8 \}$, and
   - seven actions $A = \{ \text{open}, \text{close}, \text{buckle}, \text{unbuckle}, \text{on}, \text{off}, \text{ring} \}$.

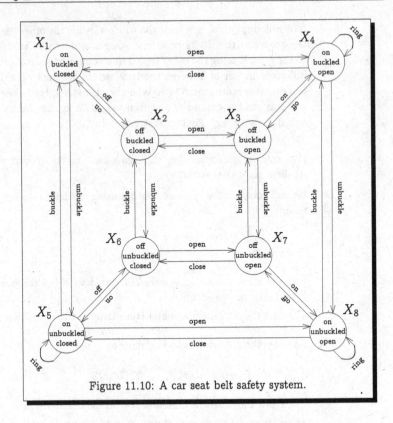

Figure 11.10: A car seat belt safety system.

For example, in state X_4, the ignition is on, the seat belt is buckled, the door is open, and the alarm is ringing.

(a) The eight states in S can be given process definitions, such as

$$X_1 \stackrel{\text{def}}{=} \text{off}.X_2 + \text{open}.X_4 + \text{unbuckle}.X_5$$

Give such a definition for each of the state variables in S.

(b) Let $D(x)$, $B(x)$, $M(x)$ and $R(x)$ be predicates defined over the states S as follows:

$$D(x) = \text{"the door is open in state } x\text{."}$$
$$B(x) = \text{"the seat belt is buckled in state } x\text{."}$$
$$M(x) = \text{"the ignition is on in state } x\text{."}$$
$$R(x) = \text{"the bell is ringing in state } x\text{."}$$

For each of these four predicates, indicate the states for which they are true.

12. Adapt the elevator system from Example 11.8 (page 290) so that it serves four floors rather than three.

13. Adapt the elevator system from Example 11.8 (page 290) so that it models two elevators operating side-by-side, which are called using the same call buttons on each floor.

14. Give a process definition for the behaviour of the simple calculator of Figure 11.1 (page 280).

15. Justify the following equalities from Section 11.5. (The first one was demonstrated in Example 11.17, page 302.)

(S_1) $E + 0 = E$.
(S_2) $E + E = E$.
(S_3) $E + F = F + E$.
(S_4) $(E + F) + G = E + (F + G)$.
(S_5) If $X \stackrel{\text{def}}{=} E$ then $X = E$.

16. Justify the following equalities from Section 11.5.

(C_1) If $E = F$ then $E + G = F + G$.
(C_2) If $E = F$ then $a.E = a.F$.

Chapter 12

Distinguishing Between Processes

Satire is a lesson, parody is a game.
　　　　　　　　　　　　- Vladimir Nabokov, *Strong Opinions.*

If we consider the properties which we used to distinguish between the vending machines from Section 11.4, we quickly notice an analogy with the way in which strategies for the two-player games of Chapter 10 were discussed; they both rely heavily on the use of modal verbs such as *may* and *must* to describe capabilities. In this chapter, we shall exploit this analogy by devising a two-player game for distinguishing between two given processes. In this game,

- the first player will aim to show that the two processes are *different*, by looking for an action that one process can do which the other cannot;

- the second player will aim to show that the two processes are *the same*, by showing that each process can copy every action made by the other.

In this game, one of the two players will always have a winning strategy (draws will not be possible); the two processes will be declared to be the same if the second player has a winning strategy, and different if the first has a winning strategy.

12.1 The Bisimulation Game

In this game we start by choosing two process states E and F (i.e., two designated states of some transition system). For example, we may consider the states X and U taken from the first of the two transition systems depicted in Figure 12.1. We may also define an *a priori* "time limit" of $n \in \mathbb{N}$ moves, or declare that the game has no time limit (i.e., take $n = \infty$). A game thus defined is represented either by $G_n(E, F)$ or $G_\infty(E, F)$. The game is played between two players, who have the following goals.

F. Moller, G. Struth, *Modelling Computing Systems*,
Undergraduate Topics in Computer Science,
DOI 10.1007/978-1-84800-322-4_13, © Springer-Verlag London 2013

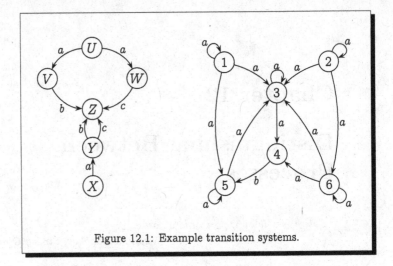

Figure 12.1: Example transition systems.

1. The first player wishes to demonstrate that the two chosen states are in some way inherently different.

2. The second player wishes to demonstrate that the two chosen states are inherently the same.

To play the game, we start by placing tokens on the two states E and F, and then proceed as follows.

1. The first player chooses one of the two tokens, and moves it forward along an arrow to another state of her choosing; if this is impossible (that is, if there are no arrows leading out of either of the states on which the tokens sit), then the second player is declared to be the winner.

2. The second player must move the *other* token forward along an arrow which has *the same label* as the arrow used by the first player; if this is impossible, then the first player is declared to be the winner.

This exchange of moves is repeated for as long as neither player gets stuck, or for a total of n exchanges of moves in the case where a finite time limit n is defined. Note that the first player gets to choose which token to move *every time it is her turn*; she does not have to keep moving the same token. If the second player succeeds in matching every move of the first player, then he is declared to be the winner. If there is no time limit, then the second player is declared to be the winner of any play of the game that goes on forever. (It may seem rather strange to declare a player to be the winner of a play which lasts forever. However, there is nothing paradoxical about this, and by doing so we ensure that there is always a winner; the game cannot end in a draw.)

Example 12.1

Suppose we start with the tokens on states X and U of the first transition system of Figure 12.1, and we assume that the time limit is (at least) 2.

1. the first player can move the token on state U along the a-labelled arrow to state V; in response the second player must move the token on state X along the a-labelled arrow to state Y.

2. The first player can then move the token on state Y along the c-labelled arrow to state Z; the second player cannot respond to this move, as there are no b-labelled arrows leading out of state V, so the first player wins.

As the second player never has any options – and thus he can never have made a bad move – this defines a winning strategy for the first player.

Example 12.2

Consider the following game played on the second transition system in Figure 12.1 with the tokens on states 1 and 2, where we assume that the time limit is (at least) 3.

1. The first player starts by moving the token on state 1 along the arrow labelled a to state 5. In response, the second player has to move the token on state 2 along an arrow labelled a; there are three ways to do this: by moving the token to state 2, to state 3, or to state 6; after some thought, he chooses to move the token to state 6.

2. The first player then moves the token on state 6 along the arrow labelled a to state 4. In response, the second player has to move the token on state 5 along an arrow labelled a; there are two ways to do this: by moving the token to state 3 or to state 5; he chooses to move the token to state 3.

3. The first player then moves the token on state 4 along the arrow labelled b to state 5. In response, the second player has to move the token on state 3 along an arrow labelled b; however, this is impossible, so the first player is declared to be the winner.

In this case, the first player was lucky: the second player had several options open to him in response to the moves of the first player, and he simply chose poorly. Had the second player responded to the opening $1 \xrightarrow{a} 5$ move of the first player by making the move $2 \xrightarrow{a} 2$, he could then have responded to all subsequent moves of the first player. In fact, the second player has a winning strategy in this game. This fact will be made evident in Section 12.4

For such a simply-defined game, the fact that there is no possibility of a draw implies that one of the two players has a winning strategy. This fact is embodied in the following.

Theorem 12.2

For any game $G_n(E, F)$ or $G_\infty(E, F)$, either the first player has a winning strategy, or the second player has a winning strategy.

Exercise 12.2   (Solution on page 470)

Prove Theorem 12.2 for finite games $G_n(E, F)$, by induction on n.

Induction cannot be used to prove the result for infinite games $G_\infty(E, F)$. Its proof is left as an exercise at the end of the chapter (Exercise 6, page 330).

Definition 12.2

We say that two process states E and F are *n-game equivalent*, written $E \sim_n F$, if, and only if, the second player has a winning strategy in the game $G_n(E, F)$. Similarly, we say that E and F are *∞-game equivalent*, written $E \sim_\infty F$, if, and only if, the second player has a winning strategy in the game $G_\infty(E, F)$.

For example, if we again consider the three vending machines from Section 11.4, we can note that their starting states are pairwise 2-game equivalent but pairwise not 3-game equivalent.

1. $V_i \sim_2 V_j$ for $i, j \in \{1, 2, 3\}$.

   The second player has a winning strategy in the game which ends after the exchange of only two moves, as all three machines start with two consecutive 10p transitions.

2. $V_1 \nsim_3 V_2$ and $V_1 \nsim_3 V_3$.

   The first player has a winning strategy in the game which lasts for three exchanges of moves, namely to play arbitrarily for the first two exchanges of moves, and then to take the transition in the V_1 process (coffee or tea) which is *not* available to the other process. The second player will be stuck at this point and lose the game.

3. $V_2 \nsim_3 V_3$.

   The first player has a winning strategy in the game which lasts for three exchanges of moves, namely to open with the transition in the

V_3 process towards the coffee transition, and then in the second move to take the transition in the V_2 process towards the tea transition. The first player can then take the tea transition in the V_2 process, which the second player cannot respond to.

Exercise 12.3   (Solution on page 471)

Recall the following processes from Exercise 11.16.

$$A \stackrel{\text{def}}{=} b.c.0 + b.d.0 \qquad\qquad C \stackrel{\text{def}}{=} a.B + a.A$$
$$B \stackrel{\text{def}}{=} A + b.(c.0 + d.0) \qquad\qquad D \stackrel{\text{def}}{=} a.B$$

For which n do we have that $C \sim_n D$? Justify your answer.

12.2 Properties of Game Equivalence

In this section, we explore various properties of game equivalences, beginning with the following characterisation which in particular provides an elegant inductive definition of finite game equivalences.

Theorem 12.3

1. $E \sim_0 F$ for all processes E and F.

2. $E \sim_{n+1} F$ if, and only if,

   - if $E \stackrel{a}{\to} E'$ then $F \stackrel{a}{\to} F'$ for some F' such that $E' \sim_n F'$; and
   - if $F \stackrel{a}{\to} F'$ then $E \stackrel{a}{\to} E'$ for some E' such that $E' \sim_n F'$.

3. $E \sim_\infty F$ if, and only if,

   - if $E \stackrel{a}{\to} E'$ then $F \stackrel{a}{\to} F'$ for some F' such that $E' \sim_\infty F'$; and
   - if $F \stackrel{a}{\to} F'$ then $E \stackrel{a}{\to} E'$ for some E' such that $E' \sim_\infty F'$.

Pictorially, 2. and 3. can be represented as follows:

Proof: The first result about 0-game equivalence is trivially true, as the second player is immediately declared to be the winner of any game which lasts for only 0 exchanges of moves.

For the second result, we note that the second player has a winning strategy in the game $G_{n+1}(E, F)$ if, and only if, regardless of what move the first player makes – either $E \xrightarrow{a} E'$ or $F \xrightarrow{a} F'$ – the second player can make a response – either $F \xrightarrow{a} F'$ or $E \xrightarrow{a} E'$ – in such a way that he still has a winning strategy in the game $G_n(E', F')$. But this is precisely what the statement in the theorem says.

Similarly for the third result, we note that the second player has a winning strategy in the game $G_\infty(E, F)$ if, and only if, regardless of what move the first player makes – either $E \xrightarrow{a} E'$ or $F \xrightarrow{a} F'$ – the second player can make a response – either $F \xrightarrow{a} F'$ or $E \xrightarrow{a} E'$ – in such a way that he still has a winning strategy in the game $G_\infty(E', F')$. Again this is precisely what the statement in the theorem says.   □

We can use Theorem 12.3 to prove that these game equivalence relations are indeed equivalence relations.

Theorem 12.4

The relations $\sim_n$ and $\sim_\infty$ are all equivalence relations.

Proof: To show that the relations $\sim_n$ and $\sim_\infty$ are reflexive, that is, that $E \sim_n E$ and $E \sim_\infty E$ for all E, we need to prove the following.

> *The second player has a winning strategy in any game in which the two tokens start on the same state E of some transition system.*

This is obvious, as the second player need merely copy every move of the first player; wherever the first player moves one of the tokens, the second player moves the other token to the same place.

To show that the relations $\sim_n$ and $\sim_\infty$ are symmetric, that is, that $F \sim_n E$ whenever $E \sim_n F$ and that $F \sim_\infty E$ whenever $E \sim_\infty F$, we need to prove the following.

> *If the second player has a winning strategy in a game in which the tokens start on states E and F of some transition system, then he also has a winning strategy in the same game but with the tokens starting on states F and E.*

Again this is obvious, due to the symmetry of the game. The second player need merely use (essentially) the same winning strategy.

To show that the relations $\sim_n$ and $\sim_\infty$ are transitive, that is, that $E \sim_n G$ whenever $E \sim_n F$ and $F \sim_n G$, and that $E \sim_\infty G$ whenever $E \sim_\infty F$ and $F \sim_\infty G$, we need to prove the following.

> *If the second player has a winning strategy in a game in which the tokens start on states E and F of some transition system, and*

he has a winning strategy in the same game but with the tokens starting on states F and G, then he also has a winning strategy in the same game but with the tokens starting on states E and G.

The details of this are left as an exercise.   □

Exercise 12.4   (Solution on page 471)

Prove, by induction on n, that the relation $\sim_n$ is transitive for each n.

That the relation $\sim_\infty$ is transitive cannot be proved by induction, but is proven in Exercise 12.7 (page 317).

Theorem 12.5

The relations $\sim_n$ and $\sim_\infty$ are strictly decreasing: $\sim_0 \supset \sim_1 \supset \sim_2 \supset \cdots \supset \sim_\infty$. In particular, if $E \sim_n F$ then $E \sim_k F$ for all $k \leq n$.

Proof: If the first player has a winning strategy in a game of length n, then she can use that strategy to win any game with a longer time limit (and in particular, the game with no predetermined finite time limit). Alternatively, if the second player has a winning strategy in a game of length n, or one with no predetermined finite time limit, then he can use that strategy to win any game with a shorter time limit. This demonstrates the sequence of inclusions of the relations: if a pair of states is in $\sim_i$ it will be in $\sim_j$ for all $j<i$, and hence $\sim_0 \supseteq \sim_1 \supseteq \sim_2 \supseteq \cdots \supseteq \sim_\infty$.

That these inclusions are strict can be noted by observing that for all $n \in \mathbb{N}$, $Cl_n \sim_n Cl$ but $Cl_n \not\sim_{n+1} Cl$; and that for all $n \in \mathbb{N}$, Clock $\sim_n$ Clock$_*$ but Clock $\not\sim_\infty$ Clock$_*$, where these clock processes were defined in Example 11.11 (page 296).   □

Exercise 12.5   (Solution on page 472)

Prove the above claims, that for all $n \in \mathbb{N}$, $Cl_n \sim_n Cl$ but $Cl_n \not\sim_{n+1} Cl$; and that for all $n \in \mathbb{N}$, Clock $\sim_n$ Clock$_*$ but Clock $\not\sim_\infty$ Clock$_*$.

12.3 Bisimulation Relations

We might expect $\sim_\infty$ to be the "limit" of the $\sim_n$ relations, that is, that the second player should have a winning strategy in the infinite game whenever he has a winning strategy for arbitrarily-long finite games. Alas, the above example disproves this intuition, as the two clocks Clock and Clock$_*$ are n-game equivalent for all n but they are not infinite-game equivalent.

Clearly these two clocks cannot be considered to be the same; the first one is guaranteed to stop after some indeterminate number of ticks, whereas the latter has the potential to tick forever. Infinite-game equivalence is thus the relation we wish to consider as defining equivalence between processes, and we shall henceforth generally refer to it as equivalence rather than infinite-game equivalence; that is, when we declare that two processes are equivalent, we shall mean that they are infinite-game equivalent.

If our intuition had been right, then to demonstrate that two processes were equivalent we could exploit Theorem 12.3 and use induction to prove them to be n-game equivalent for all $n \in \mathbb{N}$. However, in general we need an alternative proof strategy to induction. Motivated by Theorem 12.3(3), we define the following notion to capture the essence of a winning strategy for the second player in an infinite game.

Definition 12.5

A *bisimulation relation* is a binary relation $\mathcal{R}$ over states which satisfies the following property: if $E\mathcal{R}F$ then

- if $E \xrightarrow{a} E'$ then $F \xrightarrow{a} F'$ for some F' such that $E'\mathcal{R}F'$; and
- if $F \xrightarrow{a} F'$ then $E \xrightarrow{a} E'$ for some F' such that $E'\mathcal{R}F'$.

We can represent this situation pictorially as follows:

$$
\begin{array}{ccc}
E & \overset{\mathcal{R}}{-----} & F \\
\downarrow a & & \downarrow a \\
E' & \underset{\mathcal{R}}{-----} & F'
\end{array}
$$

As desired, a bisimulation relation $\mathcal{R}$ represents a winning strategy for the second player in an infinite game: whenever the two tokens are on states which are related by $\mathcal{R}$, the second player can match any move of the first player in such a way as to ensure that the tokens once again end up on states related by $\mathcal{R}$. In this way, the second player can repeatedly match the moves of the first player *ad infinitum*.

Theorem 12.6

The second player has a winning strategy in an infinite game with the tokens starting on states E and F if, and only if, $E\mathcal{R}F$ for some bisimulation relation $\mathcal{R}$. Hence in particular, $\mathcal{R} \subseteq \sim_\infty$ for any bisimulation relation $\mathcal{R}$.

Proof: If $E\mathcal{R}F$ for some bisimulation $\mathcal{R}$, then the second player can merely use the winning strategy represented by $\mathcal{R}$ as outlined above in order to win the infinite game with the tokens starting on states E and F.

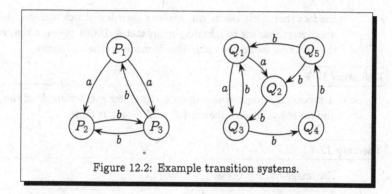

Figure 12.2: Example transition systems.

Conversely, by Theorem 12.3(3), the relation $\sim_\infty$ itself is a bisimulation relation. Hence, if the second player has a winning strategy in an infinite game with the tokens starting on states E and F, then $E\mathcal{R}F$ for the bisimulation relation $\mathcal{R} = \sim_\infty$. □

Example 12.6

Consider the two transition systems in Figure 12.2. It is straightforward to confirm, from Definition 12.5, that the following binary relation is a bisimulation relation:

$$\mathcal{R} \;=\; \big\{ (P_1, Q_1),\, (P_2, Q_2),\, (P_2, Q_4),\, (P_3, Q_3),\, (P_3, Q_5) \big\}.$$

As $(P_1, Q_1) \in \mathcal{R}$, by Theorem 12.6 we get that $P_1 \sim_\infty Q_1$.

Exercise 12.6   (Solution on page 473)

Prove that the relation $\mathcal{R}$ in Example 12.6 is a bisimulation relation.

Exercise 12.7   (Solution on page 473)

Prove that if $\mathcal{R}$ and $\mathcal{S}$ are bisimulation relations over the states of a labelled transition system, then so is $\mathcal{R} \circ \mathcal{S}$. Infer from this that $\sim_\infty$ is a transitive: that if $E \sim_\infty F$ and $F \sim_\infty G$ then $E \sim_\infty G$.

As a final observation regarding the relationship between the finite-game equivalences $\sim_n$ and the infinite-game equivalence $\sim_\infty$, we note that the reason $\sim_\infty \neq \bigcap_{n\in\mathbb{N}} \sim_n$ in the case of the two clocks – that is, that we can have Clock $\sim_n$ Clock$_\star$ for all $n \in \mathbb{N}$ but Clock $\not\sim_\infty$ Clock$_\star$ – is solely due to the fact that these clocks can perform their initial tick action in infinitely-many ways, leading to infinitely-many states. If this were not the case, then the relations would coincide. This is made precise as follows.

Definition 12.7

A process is image-finite if, and only if, for every state E of the process, and for every label a, the set $\{ F : E \xrightarrow{a} F \}$ is finite.

Theorem 12.7

For image-finite processes, $\sim_\infty = \bigcap_{n \in \mathbb{N}} \sim_n$.

Proof: Inclusion in one direction, $\sim_\infty \subseteq \bigcap_{n \in \mathbb{N}} \sim_n$, is guaranteed by Theorem 12.5: $\sim_\infty \subseteq \sim_n$ for all $n \in \mathbb{N}$, so $\sim_\infty \subseteq \bigcap_{n \in \mathbb{N}} \sim_n$.

To show inclusion in the other direction, $\bigcap_{n \in \mathbb{N}} \sim_n \subseteq \sim_\infty$, it suffices to prove that the relation $\mathcal{R} = \bigcap_{n \in \mathbb{N}} \sim_n$ is a bisimulation relation, for then by Theorem 12.6 we would have that $\mathcal{R} \subseteq \sim_\infty$ as desired.

To this end, let $E\mathcal{R}F$ be an arbitrary pair of states related by $\mathcal{R}$, that is, $E \sim_n F$ for all $n \in \mathbb{N}$. Assume first that $E \xrightarrow{a} E'$. Since $E \sim_{n+1} F$ for all $n \in \mathbb{N}$, by Theorem 12.3(2) we have that for each $n \in \mathbb{N}$, $F \xrightarrow{a} F_n$ for some F_n with $E' \sim_n F_n$. However, by image-finiteness there can be only finitely-many such F_n. Hence the same state F' must appear as F_n for infinitely-many values of n; that is, $F \xrightarrow{a} F'$ with $E' \sim_n F'$ for infinitely-many $n \in \mathbb{N}$, and hence by Theorem 12.5 for *all* $n \in \mathbb{N}$. Hence $E'\mathcal{R}F'$.

By a symmetric argument, we can show that if $F \xrightarrow{a} F'$ then $E \xrightarrow{a} E'$ for some E' with $E'\mathcal{R}F'$. Hence $\mathcal{R}$ is indeed a bisimulation. $\square$

Exercise 12.8 (Solution on page 474)

In the definition of the bisimulation game, the first player was free to move either token at each move. Suppose instead she must always move the same token with each move. For example, if for her first move she moves the token on state F, then she must always move that token in every move; at no time can she switch and move the token which started on state E. Let $E \asymp_n F$ if, and only if, the second player has a winning strategy in this new game played for at most n rounds (where n may be ∞).

1. Show that $\asymp_n$ is an equivalence relation.

2. Show that $E \sim_n F$ implies $E \asymp_n F$. That is, if the second player has a winning strategy in the bisimulation game, then he has a winning strategy in this new game.

3. Show that $E \asymp_n F$ in general does not imply that $E \sim_n F$. (Hint: consider the processes $a.b.0$ and $a.b.0 + a.0$.)

12.4   Bisimulation Colourings

Given that we cannot in general employ the inductive characterisation for finite-game equivalences to prove that two process states are (infinite-game) equivalent, we here devise an alternative approach to inferring if and when a winning strategy exists for the second player in an infinite game. The approach relies on colouring the states of the process in a particular fashion, thus partitioning the states into equivalence classes defined by colour.

Definition 12.8

A *bisimulation colouring* of a transition system is a colouring of the states which satisfies the following property:

> If *some* state with some colour C has a transition leading out of it into a state with some colour C', then *every* state coloured C has an identically-labelled transition leading out of it into a state coloured C'.

For example, if some red state has an a-transition leading to a blue state, then *every* red state has an a-transition leading to a blue state.

That is to say, if E and F have the same colour, then

- if $E \xrightarrow{a} E'$ then $F \xrightarrow{a} F'$ for some F' such that E' and F' have the same colour; and
- if $F \xrightarrow{a} F'$ then $E \xrightarrow{a} E'$ for some E' such that E' and F' have the same colour.

Two states E and F are *bisimulation equivalent* or *bisimilar*, written $E \sim F$, if they have the same colour in some *bisimulation colouring*.

As a trivial example, if we assign each state its own unique colour, then this would clearly be a bisimulation colouring. However, finding a bisimulation colouring which assigns the same colour to two different states allows us to conclude that these two states are equivalent. This fact is recorded in the following.

Theorem 12.8

$E \sim F$ if, and only if, $E \sim_\infty F$.

Proof: Given a bisimulation colouring of a transition system, the binary relation $\mathcal{R}$ which relates like-coloured states is clearly a bisimulation relation (according to Definition 12.5), and hence, by Theorem 12.6, any two like-coloured states must be infinite-game equivalent. That is, if $E \sim F$ (i.e., E and F have the same colour in some bisimulation colouring) then $E \sim_\infty F$.

Conversely, consider colouring a transition system in such a way that any two states E and F have the same colour if, and only if, $E \sim_\infty F$. By Theorem 12.3(3), this colouring is clearly a bisimulation colouring (according to Definition 12.8). Thus, if $E \sim_\infty F$ then E and F have the same colour in this bisimulation colouring, and hence $E \sim F$.   □

This new characterisation of equivalence gives rise to the following approach to demonstrating that two states of a transition system are (or are not) equivalent. We start with all states being the same colour (white, say), and refine this colouring, always maintaining the following invariant:

> **Invariant:** *If $E \sim F$ then E and F have the same colour.*

In this way, we start with a single equivalence class of states (ie, start with all states assigned the same colour), and refine this partition by subdividing the equivalence classes (by assigning some of the states in an equivalence class a new colour). This *partition refinement algorithm* can be effectively implemented to prove (or disprove) equivalences.

As an illustrative example, consider the second transition system of Figure 12.1.

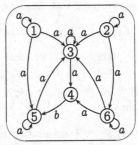

The initial all-white colouring is *not* a bisimulation colouring, as the white state 4 has a b-transition to a white state 5, whereas the other white states 1, 2, 3, 5 and 6 do not have b-transitions to white states. Hence, by the invariant, state 4 cannot be equivalent to the other white states; in any bisimulation colouring, state 4 must have a different colour from states 1, 2, 3, 5 and 6. Hence we may safely refine our colouring by making state 4 a different colour (black, say).

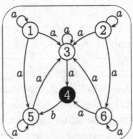

This is still not a bisimulation colouring, as the white states 3 and 6 have a-transitions to black states, whereas the other white states 1, 2 and 5 do not. Hence, by the invariant, states 3 and 6 cannot be equivalent to the other white states; in any bisimulation colouring, states 3 and 6 must have a different colour from states 1, 2 and 5. Hence we may safely refine our colouring by making states 3 and 6 a different colour (grey, say).

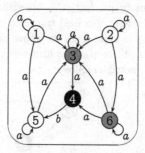

This colouring *is* a bisimulation colouring, which by construction satisfies our invariant. To confirm this, we merely enumerate the possibilities.

1. every white states has an a-labelled arrow leading into a white state, and an a-labelled arrow leading into a grey state;

2. every grey state has an a-labelled arrow leading into a grey state, and an a-labelled arrow leading into a black state; and

3. every black state has a b-labelled arrow leading into a white state.

Hence, two states in this transition system are equivalent if, and only if, they have the same colour.

For the first transition system in Figure 12.1, a little reflection reveals that no bisimulation colouring of the states of this transition system exists in which the states X and U have the same colour.

<table>
<tr><td>Exercise 12.9</td><td>(Solution on page 475)</td></tr>
</table>

Prove the above claim that the states X and U of the first transition system in Figure 12.1 cannot have the same colour in any bisimulation colouring.

This completes the outline of our algorithm for determining whether two states of a transition system are equivalent. The algorithm works by partitioning the states into equivalence classes, by starting with the trivial partition consisting of a single class containing all states, and repeatedly refining the partition by splitting one of the classes into two separate sub-classes; it does this when it discovers that none of the states of one of the new sub-classes can be equivalent to any of the states of the other. If we carry this procedure out on a transition system with n states, then clearly it can perform no more than n refinements, as each refinement gives rise to a new class and we cannot produce a partition with more than n classes. Furthermore, during each iteration we need only scan the edges of the transition system looking for a transition with which we can split a partition. Hence if there are k edges in the transition system, then this naive implementation of the algorithm would execute in time proportional to nk.

As a useful by-product, this algorithm produces a minimal-sized (in terms of the number of states) transition system which is equivalent to the original transition system. In the above example, the minimal-sized transition system has three states, which we might refer to as white, black and grey, and is depicted as follows.

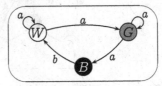

Exercise 12.10   (Solution on page 475)

Carry out the above bisimulation colouring algorithm on the first transition system of Figure 12.1, explaining each step in detail as above.

Note that the algorithm is nondeterministic; there may be several ways of splitting a set of like-coloured states. For example, starting with all states of the transition system in question white, there are three possible ways to proceed.

1. White states U and X both have an a-transition leading to a white state, while white states V, W, Y and Z do not.

2. White states V and Y both have a b-transition leading to a white state, while white states U, W, X and Z do not.

3. White states W and Y both have a c-transition leading to a white state, while white states U, V, X and Z do not.

It doesn't matter which choice you make; the end result will be the same.

★ **12.5   The Bisimulation Game Revisited: To Infinity and Beyond!**

As we observed, the relations $\sim_n$ representing n-game equivalence do not, in general, provide an adequate sequence of approximations to $\sim_\infty$, the ∞-game equivalence. This was demonstrated in Exercise 12.5 by the example of the clocks, in which $\texttt{Clock} \sim_n \texttt{Clock}_*$ for all $n \in \mathbb{N}$ but $\texttt{Clock} \not\sim_\infty$ $\texttt{Clock}_*$. All is not lost with the idea of approaching $\sim_\infty$ by a sequence of approximations. The solution – which seems very odd on first encountering it – is to take the advice of Buzz Lightyear and go to infinity and beyond. The example of the clocks shows that it is not enough just to go to infinity through the natural numbers $\sim_0, \sim_1, \sim_2, \sim_3, \ldots$. All we need to do is make sense of the idea of going beyond infinity.

Consider two young children playing the game of *"Who can name the largest number?"* in which they take turns naming larger and larger numbers. They quickly run up against the problem of what numbers come after *one-million, one-billion, one-trillion, one-quadrillion, ...*, until one of them discovers the number *googol* (10^{100}, a one followed by 100 zeros); but the other quickly responds with an even bigger number: *googol-plus-one!* The first child's argument that *"There's no such thing, googol is the biggest number!"* is of course wrong. But then eventually, one of the children names the "number" *infinity*.

It is possible to accept the idea of naming *infinity* as a number, which by definition is bigger than any natural number, and even to give it its own symbol: ω. But we will then be able to consider $\omega+1$ as a bigger number, and $\omega+2$ as an even bigger number, and even $\omega+\omega$ as a far, far greater number; these are all infinitely-big numbers, but some are just bigger than others.

We already noted in Section 6.4, when comparing the sizes (cardinalities) of sets, that *infinity* comes in different varieties; in particular, the cardinality of the set of rational numbers is the same as the cardinality of the set of natural numbers (Exercise 6.16) but strictly smaller than the cardinality of the set of reals (Example 6.16). Infinite *counting* numbers (as opposed to *measuring* numbers) also exist as mathematical objects, and are collectively known as *ordinal numbers*. These are what will allow us to approximate $\sim_\infty$.

12.5.1 Ordinal Numbers

The ordinal numbers are an extension of the natural numbers as motivated above. The initial segment of ordinals is as follows:

$$0, 1, 2, \ldots, \omega, \ \omega+1, \ \omega+2, \ \ldots, \ \omega+\omega, \ \omega+\omega+1, \ \omega+\omega+2, \ \ldots$$

Thus, after all finite ordinals have been listed (the natural numbers), the first infinite ordinal ω is listed, and we can once again list ever-bigger ordinals by successively adding one; after adding each natural number to ω we reach the ordinal $\omega+\omega$, or $\omega\times2$, from which we continue the scheme, *ad infinitum*. The collection of ordinal numbers is denoted by $\mathcal{O}$. We shall not concern ourselves with the complete theory of ordinal numbers. All we will need to know about ordinals are the following four facts:

1. Every ordinal X has a *successor* $X+1$, whose *predecessor* is X.

2. An ordinal is either: zero (i.e., 0); or a successor ordinal (i.e., $X+1$ for some ordinal X); or a *limit* ordinal which has a value which is greater than all previous ordinals, but has no predecessor.

   The first limit ordinal – which is the first infinite ordinal – is ω. It is the smallest ordinal greater than any finite ordinal (i.e., natural) number; the next limit ordinal is $\omega+\omega$ (which is also written $\omega\times2$), then $\omega+\omega+\omega$ (or $\omega\times3$), and so on.

3. Given any set S there is an ordinal $X \in \mathcal{O}$ which represents the cardinality of S; that is, there is a bijection between the set S and the set $\{Y \in \mathcal{O} : Y < X\}$.

4. In order to show that a property $P(X)$ holds for all ordinals X, it suffices to show the following:

$P(X)$ holds for X whenever $P(Y)$ holds for all $Y < X$; that is,

$$\big(\forall Y < X : P(Y)\big) \Rightarrow P(X).$$

This principle is known as *transfinite induction* and is a restatement of the principle of strong induction from Section 9.4.

For those who find this brief initiation into the world of ordinal numbers confusing, you may find it helpful to concentrate on the natural numbers, and just think of ω whenever limit ordinals are mentioned in what follows.

Example 12.10

Consider the set $\mathbb{N} \times \mathbb{N}$ of pairs of natural numbers ordered lexicographically: $(i, j) < (p, q)$ if, and only if, either $i < p$ or $i = p$ and $j < q$. Thus, we can list these out in order as follows:

$$(0,0) < (0,1) < (0,2) < \cdots < (1,0) < (1,1) < (1,2) < \cdots$$
$$< (2,0) < (2,1) < (2,2) < \cdots$$
$$< (3,0) < (3,1) < (3,2) < \cdots$$
$$< (4,0) < (4,1) < (4,2) < \cdots$$
$$< \cdots$$

This gives us a way to view the start of the list of ordinal numbers, namely by associating the pair $(i, j) \in \mathbb{N} \times \mathbb{N}$ with the ordinal number $\omega\times i + j$.

12.5.2 Ordinal Bisimulation Games

In Section 12.1 we defined the bisimulation game as either lasting for a predefined finite number n of exchanges of moves, denoting the game by

$G_n(E, F)$; or as continuing for as long as each player can make a move, denoting the game in this case by $G_\infty(E, F)$. We can refine this notion by defining the game $G_X(E, F)$ for any ordinal number X. From now on, we shall use $G_X(E, F)$ to denote both the game itself as well as the position that the game is in, with X denoting in a precise sense the length of the game.

1. From position $G_0(E, F)$, the second player is declared to be the winner.

   This reflects the idea that the second player automatically wins any game of length 0, as he need not copy any moves of the first player.

2. From position $G_{X+1}(E, F)$, the two players exchange moves once as usual, and the play continues from $G_X(E', F')$, where E' and F' are the states to which the two tokens have been moved.

   This reflects the usual idea that a game of length $X+1$ consists of a single exchange of moves followed by a game of length X.

3. From position $G_\lambda(E, F)$ where λ is a limit ordinal, the first player chooses a value $X < \lambda$, and the play continues from position $G_X(E, F)$.

   This reflects the idea that $G_\lambda(E, F)$ encompasses all games of length less than λ; that is, $G_X(E, F)$ for any $X < \lambda$. If the second player has a winning strategy in all such shorter games then he can force a win in any such game that the first player chooses, so the second player can force a win in this game $G_\lambda(E, F)$. However, if the first player has a winning strategy in some such shorter game, then she can choose that game and use her winning strategy to win the game $G_\lambda(E, F)$.

The following result corresponds to Theorem 12.2 (page 312), and is similarly proved but by transfinite induction rather than simple induction over the natural numbers.

Theorem 12.10

For any game $G_X(E, F)$, either the first player has a winning strategy, or the second player has a winning strategy.

Proof: By transfinite induction. For the case $X = 0$, the second player clearly has a winning strategy for the game $G_0(E, F)$.

Suppose that $X = Y+1$ is a successor ordinal, and that for any game $G_Y(E', F')$ one of the two players has a winning strategy.

Suppose that $X = Y+1$ is a successor ordinal, and that for any game $G_Y(E', F')$ one of the two players has a winning strategy. The argument that one of the two players has a winning strategy in the game $G_{Y+1}(E, F)$ is identical to the induction step in the proof of Theorem 12.2.

- Suppose that no matter what the first player does as her first move in the game $G_{Y+1}(E, F)$, the second player can respond in such a way that he gets into a position in which he has a winning strategy in the game of length Y. This clearly provides a winning strategy for the second player in the game $G_{Y+1}(E, F)$.

- Hence, if the second player does *not* have a winning strategy in the game $G_{Y+1}(E, F)$, then the first player can make a move in such a way that any response the second player makes results in a position from which the second player does not have a winning strategy in the game of length Y; but then by the inductive hypothesis, the first player has a winning strategy in the game of length Y from this resulting position, which means she has a winning strategy for the game $G_{Y+1}(E, F)$.

Suppose finally that X is a limit ordinal, and that for any game $G_Y(E', F')$ with $Y < X$ one of the players has a winning strategy.

- If there is some $Y < X$ such that the first player has a winning strategy in the game $G_Y(E, F)$, then she can choose this value $Y < X$ and use this winning strategy to win the game $G_X(E, F)$.

- If there is no $Y < X$ such that the first player has a winning strategy in the game $G_Y(E, F)$, then by the induction hypothesis the second player has a winning strategy for the game $G_Y(E, F)$ for each $Y < X$, and hence a winning strategy for the game $G_X(E, F)$.   □

Definition 12.10

We say that two process states E and F are *X-game equivalent*, written $E \sim_X F$, if, and only if, the second player has a winning strategy in the game $G_X(E, F)$.

Example 12.11

From Exercise 12.5 we know that $\text{Clock} \sim_\omega \text{Clock}_\star$ since $\text{Clock} \sim_n \text{Clock}_\star$ for all $n \in \mathbb{N}$ (i.e., $\text{Clock} \sim_X \text{Clock}_\star$ for all $X < \omega$).

However, $\text{Clock} \not\sim_{\omega+1} \text{Clock}_\star$, since the move $\text{Clock}_\star \xrightarrow{tick} \text{Cl}$ by the first player in the game $G_{\omega+1}(\text{Clock}_\star, \text{Clock})$ must be matched by a move $\text{Clock} \xrightarrow{tick} \text{Cl}_n$ for some $n \in \mathbb{N}$, but for *no* $n \in \mathbb{N}$ do we have that $\text{Cl} \sim_\omega \text{Cl}_n$. On the other hand, we do have that $tick.\text{Clock}_\star \sim_{\omega+1} tick.\text{Clock}$.

Exercise 12.11   (Solution on page 477)

Give process states E_n and F_n such that $E_n \sim_{\omega+n} F_n$ but $E_n \not\sim_{\omega+n+1} F_n$.

We can now extend the results of Section 12.2 about game equivalence to ordinal game equivalence. We leave most of the proofs as exercises, as they are straightforward adaptations of the proofs of the analogous results presented in Section 12.2. However, we prove the last result, which is the goal of this section: that the sequence of equivalences $\sim_X$ does indeed properly approximate $\sim_\infty$.

Theorem 12.11

1. $E \sim_0 F$ for all processes E and F.

2. $E \sim_{X+1} F$ if, and only if,

   - if $E \xrightarrow{a} E'$ then $F \xrightarrow{a} F'$ for some F' such that $E' \sim_X F'$; and
   - if $F \xrightarrow{a} F'$ then $E \xrightarrow{a} E'$ for some E' such that $E' \sim_X F'$.

3. For limit ordinals λ, $E \sim_\lambda F$ if, and only if, $E \sim_X F$ for all $X < \lambda$.

Theorem 12.12

The relations $\sim_X$ are all equivalence relations.

Theorem 12.13

The relations $\sim_X$ are strictly decreasing. That is, $\sim_X \subset \sim_Y$ whenever $X > Y$.

Specifically, if for each ordinal $X \in \mathcal{O}$ we define the process

$$E_X \stackrel{\text{def}}{=} \sum_{Y < X} a.E_Y,$$

then for all ordinals X and Y with $X < Y$, $E_X \sim_X E_Y$ but $E_X \not\sim_{X+1} E_Y$.

Theorem 12.14

$\sim_\infty = \bigcap_{X \in \mathcal{O}} \sim_X$.

Proof: Suppose that $E \sim_\infty F$; we shall show by transfinite induction that $E \sim_X F$ for all $X \in \mathcal{O}$.

- If $X = 0$ then clearly $E \sim_0 F$.
- Suppose that $X = Y + 1$ is a successor ordinal.

  - If $E \xrightarrow{a} E'$ then $F \xrightarrow{a} F'$ for some F' such that $E' \sim_\infty F'$, and hence by induction $E' \sim_Y F'$.
  - If $F \xrightarrow{a} F'$ then $E \xrightarrow{a} E'$ for some E' such that $E' \sim_\infty F'$, and hence by induction $E' \sim_Y F'$.

Thus we must have that $E \sim_{Y+1} F$.

- Suppose finally that X is a limit ordinal. Then by induction we have that $E \sim_Y F$ for all $Y < X$, and hence $E \sim_X F$.

To show inclusion in the other direction, $\bigcap_{X \in \mathcal{O}} \sim_X \subseteq \sim_\infty$, it suffices to prove that the relation $\mathcal{R} = \bigcap_{X \in \mathcal{O}} \sim_X$ is a bisimulation relation, for then by Theorem 12.6 we would have that $\mathcal{R} \subseteq \sim_\infty$ as desired.

To this end, let $E\mathcal{R}F$ be an arbitrary pair of states related by $\mathcal{R}$, that is, $E \sim_X F$ for all $X \in \mathcal{O}$. Assume first that $E \xrightarrow{a} E'$. Since $E \sim_{X+1} F$ for all $X \in \mathcal{O}$, by Theorem 12.11(2) we have that for each $X \in \mathcal{O}$, $F \xrightarrow{a} F_X$ for some F_X with $E' \sim_X F_X$. The set $\{ F_X : X \in \mathcal{O} \}$ can be no greater that the set of all states which are reachable from the state F, and there are ordinal numbers X which are arbitrarily larger than the cardinality of this set of states. Hence there must be some state F' which appears as F_X for arbitrarily-large values of X; that is, $F \xrightarrow{a} F'$ with $E' \sim_X F'$ for arbitrarily-large $X \in \mathcal{O}$, and hence by Theorem 12.13 for *all* $X \in \mathcal{O}$. Hence $E'\mathcal{R}F'$.

By a symmetric argument, we can show that if $F \xrightarrow{a} F'$ then $E \xrightarrow{a} E'$ for some E' with $E'\mathcal{R}F'$. Hence $\mathcal{R}$ is indeed a bisimulation. $\square$

12.6 Additional Exercises

1. Carry out the bisimulation colouring algorithm step-by-step on the labelled transition system defined by the following process definition, and use this to provide an equivalent system with a minimal number of states.

$$W \stackrel{\text{def}}{=} b.X + c.Z \qquad X \stackrel{\text{def}}{=} a.Y$$
$$Y \stackrel{\text{def}}{=} c.X + b.Z \qquad Z \stackrel{\text{def}}{=} a.W + a.Y$$

2. Carry out the bisimulation colouring algorithm step-by-step on the labelled transition system defined by the following process definition, and use this to provide an equivalent system with a minimal number of states.

$$X_1 \stackrel{\text{def}}{=} a.X_1 + b.X_3 \qquad X_4 \stackrel{\text{def}}{=} a.X_4 + b.X_3$$
$$X_2 \stackrel{\text{def}}{=} a.X_3 + a.X_6 + b.X_1 \qquad X_5 \stackrel{\text{def}}{=} a.X_3 + a.X_6 + b.X_1$$
$$X_3 \stackrel{\text{def}}{=} a.X_5 \qquad X_6 \stackrel{\text{def}}{=} a.X_3 + a.X_5 + b.X_4$$

3. Consider the following labelled transition system.

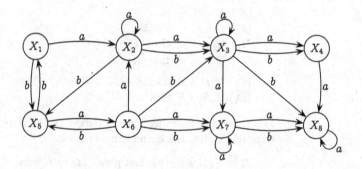

(a) Which states are 2-game equivalent to state X_6?

(b) Which states are 2-game equivalent, but *not* 3-game equivalent, to state X_6?

(c) Which states are n-game equivalent to state X_5 for all n?

4. Consider the following labelled transition system.

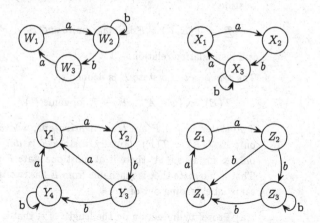

(a) For which n do we have $W_1 \sim_n X_1$? Justify your answer.

(b) For which n do we have $W_1 \sim_n Y_1$? Justify your answer.

(c) For which n do we have $W_1 \sim_n Z_1$? Justify your answer.

(d) For which n do we have $X_1 \sim_n Y_1$? Justify your answer.

(e) For which n do we have $X_1 \sim_n Z_1$? Justify your answer.

(f) For which n do we have $Y_1 \sim_n Z_1$? Justify your answer.

5. Show that the algebraic laws from Section 11.5 are true of bisimulation equivalence:

$(S_1) \ E + 0 \sim E.$

(S_2) $E + E \sim E$.

(S_3) $E + F \sim F + E$.

(S_4) $(E + F) + G \sim E + (F + G)$.

(S_5) If $X \stackrel{\text{def}}{=} E$ then $X \sim E$.

(C_1) If $E \sim F$ then $E + G \sim F + G$.

(C_2) If $E \sim F$ then $a.E \sim a.F$.

6. Prove that the following binary relation on the states of a labelled transition system is a bisimulation relation:

$$R = \{ (E, F) : \text{ the first player does } not \text{ have} $$
$$\text{a winning strategy in the game } G_\infty(E, F) \}.$$

   Conclude from this the result from Theorem 12.2 that for any game $G_\infty(E, F)$, either the first player has a winning strategy, or the second player has a winning strategy.

7. In Theorem 12.7, only one of the two processes need be image-finite in order for the conclusion to be true. Prove this by showing that the relation

$$\mathcal{R} = \{ (E, F) : F \text{ is image-finite and } E \sim_k F \text{ for all } k \in \mathbb{N} \}$$

   is a bisimulation relation.

8. The *trace set* of a state E is defined as

$$\mathcal{T}(E) = \{ s \in A^* : E \stackrel{s}{\to} F \text{ for some } F \}.$$

   Two states E and F are *trace equivalent*, written $E =_t F$, if, and only if, $\mathcal{T}(E) = \mathcal{T}(F)$. Finally, a state E is *deterministic* if, and only if, for all $s \in A^*$ there is at most one state F such that $E \stackrel{s}{\to} F$. That is, no state that is reachable from E has two transitions with the same label leading out of it.

   (a) Prove, by induction on the length of s, that if $E \sim_n F$ and $E \stackrel{s}{\to} E'$ with $k = length(s) \le n$, then $F \stackrel{s}{\to} F'$ for some F' with $E' \sim_{n-k} F'$. Deduce from this that if $E \sim F$ then $E =_t F$.

   (b) Prove that $\mathcal{R} = \{ (E, F) : E =_t F \text{ and } E, F \text{ are deterministic} \}$ is a bisimulation relation. Deduce from this that if $E =_t F$ and E and F are deterministic, then $E \sim F$.

9. A *trace bisimulation relation* is a binary relation $\mathcal{R}$ over states which satisfies the following property (where the extended transition relation $\to \subseteq S \times A^* \times S$ is defined in the previous exercise):

   If $E\mathcal{R}F$ then

   • if $E \stackrel{s}{\to} E'$ then $F \stackrel{s}{\to} F'$ for some F' such that $E'\mathcal{R}F'$; and

- if $F \xrightarrow{s} F'$ then $E \xrightarrow{s} E'$ for some E' such that $E'\mathcal{R}F'$.

In terms of the bisimulation game, this reflects a change in the rules which allows the first player to make a sequence of transitions, rather than a single transition, which the second player must copy.

Prove that $\mathcal{R}$ is a trace bisimulation relation if, and only if, $\mathcal{R}$ is a bisimulation.

10. A set $R \subseteq \Sigma$ is a *refusal set* of E if, and only if, $E \xrightarrow{a}\!\!\!\!/$ for any $a \in R$. A pair $(w, R) \in \Sigma^* \times 2^\Sigma$ is a *failure* of E if, and only if, $E \xrightarrow{w} F$ for some F such that R is a refusal set of F. E and F are *failures equivalent*, written $E =_f F$, if, and only if, they possess the same failures.

   (a) Prove that $E \sim F$ implies $E =_f F$, and that $E =_f F$ implies $E =_t F$.

   (b) Recalling the vending machines V_1, V_2 and V_3, prove that $V_1 \neq_f V_2$ but that $V_2 =_f V_3$, thus showing that the reverse implications do not hold in general.

   (c) What is the relationship between $=_f$ and $\asymp$, the simulation equivalence from Exercise 12.8?

11. Ordinal numbers, viewed as sets, can be defined as follows:

   - if S is a set of ordinals, then so is $\bigcup S$;
   - if X is an ordinal, then so is $X^+ = X \cup \{X\}$;
   - nothing is an ordinal number unless it is constructed from the above two rules.

Thus we can construct the first few ordinals as follows:

   - $0 = \bigcup \emptyset = \emptyset$
   - $1 = 0^+ = 0 \cup \{0\} = \emptyset \cup \{0\} = \{0\}$
   - $2 = 1^+ = 1 \cup \{1\} = \{0, 1\}$
   - $3 = 2^+ = 2 \cup \{2\} = \{0, 1, 2\}$

   $\vdots$

   - $n = \{0, 1, 2, \ldots, n-1\}$

   $\vdots$

   - $\omega = \bigcup\{0, 1, 2, \ldots\} = \{0, 1, 2, \ldots\}$
   - $\omega+1 = \omega^+ = \omega \cup \{\omega\} = \{0, 1, 2, \ldots, \omega\}$

   $\vdots$

Intuitively, an ordinal is the set of ordinals less than it; and the less-than relation corresponds to membership: $X < Y$ if, and only if, $X \in Y$.

Prove the following facts about ordinal numbers X, Y and Z as defined above.

(a) Every element of an ordinal X is itself an ordinal.
   (Proof: By induction on X.)

(b) If $X \in Y$ and $Y \in Z$ then $X \in Z$; that is, $\in$ is transitive.
   (Proof: By induction on Z.)

(c) If $X \in Y$ then $X \subseteq Y$.
   (Proof: Follows directly from previous result.)

(d) $X \notin X$.
   (Proof: By induction on X.)

(e) $X \cap Y$ is an ordinal.
   (Proof: By induction on X.)

12. Prove Theorem 12.11. (page 327).

13. Prove Theorem 12.12 (page 327).

14. Prove Theorem 12.13 (page 327).

Chapter 13

Logical Properties of Processes

I summed up all systems in a phrase, and all existence in an epigram.

- Oscar Wilde.

Thus far in Part II of the book, we have developed the understanding of what a process is, namely a labelled transition system, as well as the means for describing processes formally with a simple process language. We have also defined when two processes are equivalent – namely when they are bisimilar, which we presented as game equivalent – as well as a procedure for determining if two processes are equivalent.

Determining equivalence between processes is instrumental for finding out if a proposed implementation of a computing system matches its specification. However, we are often not interested in the complete behaviour of a system, but rather only in certain aspects. For example, we may not be interested – for the moment – in what actions a certain system does, but rather we might only want to know if it can ever *deadlock*, that is, evolve into a state in which it can perform no actions. This would be very useful in the analysis of systems which are expected to be perpetual, such as operating systems (particularly those running on critical systems). In other instances we may be interested only in knowing if a given system may or will ever perform a particular action, for example service a request such as printing a document that has been sent to the printer queue.

In this chapter we shall consider a simple logic for expressing properties of systems, as well as the means for determining whether or not a given process satisfies such properties. The properties which the logic can express will be dynamic (behavioural) properties which describe what actions a process can or cannot do, rather than static properties such as how many states a process has which are irrelevant for its correct functioning.

A given property will potentially hold of many different systems, and fail to hold of many others. However, the properties that we express should respect our understanding of equivalence: if a given property holds of a particular process, then it should hold of any other equivalent process. Con-

F. Moller, G. Struth, *Modelling Computing Systems*,
Undergraduate Topics in Computer Science,
DOI 10.1007/978-1-84800-322-4_14, © Springer-Verlag London 2013

versely, if two processes are not equivalent, then you should be able to express a property which distinguishes between these two processes; that is, a property which holds of one of the processes but not the other. The logic which we describe in this chapter is of this nature.

Example 13.1

Consider the following two statements about a particular computer:

1. "The computer consists of three parts: a CPU, a memory unit, and a bus for communicating with the environment."

2. "CONTROL-ALT-DELETE can be pressed; this will shut down the computer, which will then not do anything further."

The first statement does not refer to the (dynamic) behaviour of the computer, but rather to its (static) structure. As such it cannot be used to distinguish between the behaviour of this and any other computer. Another computer may (and likely will) consist of the same three parts yet behave completely differently; while yet another may behave identically to the computer in question despite being built completely differently.

By contrast, the second statement describes one particular aspect of the behaviour of the computer which we may want our computer to demonstrate. Another computer built from the same three parts may be unacceptable if it does not behave the same when you press the CONTROL-ALT-DELETE combination of keystrokes.

13.1 The Mays and Musts of Processes

In trying to understand the differences between the behaviours of the various vending machines in Section 11.4, we were led to making statements such as the following two:

1. We _may_ do a '10p' action and end up in a state in which
   we _may_ do a '10p' action and end up in a state in which
   we _cannot_ do a 'coffee' action.

2. We _may_ do a '10p' action and end up in a state in which
   no matter how we do a '10p' action
   we _must_ end up in a state in which
   we _cannot_ do a 'tea' action.

Thus we are expressing capabilities (and inabilities) of a process using the two auxiliary verbs _may_ and _must_, which are known as _modal helping verbs_ as they help set the modality – necessity or possibility – of the main verb. In fact, we are using these verbs in a very strict manner, namely in the following two contexts:

$\langle a \rangle P$:   *we __may__ do an 'a' action and*
end up in a state in which P is true;

$[a]P$:   *no matter how we do an 'a' action*
we __must__ end up in a state in which P is true.

We will use the above notation, $\langle a \rangle P$ (pronounced "diamond-a" P) and $[a]P$ (pronounced "box-a" P), for writing down such statements.

How, then, can we express the simple property that we may do a 'coffee' action? It doesn't suffice to simply write:

$\langle \texttt{coffee} \rangle$

as – following the translations given above – this reads in English as:

we __may__ do a 'coffee' action and end up in a state in which.

This is not grammatically correct. In order to complete the sentence, we must indicate a property that we require to be true of the state into which the process evolves after doing the 'coffee' action. Every modality "$\langle a \rangle$" and "$[a]$" has to be followed by some property P.

In this case, however, we don't require anything in particular to be true in the state we get into after doing the 'coffee' action; we are only content that we *can* evolve into such a state. To solve this problem, we can use the property true, which of course is itself true of any process state. Thus, to express the property that we may do a 'coffee' action, we would write:

$\langle \texttt{coffee} \rangle \texttt{true}$

which more fully says

we __may__ do a 'coffee' action and
end up in a state in which __true__ is true.

Although the final clause is redundant, as true is always true (that is, true is true in every state), it is nonetheless necessary in order to turn the expression into a complete logical statement.

We may now express the two properties of our vending machines:

1. $\langle 10p \rangle \langle 10p \rangle \neg \langle \texttt{coffee} \rangle \texttt{true}$

2. $\langle 10p \rangle [10p] \neg \langle \texttt{tea} \rangle \texttt{true}$

If we read these two lines as English statements following the translations given above for the new notation – as well as reading the negation of a property, $\neg P$, as *"it is not the case that P"* – we arrive at the following:

1. *We may do a '10p' action and end up in a state in which*
      we may do a '10p' action and end up in a state in which
         it is not the case that
            we may do a 'coffee' action and
            end up in a state in which true is true.

2. *We may do a '10p' action and end up in a state in which*
      no matter how we do a '10p' action
         we must end up in a state in which
            it is not the case that
               we may do a 'tea' action and
               end up in a state in which true is true.

Exercise 13.1   (Solution on page 477)

Explain what each of the following properties expresses.

1. ⟨coffee⟩true
2. ⟨coffee⟩false
3. [coffee]true
4. [coffee]false

Exercise 13.2   (Solution on page 477)

How can we express the property that we cannot do two '*a*' actions in a row? Give your answer using the above notation, and write out your property in English.

Exercise 13.3   (Solution on page 478)

How can we express a property that distinguishes between the clock C1 from Example 11.7 which ticks forever, and the clock $C1_*$ from Exercise 11.8 which may tick forever or may stop ticking after any tick? That is, give a property using the above notation which is true of C1 but false of $C1_*$. Write out your property in English as well.

13.2   A Modal Logic for Properties

In the previous section we presented the core of a simple logical language for expressing properties which may be true or false of a given process.

In this section we complete the description of this simple logic, which we shall simply call **HML** (for Hennessy-Milner Logic, after its inventors). This language consists essentially of propositional logic with the additional two modal connectives $\langle a \rangle P$ ("diamond-a" P) and $[a]P$ ("box-a" P):

$$P, Q \ ::= \ \text{true} \ | \ \text{false} \ | \ \neg P \ | \ P \wedge Q \ | \ P \vee Q \ | \ \langle a \rangle P \ | \ [a]P.$$

A formula P of **HML** represents a property which may or may not be true in a given state E of a process. If it is true in that state, we shall write $E \models P$ and say that the state E *satisfies* the property P; otherwise we will write $E \not\models P$ and say that the state E does *not* satisfy the property P; that is, by $E \not\models P$ we mean $\neg(E \models P)$. If a property is true in *some* state, then we say that the property is *satisfiable*; and if it is true in *every* state, then we say that it is *valid*.

Whether or not a property is true in a given state is defined inductively on the structure of the formula P as follows:

- $E \models \text{true}$ for all E.

  The property true is true in *every* state.

- $E \not\models \text{false}$ for all E.

  The property false is *not* true in *any* state.

- $E \models \neg P$ if, and only if, $E \not\models P$.

  The property $\neg P$ is true in a state if, and only if, P is *not* true in that state.

- $E \models P \wedge Q$ if, and only if, $E \models P$ *and* $E \models Q$.

  The property $P \wedge Q$ is true in a state if, and only if, both P and Q are true in that state.

- $E \models P \vee Q$ if, and only if, $E \models P$ *or* $E \models Q$.

  The property $P \vee Q$ is true in a state if, and only if, either P or Q (or both) is true in that state.

- $E \models \langle a \rangle P$ if, and only if, $F \models P$ for *some* state F such that $E \xrightarrow{a} F$.

  The property $\langle a \rangle P$ is true in a state if, and only if, you can do an 'a' transition from that state to a state in which the property P is true.

- $E \models [a]P$ if, and only if, $F \models P$ for *all* F such that $E \xrightarrow{a} F$.

  The property $[a]P$ is true in a state if, and only if, the property P is true in *every* state that you can get to by doing an 'a' transition from that state.

The syntax and semantics of the logic **HML** is summarised in Figure 13.1.

We shall make use of the following shorthand abbreviations:

$E \models \text{true}$ for *all* E.

$E \models \text{false}$ for *no* E.

$E \models \neg P$ if, and only if, $E \not\models P$.

$E \models P \wedge Q$ if, and only if, $E \models P$ *and* $E \models Q$.

$E \models P \vee Q$ if, and only if, $E \models P$ *or* $E \models Q$.

$E \models \langle a \rangle P$ if, and only if, $F \models P$ for *some* F such that $E \xrightarrow{a} F$.

$E \models [a] P$ if, and only if, $F \models P$ for *all* F such that $E \xrightarrow{a} F$.

Figure 13.1: Syntax and semantics of the modal logic **HML**.

$\langle - \rangle P = \bigvee_a \langle a \rangle P$ where the disjunction ranges over the whole set of actions of a process.

This property is true in a state if, and only if, you can do *some* transition from that state to a state in which the property P is true.

$[-]P = \bigwedge_a [a]P$ where the conjunction ranges over the whole set of actions of a process.

This property is true in a state if, and only if, the property P is true in *every* state that you can get to by doing a transition from that state.

$\langle K \rangle P = \bigvee_{a \in K} \langle a \rangle P$ where K is a set of actions (typically written without set braces, as in $\langle a, b, c \rangle P$).

This property is true in a state if, and only if, you can do an 'a' transition from that state, for some $a \in K$, to a state in which the property P is true. This is the same as $\langle a \rangle P$ when $K = \{a\}$; the same as $\langle - \rangle P$ when K is the set of all actions of a process; and the same as false when $K = \emptyset$.

$[K]P = \bigwedge_{a \in K} [a]P$ where K is a set of actions (typically written without set braces, as in $[a, b, c]P$).

This property is true in a state if, and only if, the property P is true in *every* state that you can get to by doing an 'a' transition from that state, for some $a \in K$. This is the same as $[a]P$ when $K = \{a\}$; the same as $[-]P$ when K is the set of all actions of a process; and the same as true when $K = \emptyset$.

$\langle -K \rangle P = \langle \overline{K} \rangle P$ where K is a set of actions (typically written without set braces, as in $\langle -a, b, c \rangle P$).

This property is true in a state if, and only if, you can do an 'a' transition from that state, for some $a \notin K$ (i.e., for some $a \in \overline{K}$), to a state in which the property P is true.

$[-K]P = [\overline{K}]P$ where K is a set of actions (typically written without set braces, as in $[-a, b, c]P$).

This property is true in a state if, and only if, the property P is true in *every* state that you can get to by doing an 'a' transition from that state, for some $a \notin K$ (i.e., for some $a \in \overline{K}$).

Note that in all of the above shorthand formulæ we assume that the number of possible actions is finite; the logic **HML** does not have infinite conjunction or disjunction.

Example 13.3

Consider the following two simple processes:

$$E \stackrel{\text{def}}{=} a.a.0 \qquad\qquad F \stackrel{\text{def}}{=} a.a.0 + a.0$$

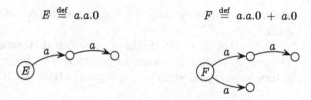

These differ in that process F may deadlock immediately after performing the first 'a' action, whereas process E is guaranteed to be able to perform a second 'a' action after performing the first 'a' action. These can be rendered in modal logic as follows:

- $F \models \langle a \rangle \neg \langle a \rangle \text{true}$ whereas $E \not\models \langle a \rangle \neg \langle a \rangle \text{true}$.

  We may do an 'a' action and end up in a state in which we cannot do another 'a' action.

  This is true in state F but not true in state E.

- $E \models [a] \langle a \rangle \text{true}$ whereas $F \not\models [a] \langle a \rangle \text{true}$.

  No matter how we do an 'a' action, we must end up in a state in which we may do another 'a' action.

  This is true in state E but not true in state F.

When first learning to think logically with the modal verbs "*may*" and "*must*," it is easy to misinterpret properties, particularly when expressing them in the language of **HML**. A common mistake arises when wanting to express the property

I must do an 'a' action.

This property is *not* captured by the formula $\langle a \rangle$true which expresses the property

I may do an 'a' action

as this allows the possibility of doing something other than an 'a' action; if, for example, I could also do a 'b' action, then it would clearly not be the case that I *must* do an 'a' action.

The next misconception is that – being a "*must*" property – we would express the desired property (that an 'a' action *must* happen) as [a]true. However, this formula only expresses what must be true *if and when you do an 'a' action*; it doesn't even assert that an 'a' action is even possible! More precisely, it asserts that:

no matter how we do an 'a' action
we must end up in a state in which true is true

which is true of *every* state of a system whether or not it can do an 'a' action!

So how then do we express the property that an 'a' action must happen? The answer is: precisely when an 'a' action *may* happen and *no other action* may happen, which we can express in **HML** as follows:

$$\langle a \rangle \text{true} \ \wedge \ \bigwedge_{b \neq a} \neg \langle b \rangle \text{true}$$

or more simply using our shorthand as follows:

$$\langle a \rangle \text{true} \ \wedge \ \neg \langle -a \rangle \text{true}$$

Exercise 13.4   (Solution on page 478)

Consider the following transition system:

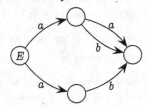

Which of the following are correct?

1. $E \models \langle a \rangle$true   5. $E \models \langle a \rangle \langle a \rangle$true   9. $E \models [a]\langle a \rangle$true

2. $E \models \langle b \rangle$true   6. $E \models \langle a \rangle \langle b \rangle$true   10. $E \models [a]\langle b \rangle$true

3. $E \models [a]$false   7. $E \models \langle a \rangle [a]$false   11. $E \models [a][a]$false

4. $E \models [b]$false   8. $E \models \langle a \rangle [b]$false   12. $E \models [a][b]$false

Exercise 13.5   (Solution on page 478)

Express the following properties regarding the lamp process from Section 11.2 pictured in Figure 11.5 (page 289). In each case, indicate which of the three states of the process (OFF, ON, BROKEN) satisfy the property in question.

1. I may do two '*pull*' actions in a row followed by a '*break*' action.

2. I may do two '*pull*' actions in a row followed by a '*reset*' action.

3. I cannot do a '*pull*' action.

4. I can only do a '*pull*' action (that is, I must do a '*pull*' action).

13.3   Negation Is Definable

In Section 11.4 we observed that the negation of a *must* property is equivalent to a *may* property, and vice versa. This should have become apparent as well from Exercise 13.4.

More precisely, we consider two formulæ of our modal logic to be equivalent if, and only if, they are true in the same states: $P \Leftrightarrow Q$ if, and only if, for all states E: $E \models P \Leftrightarrow E \models Q$. Our observations about negating modal properties are then expressed as follows:

$$\neg \langle a \rangle P \;\Leftrightarrow\; [a]\neg P \qquad and$$

$$\neg [a] P \;\Leftrightarrow\; \langle a \rangle \neg P$$

In words these say the following: the property which states that

> we <u>cannot</u> do an '*a*' action and
> end up in a state in which P is true

is equivalent to

> no matter how we do an '*a*' action
> we <u>must</u> end up in a state in which $\neg P$ is true;

and the property which states that

> it is <u>not</u> true that no matter how we do an '*a*' action
> we <u>must</u> end up in a state in which P is true

is equivalent to

> we <u>may</u> do an '*a*' action and
> end up in a state in which $\neg P$ is true.

We can motivate this correspondence by expressing the meaning of the modal connectives in predicate logic. A *may* property says something about *some* state to which you can go, whereas a *must* property says something about *all* states to which you can go:

$$E \models \langle a \rangle P \text{ if, and only if, } \exists F (E \xrightarrow{a} F \wedge F \models P)$$

$$E \models [a]P \text{ if, and only if, } \forall F (E \xrightarrow{a} F \Rightarrow F \models P)$$

We can then reason about these properties using the rules for quantification from Section 4.3:

$$\neg \forall x \, P(x) \Leftrightarrow \exists x \, \neg P(x) \quad \text{and} \quad \neg \exists x \, P(x) \Leftrightarrow \forall x \, \neg P(x).$$

For example, we can show the equivalence $\neg \langle a \rangle P \Leftrightarrow [a]\neg P$ as follows:

$$
\begin{aligned}
E \models \neg \langle a \rangle P &\Leftrightarrow E \not\models \langle a \rangle P \\
&\Leftrightarrow \neg \exists F (E \xrightarrow{a} F \wedge F \models P) \\
&\Leftrightarrow \forall F \neg (E \xrightarrow{a} F \wedge F \models P) \\
&\Leftrightarrow \forall F (E \xrightarrow{a} F \Rightarrow F \not\models P) \\
&\Leftrightarrow \forall F (E \xrightarrow{a} F \Rightarrow F \models \neg P) \\
&\Leftrightarrow E \models [a]\neg P
\end{aligned}
$$

Exercise 13.6   (Solution on page 478)

Show the equivalence $\neg [a]P \Leftrightarrow \langle a \rangle \neg P$ by using the rules for quantification to prove that $E \models \neg [a]P \Leftrightarrow E \models \langle a \rangle \neg P$.

Example 13.6

In order to express that we cannot do an '*a*' action, we can write

$\neg \langle a \rangle$true   (*It is not the case that we can do an '*a*' action.*)

By the above observation, since $\neg$true $=$ false this is equivalent to the expression

[*a*]false   (*No matter how we do an '*a*' action*
        we must end up in a state in which false is true.)

Since false cannot be true in *any* state, this means that there must be no possibility of doing an '*a*' action, as if we could do an '*a*' action we would have to end up in a state in which false is true.

Although **HML** includes negation, we can show that any property that can be expressed in **HML** can be expressed without using negation. That is, any formula P of **HML** can be transformed into a formula $pos(P)$ which contains no negation symbol and is semantically equivalent to P in the sense that $E \models pos(P)$ if, and only if, $E \models P$. This transformation is defined together with a dual transformation $neg(P)$ which transforms the formula P into one which contains no negation symbols yet is semantically equivalent to $\neg P$ in that $E \models neg(P)$ if, and only if, $E \not\models P$. Both transformations involve pushing negations into formulas using De Morgan's Laws, and are defined inductively on the structure of the formula P as follows:

$$pos(\text{true}) = \text{true} \qquad\qquad neg(\text{true}) = \text{false}$$
$$pos(\text{false}) = \text{false} \qquad\qquad neg(\text{false}) = \text{true}$$
$$pos(\neg P) = neg(P) \qquad\qquad neg(\neg P) = pos(P)$$
$$pos(P \wedge Q) = pos(P) \wedge pos(Q) \qquad neg(P \wedge Q) = neg(P) \vee neg(Q)$$
$$pos(P \vee Q) = pos(P) \vee pos(Q) \qquad neg(P \vee Q) = neg(P) \wedge neg(Q)$$
$$pos(\langle a \rangle P) = \langle a \rangle pos(P) \qquad\qquad neg(\langle a \rangle P) = [a]neg(P)$$
$$pos([a]P) = [a]pos(P) \qquad\qquad neg([a]P) = \langle a \rangle neg(P)$$

It is immediately clear that $pos(P)$ and $neg(P)$ are negation-free terms, as negation does not appear on the right-hand side of any of the defining equations.

Theorem 13.6

For any process E and any formula P of **HML**:

1. $E \models pos(P)$ if, and only if, $E \models P$; and

2. $E \models neg(P)$ if, and only if, $E \not\models P$.

Proof: By induction on the structure of P. The details are left as an exercise.

Exercise 13.7   (Solution on page 478)

Prove Theorem 13.6

Exercise 13.8   (Solution on page 482)

Prove, by induction on the structure of P, that $neg(neg(P)) = P$.

13.4 The Vending Machines Revisited

We can now express precisely the differences between the three vending machines V_1, V_2 and V_3 introduced in Section 11.4, by writing down formulæ of the logic **HML** which distinguish between them. Specifically, we shall produce three formulæ P_1, P_2 and P_3 of **HML** such that $V_i \models P_i$ for each i, but $V_i \not\models P_j$ whenever $i \neq j$. That is, formula P_i will distinguish the machine V_i from the other two machines by expressing a property which is true of machine V_i but not true of the others.

1. $P_1 = [\text{10p}][\text{10p}]\langle\text{tea}\rangle\text{true}$

   This formula expresses the property that after doing two consecutive '10p' actions, we must be in a state in which we can do a tea move. This is true of V_1 as there is only one state in which we can be after doing the two '10p' actions, namely the state

   $$\texttt{coffee.collect.}V_1 \texttt{ + tea.collect.}V_1$$

   and it is certainly the case that we may do a tea move from this state.

   However, this is neither true of V_2 nor of V_3; in both of these cases it is possible to do two consecutive '10p' actions and end up in a state where a 'tea' action is *not* possible (only a 'coffee' action). That is, V_2 and V_3 satisfy the formula

   $$P_1' = \langle\text{10p}\rangle\langle\text{10p}\rangle[\text{tea}]\text{false}$$

   while V_1 does not. (This formula is the negation of the one in question.)

2. $P_2 = [\text{10p}]\langle\text{10p}\rangle[\text{tea}]\text{false}$

   This formula expresses the property that after doing a '10p' action, we will be able to do a further '10p' action and end up in a state where we cannot do a 'tea' action. This is true of V_2 as there is only one state in which we can be after doing the first '10p' action, namely the state

   $$\texttt{10p.coffee.collect.}V_1 \texttt{ + 10p.tea.collect.}V_1$$

   and we can indeed do a further '10p' action, getting to the state

   $$\texttt{coffee.collect.}V_1$$

   in which we cannot do a 'tea' action.

   However, this is neither true of V_1 nor of V_3; in these cases it is possible to do the following '10p' actions:

   - $V_1 \xrightarrow{\text{10p}} \texttt{10p.(coffee.collect.}V_1 \texttt{ + tea.collect.}V_1\texttt{)}$

$\bullet\ V_3 \xrightarrow{\ 10\text{p}\ } 10\text{p}.\texttt{tea}.\texttt{collect}.V_3$

In both cases we end up in a state from which, after doing a further '10p' action, we can do a 'tea' action. V_1 and V_3 thus satisfy the formula

$$P_2' = \langle 10\text{p}\rangle[10\text{p}]\langle\text{tea}\rangle\text{true}$$

while V_2 does not. (This formula is the negation of the one in question.)

3. $P_3 = \langle 10\text{p}\rangle[10\text{p}][\text{tea}]\text{false}$

This formula expresses the property that it is possible to do a '10p' action and end up in a state from which we cannot do a further '10p' action followed by a 'tea' action. This is true of V_3 as we can make the transition

$$V_3 \xrightarrow{\ 10\text{p}\ } 10\text{p}.\texttt{coffee}.\texttt{collect}.V_3$$

and indeed find ourselves in a state from which we cannot do a further '10p' action followed by a 'tea' action.

However, this is neither true of V_1 nor of V_2; in each of these cases there is only one 10p transition possible, namely:

$\bullet\ V_1 \xrightarrow{\ 10\text{p}\ } 10\text{p}.(\texttt{coffee}.\texttt{collect}.V_1 + \texttt{tea}.\texttt{collect}.V_1)$

$\bullet\ V_2 \xrightarrow{\ 10\text{p}\ } 10\text{p}.\texttt{coffee}.\texttt{collect}.V_2 + 10\text{p}.\texttt{tea}.\texttt{collect}.V_2$

In both cases we end up in a state from which we can do a further '10p' action followed by a 'tea' action. V_1 and V_2 thus satisfy the formula

$$P_3' = [10\text{p}]\langle 10\text{p}\rangle\langle\text{tea}\rangle\text{true}$$

while V_3 does not. (This formula is the negation of the one in question.)

Exercise 13.9   (Solution on page 482)

Recall the following processes from Exercise 11.16.

$$A \stackrel{\text{def}}{=} b.c.0 + b.d.0 \qquad\qquad C \stackrel{\text{def}}{=} a.B + a.A$$
$$B \stackrel{\text{def}}{=} A + b.(c.0 + d.0) \qquad\qquad D \stackrel{\text{def}}{=} a.B$$

Give two formulæ of **HML** which distinguish between C and D: one formula which is true of C but not true of D; and one formula which is true of D but not true of C.

13.5 Modal Properties and Bisimulation

We have now developed two methods for distinguishing between processes.

1. In Chapter 12 we explicitly defined what it means for two processes to be equivalent, in terms of winning strategies in games.

2. In this chapter we defined a modal logic for expressing properties of processes with which we can distinguish between two processes.

We may well wonder if these two methods give the same results.

1. We should be disturbed if two equivalent processes could be differentiated by some formula of the modal logic. This would question the usefulness of the logic as a tool for reasoning about the behaviour of processes.

2. It would also be disappointing, though less of a concern, if the modal logic could not distinguish between some pair of non-equivalent processes. This would mean simply that the logic is too weak to express all aspects of the behaviour of a process.

However, we devised the equivalence based on a consideration of the capabilities of the processes as expressed using precisely the types of modal verbs which form the basis of our logic **HML**. Hence our intuition suggests that the distinguishing power of the modal logic should coincide with the equivalence. In this section we explore and confirm this intuition.

To determine if two processes are n-game equivalent, we need to explore only the first n transitions of the processes; the behaviour of the processes after n transitions is irrelevant. In the same way, in order to determine whether or not some formula of the modal logic is true of some process, we need only to explore the initial behaviour of the process; exactly how deeply we need explore the process depends on the complexity of the formula, as defined by its modal depth.

Definition 13.9

The *modal depth* $md(P)$ of a formula P of **HML** is defined inductively as follows.

$$md\,(true) = 0 \qquad md\,(P \wedge Q) = \max(md\,(P),\, md\,(Q))$$
$$md\,(false) = 0 \qquad md\,(P \vee Q) = \max(md\,(P),\, md\,(Q))$$

$$md\,(\neg P) = md\,(P) \qquad md\,(\langle a \rangle P) = 1 + md\,(P)$$
$$md\,([a]P) = 1 + md\,(P)$$

The modal depth simply counts the maximum number of modal operators along any path in the syntax tree of an **HML** formula. For example, the formula $\langle a \rangle ([b]\text{false} \wedge [a]\langle b \rangle \text{true})$ has a modal depth of 3, as evidenced by the following syntax tree for the formula.

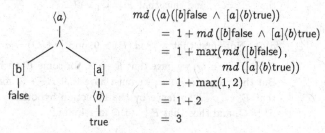

$$md\left(\langle a \rangle ([b]\text{false} \wedge [a]\langle b \rangle \text{true})\right)$$
$$= 1 + md\left([b]\text{false} \wedge [a]\langle b \rangle \text{true}\right)$$
$$= 1 + \max(md\left([b]\text{false}\right),$$
$$md\left([a]\langle b \rangle \text{true}\right))$$
$$= 1 + \max(1, 2)$$
$$= 1 + 2$$
$$= 3$$

The following theorem demonstrates that no formula of modal depth n can distinguish between two processes which are n-game equivalent. The immediate corollary to this is our first desired result: that we cannot use the logic to distinguish between equivalent processes.

Theorem 13.9

If $E \models P$ and $E \sim_n F$ where $n = md\,(P)$, then $F \models P$. That is, no formula of modal depth n can distinguish between two n-game equivalent processes.

Proof: By induction on the structure of P, arguing by cases on the structure of P:

$\underline{P = \textbf{true}}$: The result is immediately true in this case, as the conclusion $F \models \text{true}$ is always true.

$\underline{P = \textbf{false}}$: The result is vacuously true in this case, as the premise $E \models \text{false}$ is false.

$\underline{P = \neg Q}$: Since $E \models \neg Q$, we have $E \not\models Q$, and hence $F \not\models Q$ by induction, so $F \models \neg Q$.

$\underline{P = Q_1 \wedge Q_2}$: Note that $n_1 = md\,(Q_1) \leq n$ and $n_2 = md\,(Q_2) \leq n$; hence $E \sim_{n_1} F$ and $E \sim_{n_2} F$.

Since $E \models Q_1 \wedge Q_2$, we have that $E \models Q_1$ and that $E \models Q_2$. Hence by the induction hypothesis (applied twice), we have that $F \models Q_1$ and that $F \models Q_2$, and thus that $F \models Q_1 \wedge Q_2$.

$\underline{P = Q_1 \vee Q_2}$: Note that $n_1 = md\,(Q_1) \leq n$ and $n_2 = md\,(Q_2) \leq n$; hence $E \sim_{n_1} F$ and $E \sim_{n_2} F$.

Since $E \models Q_1 \vee Q_2$, we have that $E \models Q_1$ or that $E \models Q_2$. Hence by the induction hypothesis (applied twice), we have that $F \models Q_1$ or that $F \models Q_2$, and thus that $F \models Q_1 \vee Q_2$.

$\underline{P = \langle a \rangle Q}$: Note first that $n = md\,(P) > 0$, and $md\,(Q) = n-1$.

Since $E \models \langle a \rangle Q$, we have that $E \xrightarrow{a} E'$ for some E' such that $E' \models Q$. But then, since $E \sim_n F$, we must have that $F \xrightarrow{a} F'$ for some F' such that $E' \sim_{n-1} F'$. Hence by the induction hypothesis, we have that $F' \models Q$, and thus that $F \models \langle a \rangle Q$ as required.

$\underline{P = [a]Q}$: Note first that $n = md\,(P) > 0$, and $md\,(Q) = n-1$.

To show that $F \models [a]Q$, we need to show that $F' \models Q$ whenever $F \xrightarrow{a} F'$.

Suppose then that $F \xrightarrow{a} F'$. Since $E \sim_n F$ we must have that $E \xrightarrow{a} E'$ for some E' such that $E' \sim_{n-1} F'$; furthermore, since $E \models [a]Q$, we must have that $E' \models Q$. Thus by the induction hypothesis, we must have that $F' \models Q$ as required. $\qquad\qquad\square$

We can express this result more succinctly if we first formulate the notion of logical equivalence with respect to the formulæ of **HML** of a fixed bounded modal depth.

(**Definition 13.10**)

Let

$$\mathbf{HML}_n = \{\, P \in \mathbf{HML} \,:\, md\,(P) \leq n \,\}$$

*be the subset of **HML** consisting of all formulæ of modal depth at most n.*

1. *Two processes E and F are n-logically equivalent, written $E \equiv_n F$, if, and only if, the following holds.*

   For all $P \in \mathbf{HML}_n$: $E \models P$ if, and only if, $F \models P$.

   That is, no formula of modal depth n (or less) can distinguish between them.

2. *The processes E and F are logically equivalent, written $E \equiv F$ if, and only if, the following holds.*

   for all $P \in \mathbf{HML}$: $E \models P$ if, and only if, $F \models P$.

   That is, no formula (of any modal depth) can distinguish between them.

Theorem 13.9 then states simply that $E \equiv_n F$ whenever $E \sim_n F$.

Corollary 13.10

If $E \sim F$ then $E \equiv F$, that is, no formula of the logic HML can distinguish between two equivalent processes E and F.

Proof: If E and F *could* be differentiated by a formula P of HML, then by the above we would have that $E \not\sim_n F$, where $n = md(P)$, and hence we would have that $E \not\sim F$. $\qquad\square$

The converse result, that two processes which cannot be distinguished by any property of HML must be equivalent, is not completely attainable. This is due to the fact that equivalence is not the limit of the n-game equivalences, while the logic HML is the limit of the bounded logics HML_n. However, as was the case with relating the finite-game equivalences to bisimulation equivalence, this result holds when we restrict ourselves to image-finite processes.

Theorem 13.10

For image-finite processes E and F, if $E \equiv_n F$ then $E \sim_n F$.

Proof: We shall prove, by induction on n, the equivalent contrapositive statement that if $E \not\sim_n F$ then there is a formula P of modal depth n such that $E \models P$ but $F \not\models P$.

The base case ($n=0$) is vacuously true, as the premise $E \not\sim_0 F$ cannot hold.

For the induction step, suppose that $E \not\sim_{n+1} F$, and assume (without loss of generality) that $E \xrightarrow{a} E'$ for some E' such that $E' \not\sim_n F'$ whenever $F \xrightarrow{a} F'$. Let

$$F \xrightarrow{a} F_1 \qquad F \xrightarrow{a} F_2 \qquad \cdots \qquad F \xrightarrow{a} F_k$$

be all of the (finitely-many) a-transitions possible from F. Then $E' \not\sim_n F_i$ for each $i = 1, 2, \ldots, k$, and hence by the induction hypothesis there are properties $P_1, P_2, \ldots, P_k$ of modal depth n such that $E' \models P_i$ but $F_i \not\models P_i$ for each $i = 1, 2, \ldots, k$. The property P we seek is then

$$P = \langle a \rangle (P_1 \wedge P_2 \wedge \cdots \wedge P_k).$$

Clearly $E \models \langle a \rangle (P_1 \wedge P_2 \wedge \cdots \wedge P_k)$ but $F \not\models \langle a \rangle (P_1 \wedge P_2 \wedge \cdots \wedge P_k)$. $\qquad\square$

Corollary 13.11

For image-finite processes E and F, if $E \equiv F$ then $E \sim F$.

Proof: If $E \equiv F$ then $E \equiv_n F$ for all n, and hence by the above, $E \sim_n F$ for all n. Thus, since E and F are image-finite, $E \sim F$. $\qquad\square$

The clock processes Clock and Clock$_\star$ from Example 11.11 pictured in Figure 11.7 (page 298) provide the counter-example to this Corollary in the case of infinite-branching processes. In this case, Clock $\sim_n$ Clock$_\star$ for all $n \in \mathbb{N}$, and hence Clock $\equiv_n$ Clock$_\star$ for all $n \in \mathbb{N}$, meaning that Clock $\equiv$ Clock$_\star$; however, Clock $\not\sim$ Clock$_\star$.

★ (13.6) Characteristic Formulæ

Given a process state E, a formula $cf(E)$ of the logic **HML** is called a *characteristic formula* for E if, and only if, for all processes F:

$$F \models cf(E) \text{ if, and only if, } F \sim E.$$

For example, the characteristic formula for 0 is

$$cf(0) = [-]\mathsf{false}$$

as this formula specifies that there are no transitions possible from the state in question.

Exercise 13.11 (Solution on page 483)

1. Argue that the characteristic formula for $a.0$ is

$$cf(a.0) = \langle a \rangle \mathsf{true} \wedge [-a]\mathsf{false} \wedge [-][-]\mathsf{false}.$$

2. Give a characteristic formula for $a.(b.0 + c.0)$.

The existence of characteristic formulæ further cements the close connection between modal properties and bisimilarity. However, the exact relationship as presented in the following theorem takes into account the finite limitation of modal formulæ: that they can reason only about the first steps of a process up to a number of steps equal to the modal depth of the formula.

Theorem 13.11

For every $n \in \mathbb{N}$ and every state E of an LTS defined over a finite set of actions, there is a formula $cf_n(E) \in \mathbf{HML}_n$ such that, for all states F,

$$F \models cf_n(E) \quad \text{iff} \quad E \sim_n F.$$

Furthermore, for every $n \in \mathbb{N}$ there are only finitely-many such formulæ $cf_n(E)$.

Proof: By induction on n.

For the base case we can take $cf_n(E) = \text{true}$, since $F \models \text{true}$ and $E \sim_0 F$ for every $F \in$ States. Clearly there are only finitely-many (namely, one) such formulæ.

For the induction step, let

$$cf_{n+1}(E) = \bigwedge \left\{ \langle a \rangle\, cf_n(E') : E \xrightarrow{a} E' \right\}$$
$$\wedge \bigwedge_{a \in A} [a] \bigvee \left\{ cf_n(E') : E \xrightarrow{a} E' \right\}.$$

There are two parts to this formula:

- The first conjunction of subformulæ characterises what transition are possible: for each transition $E \xrightarrow{a} E'$, it must be possible to do an a transition into a state characterised by the formula $cf_n(E')$.

- The second conjunction of subformulæ characterises the states into which such a transition must evolve: upon performing an a transition, the process must evolve into a state characterised by the formula $cf_n(E')$ for some E' such that $E \xrightarrow{a} E'$.

Recalling the assumption that A is finite, we can note that – even though there may be infinitely-many transitions $E \xrightarrow{a} E'$ – the two sets of subformulæ are, by induction, finite; hence, this is a well-formed formula (ie, it is of finite size), and there can only be finitely-many such formulæ.

Suppose that $F \models cf_{n+1}(E)$.

- If $E \xrightarrow{a} E'$ then, since $F \models \langle a \rangle\, cf_n(E')$,

  $F \xrightarrow{a} F'$ for some F' such that $F' \models cf_n(E')$,

  and thus by induction $E' \sim_n F'$.

- If $F \xrightarrow{a} F'$ then, since $F \models [a] \bigvee_{E \xrightarrow{a} E'} cf_n(E')$,

  $E \xrightarrow{a} E'$ for some E' such that $F' \models cf_n(E')$,

  and thus by induction $E' \sim_n F'$.

Hence, by the above Lemma, $E \sim_{n+1} F$.

Now suppose that $E \sim_{n+1} F$.

- If $E \xrightarrow{a} E'$ then, by the above Lemma, $F \xrightarrow{a} F'$ for some F' such that $E' \sim_n F'$, and thus by induction $F' \models cf_n(E')$. As this is true of all $a \in A$ and all E' such that $E \xrightarrow{a} E'$,

$$\|\text{true}\| = \text{States}$$
$$\|\text{false}\| = \emptyset$$
$$\|\neg P\| = \overline{\|P\|}$$
$$\|P \wedge Q\| = \|P\| \cap \|Q\|$$
$$\|P \vee Q\| = \|P\| \cup \|Q\|$$
$$\|\langle a\rangle P\| = \{E \in \text{States} : E \xrightarrow{a} E' \text{ for some } E' \in \|P\|\}$$
$$\|[a]P\| = \{E \in \text{States} : E \xrightarrow{a} E' \text{ implies } E' \in \|P\|\}$$

Figure 13.2: Global semantics of the modal logic **HML**.

$$F \models \bigwedge_{E \xrightarrow{a} E'} \langle a\rangle \, cf_n(E')$$

- If $F \xrightarrow{a} F'$ then, by the above Lemma, $E \xrightarrow{a} E'$ for some E' such that $E' \sim_n F'$, and thus by induction $F' \models cf_n(E')$. As this is true of all $a \in A$ and all F' such that $F \xrightarrow{a} F'$,

$$F \models \bigwedge_{a \in A} [a] \bigvee_{E \xrightarrow{a} E'} cf_n(E')$$

Hence $F \models cf_{n+1}(E)$. $\square$

★ (13.7) ## Global Semantics

An alternative way to define the semantics of properties of **HML** is by associating to each property $P \in$ **HML** the set $\|P\|$ of states which satisfy the property P. Determining whether or not $E \models P$ then would correspond to determining if $E \in \|P\|$.

An inductive definition of the semantic function $\|P\|$ is given in Figure 13.2, where the set States represents the set of all states of the underlying transition system. With this definition, we get the following result.

Theorem 13.12

$E \models P$ if, and only if, $E \in \|P\|$.

Proof: By induction on the structure of P, arguing by cases on the structure of P.

$\underline{P = \textbf{true}}$: $E \models \text{true} \Leftrightarrow E \in \text{States} \Leftrightarrow E \in \|\text{true}\|$.

$\underline{P = \text{false}}$: $E \models \text{false} \Leftrightarrow E \in \emptyset \Leftrightarrow E \in \|\text{false}\|$

$\underline{P = \neg P}$: $E \models \neg P \Leftrightarrow E \not\models P \Leftrightarrow E \notin \|P\| \Leftrightarrow E \in \overline{\|P\|} \Leftrightarrow E \in \|\neg P\|$

$\underline{P = Q_1 \land Q_2}$: $E \models Q_1 \land Q_2 \Leftrightarrow E \models Q_1$ and $E \models Q_2$

$\qquad\qquad\qquad \Leftrightarrow E \in \|Q_1\| \cap \|Q_2\| \Leftrightarrow E \in \|Q_1 \land Q_2\|$

$\underline{P = Q_1 \lor Q_2}$: $E \models Q_1 \lor Q_2 \Leftrightarrow E \models Q_1$ or $E \models Q_2$

$\qquad\qquad\qquad \Leftrightarrow E \in \|Q_1\| \cup \|Q_2\| \Leftrightarrow E \in \|Q_1 \lor Q_2\|$

$\underline{P = \langle a \rangle Q}$: $E \models \langle a \rangle Q \Leftrightarrow E \xrightarrow{a} E'$ such that $E' \models Q$

$\qquad\qquad \Leftrightarrow E \xrightarrow{a} E'$ such that $E' \in \|Q\| \Leftrightarrow E \in \|\langle a \rangle Q\|$

$\underline{P = [a]Q}$: $E \models [a]Q \Leftrightarrow E \xrightarrow{a} E'$ implies $E' \models Q$

$\qquad\qquad \Leftrightarrow E \xrightarrow{a} E'$ implies $E' \in \|Q\| \Leftrightarrow E \in \|[a]Q\| \qquad \square$

Exercise 13.12 (Solution on page 483)

Consider the following transition system:

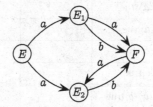

Compute the following sets:

1. $\|\langle a \rangle \text{true}\|$ 3. $\|\langle a \rangle \langle a \rangle \text{true}\|$ 5. $\|\langle a \rangle [a] \text{false}\|$

2. $\|\langle b \rangle \text{true}\|$ 4. $\|\langle b \rangle \langle b \rangle \text{true}\|$ 6. $\|[b] \langle a \rangle \text{true}\|$

13.8 Additional Exercises

1. Give properties of the modal logic **HML** which distinguish between the clocks Cl_n of Example 11.7. That is, for each $n \in \mathbb{N}$, give a formula P_n of **HML** which is true of Cl_n but false of Cl_k for every $k \neq n$.

2. What does the property $\langle a \rangle \text{false}$ say? Can you give an example process which satisfies this property?

3. Express the negation of each of the following properties without using negation operator ¬. In each case, write out each property and its negation in English.

   (a) $[a](\langle b\rangle\text{true} \wedge \langle c\rangle\text{true})$.

   (b) $[a]\langle b\rangle(\langle a\rangle\text{true} \vee \langle b\rangle[a]\text{false})$.

4. Consider the following 4-state transition system.

$$X \stackrel{\text{def}}{=} a.0 + Y$$

$$Y \stackrel{\text{def}}{=} a.Z$$

$$Z \stackrel{\text{def}}{=} b.0$$

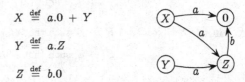

   Fill in the following table with the states satisfying the relevant properties. (The first line has been filled in to get you started.)

property P	states satisfying P	negation $\neg P$	states satisfying $\neg P$
$\langle a\rangle$true	X, Y	$[a]$false	$Z, 0$
$[a]$true			
$\langle b\rangle$true			
$[b]$true			
$\langle a\rangle\langle b\rangle$true			
$\langle a\rangle[b]$true			
$[a]\langle b\rangle$true			
$[a][b]$true			

5. Consider the following labelled transition system.

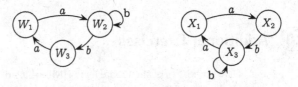

   Give a modal logic formula which distinguishes between W_1 and X_1. Argue why no formula of smaller modal depth can distinguish between these two states.

6. Give a labelled transition system with a state s which satisfies all of the following:

   - $\langle a \rangle(\langle a \rangle \text{true} \ \wedge \ \langle b \rangle \langle a \rangle \text{true})$
   - $\langle a \rangle \langle b \rangle(\langle b \rangle \text{true} \ \wedge \ [a]\text{false})$
   - $\langle a \rangle \langle b \rangle([a]\text{false} \ \wedge \ [b]\text{false})$

7. Recall the specification of the car safety system from Exercise 11 on page 305.

   (a) Express $R(x)$ in the modal logic M in two ways:
       i. óne way involving only the action "ring"; and
       ii. another way *not* involving the action "ring".

       (Hint: First express $R(x)$ in terms of $D(x)$, $B(x)$ and $M(x)$.)

   (b) Which states satisfy the following formulæ?
       i. $\langle \text{buckle} \rangle \text{true} \ \wedge \ \langle \text{close} \rangle \text{true}$
       ii. $\langle \text{buckle} \rangle \text{true} \ \wedge \ [\text{close}]\text{false}$
       iii. $\langle \text{on} \rangle \langle \text{ring} \rangle \text{true}$
       iv. $[\text{on}]\langle \text{ring} \rangle \text{true}$
       v. $\langle \text{open} \rangle \big(\langle \text{buckle} \rangle \text{true} \ \wedge \ \langle \text{off} \rangle \text{true}\big)$
       vi. $\langle \text{open} \rangle \big(\langle \text{buckle} \rangle \text{true} \ \vee \ \langle \text{off} \rangle \text{true}\big)$

8. Prove or disprove the following. (Here, equality between formulæ means that the formulæ express the same properties.)

   (a) $\langle a \rangle(P \wedge Q) = \langle a \rangle P \wedge \langle a \rangle Q$.
   (b) $\langle a \rangle(P \vee Q) = \langle a \rangle P \vee \langle a \rangle Q$.
   (c) $[a](P \wedge Q) = [a]P \wedge [a]Q$
   (d) $[a](P \vee Q) = [a]P \vee [a]Q$.

9. As defined, the modal logic **HML** involves only binary conjunctions and disjunctions, $P \wedge Q$ and $P \vee Q$, and thus by extension finite conjunctions and disjunctions, $\bigwedge \mathcal{F}$ and $\bigvee \mathcal{F}$ for finite sets of formulæ $\mathcal{F}$.

   Prove that if we allow infinite conjunctions and disjunctions, then the logical characterisation of bisimulation equivalence is tight: that $E \sim F$ if, and only if, E and F satisfy the same (infinitary) modal logic formulæ.

10. Let **HML**$_t$ be the subset of **HML** formulæ generated by the following BNF equation:

    $$P \ ::= \ \text{true} \ | \ \langle a \rangle P$$

    Show that **HML**$_t$ characterises trace equivalence $=_t$ from Exercise 8 on page 330, in the sense that $E =_t F$ if, and only if, E and F satisfy the same formulæ of **HML**$_t$.

11. Let **HML**$_\asymp$ be the subset of **HML** formulæ generated by the following BNF equation:

$$P, Q \ ::= \ \text{true} \ \mid \ \langle a \rangle P \ \mid \ P \wedge Q$$

Show that **HML**$_\asymp$ characterises simulation equivalence $\asymp$ from Exercise 12.8 on page 318, in the sense that $E \asymp F$ if, and only if, E and F satisfy the same formulæ of **HML**$_\asymp$.

12. Let **HML**$_f$ be the subset of **HML** formulæ generated by the following BNF equation:

$$P \ ::= \ [K]\text{false} \ \mid \ \langle a \rangle P$$

where $K \subseteq \Sigma$ is a set of actions. Show that **HML**$_f$ characterises failures equivalence $=_f$ from Exercise 10 on page 331, in the sense that $E =_f F$ if, and only if, E and F satisfy the same formulæ of **HML**$_f$.

Chapter 14

Concurrent Processes

Many hands make light work.

- John Heywood.

Thus far the systems that we have considered have been simple sequential processes, and have deviated from the standard (deterministic) notion of a sequential program only by the presence of (nondeterministic) choice. Of course the real interest in the study of systems arises when we permit processes to run in parallel and interact with one another. There are a variety of ways in which one might introduce operators into the language to permit such concurrent process behaviour. In this chapter we introduce a relatively simple operator, referred to as *synchronisation merge*, and demonstrate its use in a variety of example applications.

14.1 Synchronisation Merge

In this section, we introduce a parallel composition operator $\parallel$ which allows two processes E and F to execute in parallel. The precise fashion in which this concurrent execution takes place must be defined; in particular, we must clearly stipulate in what fashion such concurrent processes may interact with one another. To motivate our study, we start with a simple example.

Example 14.1

Suppose we have a very simple factory employing two workers. The first worker takes in jobs one at a time and, after carrying out some work on a job, passes it on to the second worker (assuming the second worker isn't still working on an earlier job). The second worker takes jobs one at a time from the first worker and, after carrying out some work on a job, sends it out of the factory.

The two workers can be represented by the following two simple processes P and Q:

F. Moller, G. Struth, *Modelling Computing Systems*,
Undergraduate Topics in Computer Science,
DOI 10.1007/978-1-84800-322-4_15, © Springer-Verlag London 2013

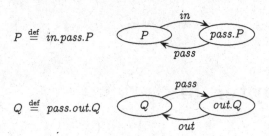

$$P \stackrel{\text{def}}{=} in.pass.P$$

$$Q \stackrel{\text{def}}{=} pass.out.Q$$

When the two workers start their work day, the first can start working immediately by taking in the first job, represented by the transition

$$P \stackrel{in}{\longrightarrow} pass.P$$

However, the second worker has to wait until the first worker has completed working on the first job and passes it on; that is, the transition

$$Q \stackrel{pass}{\longrightarrow} out.Q$$

cannot take place in reality until the associated transition

$$pass.P \stackrel{pass}{\longrightarrow} P$$

takes place. The two workers *synchronise* on the *pass* action; they must do this action together.

Consider what we would see if we were to watch these two workers. The behaviour of these two processes P and Q running together would be represented by the following process, where we represent the two relevant process states side-by-side separated by parallel lines $\|$:

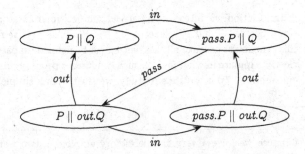

Note that having passed a job on, the first worker can take in another job; however, this job cannot be passed on until the second worker has sent out the previous job.

In the above example, two processes P and Q are made to run in parallel. This parallel composition is written as $P \| Q$, and each process is allowed to perform certain of its actions independent of the other, but is forced to synchronise with the other process on certain other actions. With this understanding, we are now prepared to explain the formal definition of the parallel composition operator ‖.

We first require each process state E to have a well-defined *synchronisation sort* Sort(E), denoting the subset of actions of the process on which it synchronises with other processes; every state of a given process will possess the same sort. The synchronisation sort of a process identifies those actions which are, in essence, external to the process, and represent those actions through which the process communicates with other processes via synchronisation. They are the actions by which processes are interconnected.

Example 14.2

Suppose Sort(P) $= \{a, b\}$, Sort(Q) $= \{a, b, c\}$ and Sort(R) $= \{b, c\}$. Composing these processes in parallel gives a system $P \parallel Q \parallel R$ in which the individual components P, Q and R are directly connected through the actions of their respective synchronisation sorts. The composed system can thus be viewed schematically as follows:

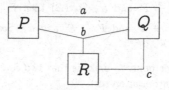

This depicts the whole system as consisting of the three physical processes P, Q and R all operating independently. The behaviour of these three processes is not depicted in the diagram, but they are inter-connected through the three actions a, b and c, which may be thought of as physical ports. The result is a process $P \parallel Q \parallel R$, which can itself be composed in parallel with further processes, with the synchronisation sort Sort($P \parallel Q \parallel R$) $= \{a, b, c\}$:

The intention of the synchronisation sort of a process is to define which actions are of importance when it comes to interaction; the individual actions of E may take place in the composition $E \parallel F$ so long as they are not in the sort of the process F. However, E must synchronise with F on any action from the sort of F which E is prepared to do. That is, E cannot do an action $a \in$ Sort(F) unless F itself is prepared to do this action, in which case E and F can perform this action in synchrony. Note that when we compose two processes E and F, the sort of the resulting processes is the union of the sorts of the components:

$$\text{Sort}(E \parallel F) \;=\; \text{Sort}(E) \cup \text{Sort}(F).$$

With this understanding in place, we may give the formal semantic definition of the *synchronisation merge $E \parallel F$* of processes E and F. There are three rules governing the behaviour of $E \parallel F$:

1. one which stipulates that $E \parallel F$ may perform a transition of E as long as it does not involve an action from the synchronisation sort of F;

2. one which stipulates that $E \parallel F$ may perform a transition of F as long as it does not involve an action from the synchronisation sort of E; and

3. one which stipulates that $E \parallel F$ may synchronise on the performance, by E and F together, of an action in either (or both) of their synchronisation sorts.

Formally, these rules are as follows.

1. If $E \xrightarrow{a} E'$ and $a \notin \text{Sort}(F)$ then $E \parallel F \xrightarrow{a} E' \parallel F$.

2. If $F \xrightarrow{a} F'$ and $a \notin \text{Sort}(E)$ then $E \parallel F \xrightarrow{a} E \parallel F'$.

3. If $E \xrightarrow{a} E'$ and $F \xrightarrow{a} F'$ and $a \in \text{Sort}(E) \cup \text{Sort}(F)$ then $E \parallel F \xrightarrow{a} E' \parallel F'$.

One further point to make is that two equivalent processes must have the same synchronisation sort.

Exercise 14.2   (Solution on page 483)

Why is it important that equivalent processes have the same sort?

Hint: We wish to make sure that if $A \sim B$ then $A \parallel X \sim B \parallel X$, that is, there should be no effect in the functioning of a system if we replace one component A with an equivalent component B. What might happen, though, if A and B have the same behaviour but different synchronisation sorts?

14.2   Counters

For any integer $k>0$, a *k-counter* is a system which stores an integer value between 0 and k (inclusively). The k-counter can be:

- *incremented*, as long as its value is less than k;

- *decremented*, as long as its value is greater than 0; and

- *tested* if its value is zero.

For example, we can define a 1-counter C by:

$$C \overset{\text{def}}{=} iszero.C + inc.dec.C$$

which defines the following transition system:

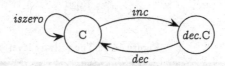

Almost as simply, we can define a 2-counter by:

$$C_2 \overset{\text{def}}{=} iszero.C_2 + inc.C_2'$$
$$C_2' \overset{\text{def}}{=} inc.C_2'' + dec.C_2$$
$$C_2'' \overset{\text{def}}{=} dec.C_2'$$

which defines the following transition system:

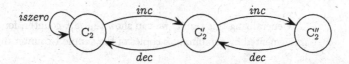

We can use two copies of the simple 1-counter to "implement" a 2-counter. Assuming that Sort(C) = {*iszero*}, the transition system C ‖ C is depicted as follows:

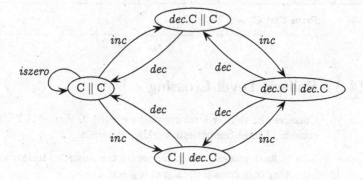

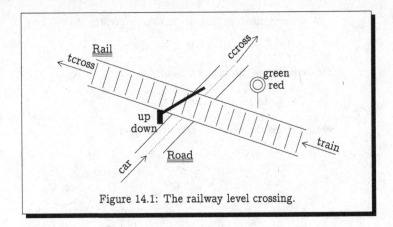

Figure 14.1: The railway level crossing.

Here, the initial state C ‖ C can do an increment action *inc* in two ways, either by allowing the left-hand 1-counter C perform this action, or by allowing the right-hand 1-counter C perform it; as this action is not in the synchronisation sort of C, neither process will block the other from performing this action.

On the other hand, the *iszero* action is only possible in the initial state C ‖ C; as the action *iszero* is in the synchronisation sort of C, both components of the parallel composition must participate in this action.

Generalising this result, we can show that a k-counter, for any k, can be implemented by combining k copies of the simple 1-counter in parallel; that is,

$$C_k \sim \underbrace{C \parallel C \parallel \cdots \parallel C.}_{k \text{ copies}}$$

Exercise 14.3 (Solution on page 484)

Prove that $C_2 \sim C \parallel C$.

14.3 Railway Level Crossing

Consider the railway level crossing depicted in Figure 14.1. This system consists of three components working in parallel.

- A Rail process, which represents the arrival of trains, assuring that they only cross if the signal is green.

- A Road process, which represents the arrival of cars, assuring that they only cross if the barrier is up.

- A Controller process, which regulates the signal and barrier, assuring that the barrier is never up at the same time that the signal is green.

This is a typical example of a control system, in which a controller process is regulating the behaviour of other processes in order to prevent undesirable behaviours. The desirable properties which the controller would like to attain are of two kinds.

1. *Safety Properties*: (*No crashes*)

    - A car may not cross at the same time as a train.

2. *Liveness Properties*: (*Eventual service*)

    - If a car arrives, eventually the barrier goes up.
    - If a train arrives, eventually the signal turns green.

We shall now describe the behaviour of the three component processes.

1. Road $\stackrel{\text{def}}{=}$ car.up.ccross.down.Road,
   with Sort(Road) = {up, down}.

   The Road process repeatedly carries out the following events.

   (a) signals the arrival of a car at the crossing (the car action);
   (b) witnesses the raising of the barrier (the up action);
   (c) signals the crossing of the car (the ccross action); and finally
   (d) witnesses the lowering of the barrier (the down action).

   The Road process is thus depicted by the following transition system.

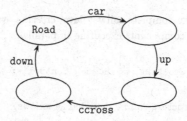

2. Rail $\stackrel{\text{def}}{=}$ train.green.tcross.red.Rail,
   with Sort(Rail) = {green, red}.

   Analogous to the Road process, the Rail process repeatedly carries out the following events.

(a) signals the arrival of a train at the crossing (the train action);

(b) witnesses the signal turning green (the green action);

(c) signals the crossing of the train (the tcross action); and finally

(d) witnesses the signal turning red (the red action).

The Rail process is thus depicted by the following transition system.

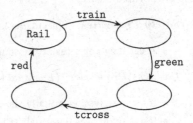

3. Controller $\stackrel{\text{def}}{=}$ green.red.Controller + up.down.Controller, with Sort(Controller) = {up, down, green, red}.

The Controller process is thus depicted by the following transition system.

The complete railway level crossing system then consists of the above three components executing in parallel:

Crossing $\stackrel{\text{def}}{=}$ Road ‖ Controller ‖ Rail

Its structure can be depicted as follows.

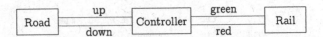

Its behaviour is thus given by the following transition system.

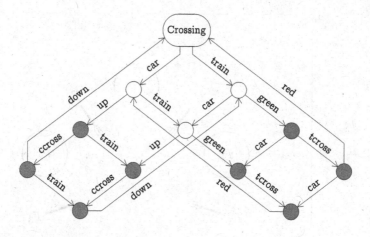

Exercise 14.4   (Solution on page 484)

Do the desired safety and liveness properties mentioned above hold? Explain why or why not. If any of these properties fail,

- can you propose a weaker yet acceptable property which does hold?
- can you propose a way to alter the definitions of the components of the system so that the property does hold?

14.4 Mutual Exclusion

When two tasks are being carried out together, problems can occur if they want to access some shared resource at the same time. A striking illustration of this is the Clayton Tunnel Accident (page 2), wherein one train was allowed to enter the tunnel which was currently occupied by another train. *Mutual exclusion* refers to the problem of ensuring that two processes can never be in their *critical section* – ie, using a shared resource such as a shared memory or printer – at the same time. If a process is granted use of such a shared resource, it must be allowed to maintain exclusive use of this resource until it has completed its use and released it. This is a ubiquitous problem in the design of concurrency systems.

14.4.1 Dining Philosophers

The problem of mutual exclusion was first identified and solved in 1965 by Edsger W. Dijkstra who also proposed the following illustration of how a system can deadlock if due concern is not taken to the use of shared

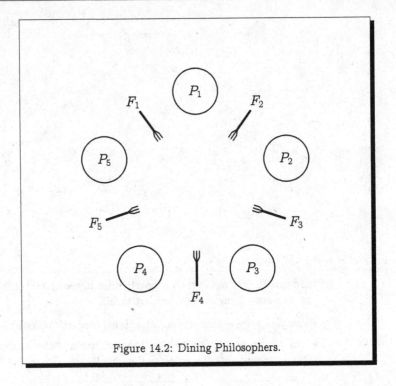

Figure 14.2: Dining Philosophers.

resources. It is a very simple problem to consider, yet offers a wealth of insight into the challenges posed by synchronisation.

Imagine there are five philosophers sitting at a round dining table thinking. Each philosopher has a plate of spaghetti in front of them, and there is a fork between every pair of plates. Figure 14.2 depicts the situation. The spaghetti is hopelessly tangled, meaning that a philosopher must use two forks together to eat it[†]. A philosopher may only use the two forks that are on either side of their own plate, which they pick up one at a time in either order. After taking a mouthful of spaghetti, the philosopher then replaces the two forks, in either order, to where they were lifted from the table. As the philosophers are deep in their own thoughts, at no point do they communicate with one another.

Our task is to design a protocol – that is, model the interactions between the philosophers and the forks – which satisfies the following correctness properties:

1. A philosopher eats only when holding two forks.

[†]The story is often described with rice rather than spaghetti, and chopsticks in place of forks, making it more immediate that two utensils are needed to eat.

2. No two philosophers may hold the same fork simultaneously (mutual exclusion).

3. The philosophers never get stuck, with every one of them forever waiting for some fork to become available (deadlock freedom).

To model this problem, we introduce the following actions:

- eat_i: philosopher i takes a bite to eat.
- $lift_{ij}$: philosopher i picks up fork j.
- $drop_{ij}$: philosopher i puts down fork j.

The behaviour of the forks is easy to describe:

$$F_1 \stackrel{\text{def}}{=} lift_{51} . drop_{51} . F_1 + lift_{11} . drop_{11} . F_1$$
$$F_2 \stackrel{\text{def}}{=} lift_{12} . drop_{12} . F_2 + lift_{22} . drop_{22} . F_2$$
$$F_3 \stackrel{\text{def}}{=} lift_{23} . drop_{23} . F_3 + lift_{33} . drop_{33} . F_3$$
$$F_4 \stackrel{\text{def}}{=} lift_{34} . drop_{34} . F_4 + lift_{44} . drop_{44} . F_4$$
$$F_5 \stackrel{\text{def}}{=} lift_{45} . drop_{45} . F_5 + lift_{55} . drop_{55} . F_5$$

That is to say, a fork is picked up by one of the two philosophers nearest to it at the table, and is subsequently placed back on the table by that same philosopher.

We have some freedom in how to define the behaviour of a philosopher, in that it is unspecified which order they pick up and set down their forks. As a first attempt, we assume that they each pick up the fork to their right first (as well as set this one down first):

$$P_1 \stackrel{\text{def}}{=} lift_{11} . lift_{12} . eat_1 . drop_{11} . drop_{12} . P_1$$
$$P_2 \stackrel{\text{def}}{=} lift_{22} . lift_{23} . eat_2 . drop_{22} . drop_{23} . P_2$$
$$P_3 \stackrel{\text{def}}{=} lift_{33} . lift_{34} . eat_3 . drop_{33} . drop_{34} . P_3$$
$$P_4 \stackrel{\text{def}}{=} lift_{44} . lift_{45} . eat_4 . drop_{44} . drop_{45} . P_4$$
$$P_5 \stackrel{\text{def}}{=} lift_{55} . lift_{51} . eat_5 . drop_{55} . drop_{51} . P_5$$

The synchronisation sorts for forks and philosopher processes are defined naturally as follows:

$$\text{Sort}(P_i) = \{ lift_{ij}, drop_{ij} : 1 \leq j \leq 5 \}.$$
$$\text{Sort}(F_j) = \{ lift_{ij}, drop_{ij} : 1 \leq i \leq 5 \}.$$

Unfortunately this protocol has the possibility of deadlocking: every philosopher may pick up a fork with their right hand before any one of them picks up the fork to their left, at which time they will all be waiting forever for their left-hand neighbour to return the fork to the table.

We can resolve this problem by changing the definition of the first (and only the first) philosopher, who is required to pick up the fork to their *left* first:

$$P_1 \stackrel{\text{def}}{=} lift_{12} . lift_{11} . eat_1 . drop_{12} . drop_{11} . P_1$$

With some thought, it becomes apparent that this refined protocol cannot deadlock.

Exercise 14.5   (Solution on page 484)

Argue that the refined protocol, in which the first philosopher picks up the left fork first, cannot deadlock.

14.4.2   Peterson's Algorithm

There have been various solutions proposed for dealing with the mutual exclusion problem. Here we examine an elegant solution proposed by Gary L. Peterson in 1981.

We consider two processes, P_1 and P_2, both of which wanting at times to enter some critical section. There are two Boolean variables: b1 which is true if P_1 wants to enter (or is in) the critical section, and b2 which is true if P_2 wants to enter (or is in) the critical section; and a variable k which has value 1 or 2 indicating which process has "ownership" of (ie, the authority to enter) the critical section. The Boolean variables b1 and b2 are initially set to false, while the initial value of k is arbitrary.

The two processes are then defined as follows (where the actual details of the critical and noncritical sections are left unspecified).

P_1: while true do	P_2: while true do
$\cdots$*noncritical section*$\cdots$	$\cdots$*noncritical section*$\cdots$
b1 := true;	b2 := true;
k := 2;	k := 1;
while (b2 and k=2) do	while (b1 and k=1) do
skip;	skip;
$\cdots$*critical section*$\cdots$	$\cdots$*critical section*$\cdots$
b1 := false	b2 := false

When process P_1 wishes to enter the critical section, it indicates this by setting b1 to true, but also sets k to 2 granting authority to the other process P_2 to enter the critical section. It then waits until either the other process P_2 does not wish to enter the critical region (ie, b2 is false) or the other process grants it authority to enter the critical region (ie, k has value 1), at which time it enters the critical region; when it exits the critical region it indicates this by setting b1 to false. Process P_2 is defined in an identical fashion.

To model these processes as labelled transition systems, we first need to represent the variables b1, b2 and k themselves as processes which interact with the processes P_1 and P_2. The variable b1 is represented by the following two-state system:

$$B_1f \stackrel{\text{def}}{=} b_1rf.B_1f + b_1wf.B_1f + b_1wt.B_1t$$
$$B_1t \stackrel{\text{def}}{=} b_1rt.B_1t + b_1wt.B_1t + b_1wf.B_1f$$

$$\text{Sort}(B_1f) = \{\, b_1rf,\ b_1rt,\ b_1wf,\ b_1wt \,\}$$

The state B_1f ("f" for "false") signifies that the variable b1 has the value false, while the state B_1t ("t" for "true") signifies that the variable b1 has the value true.

- Processes read the value of the variable by synchronising with the process on the actions b_1rf and b_1rt ("r" for "read"): the action b_1rf represents the process telling the environment that the value of b1 is false, while the action b_1rt represents the process telling the environment that the value of b1 is true. The state of the variable process does not change on these actions.

- Processes write a value to the variable by synchronising with the process on the actions b_1wf and b_iwt ("w" for "write"): the action b_1wf ("w" for "write") represents the value of the variable b1 being set to false, while the action b_1wt represents the value of the variable b1 being set to true. The state of the variable process changes as appropriate on these actions.

All of the reading and writing actions are included in the sort of the process, as clearly these actions must be done in synchrony with this process.

The variable b2 has an analogous definition:

$$B_2f \stackrel{\text{def}}{=} b_2rf.B_2f + b_2wf.B_2f + b_2wt.B_2t$$
$$B_2t \stackrel{\text{def}}{=} b_2rt.B_2t + b_2wt.B_2t + b_2wf.B_2f$$

$$\text{Sort}(B_2f) = \{\, b_2rf,\ b_2rt,\ b_2wf,\ b_2wt \,\}$$

Finally, the variable k is similarly defined:

$$K_1 \stackrel{\text{def}}{=} kr1.K_1 + kw1.K_1 + kw2.K_2$$
$$K_2 \stackrel{\text{def}}{=} kr2.K_2 + kw2.K_2 + kw1.K_1$$

$$\text{Sort}(K_1) = \{\, kr1,\ kr2,\ kw1,\ kw2 \,\}$$

Again, the variable k is represented by a two-state process, representing the two values that the variable k can hold (either 1 or 2); and there are actions representing the reading and writing of these two values.

We now turn our attention to defining the processes P_1 and P_2. As the behaviour of the processes within the noncritical and critical sections are irrelevant for our study – we are only interested in ensuring that mutual exclusion is attained – we ignore these completely. The behaviour of the process P_1 thus starts with setting the value of b1 to true and the value of k to 2:

$$P_1 \overset{\text{def}}{=} \text{b}_1\text{wt}.\text{kw2}.W_1$$

The process W_1 represents the process at the point of executing the while loop waiting to enter the critical section:

```
while (b2 and k=2) do skip
```

The process will stay in the state W_1 for as long as the value of b2 is true (ie, the action b_2rt can occur) and the value of k is 2 (ie, the action kr2 can occur). However, if either of these is false, that is if the value of b2 is false (ie, the action b_2rf can occur) or the value of k is 1 (ie, the action kr1 can occur), then the process will move into a new state R_1 signifying that the process is ready to enter the critical section:

$$W_1 \overset{\text{def}}{=} \text{b}_2\text{rt}.W_1 + \text{kr2}.W_1 + \text{b}_2\text{rf}.R_1 + \text{kr1}.R_1$$

Finally, in the state R_1 the process will enter the critical section, then (ultimately) exit it, and set the value of the variable b1 to be false, before returning then to the initial state:

$$R_1 \overset{\text{def}}{=} enter.exit.\text{b}_1\text{wf}.P_1$$

The synchronisation sort of the process P_1 contains the three relevant writing events:

$$\text{Sort}(P_1) = \{\,\text{b}_1\text{wt},\ \text{b}_1\text{wf},\ \text{kw2}\,\},$$

as the variables can only change value if they are written to.

The process P_2 is defined analogously to process P_1:

$$P_2 \overset{\text{def}}{=} \text{b}_2\text{wt}.\text{kw1}.W_2$$
$$W_2 \overset{\text{def}}{=} \text{b}_1\text{rt}.W_2 + \text{kr1}.W_2 + \text{b}_1\text{rf}.R_2 + \text{kr2}.R_2$$
$$R_2 \overset{\text{def}}{=} enter.exit.\text{b}_2\text{wf}.P_2$$
$$\text{Sort}(P_2) = \{\,\text{b}_2\text{wt},\ \text{b}_2\text{wf},\ \text{kw1}\,\}$$

The two processes P_1 and P_2 are then depicted by the following labelled transition systems:

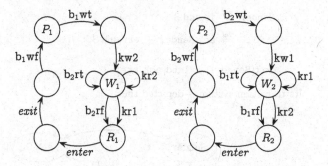

The whole system is then the concurrent composition of the processes P_1 and P_2 with the variable processes:

$$\text{PETERSON} \stackrel{\text{def}}{=} P_1 \parallel P_2 \parallel B_1f \parallel B_2f \parallel K_1.$$

Exercise 14.6   (Solution on page 485)

Argue that the two processes P_1 and P_2 can never both be in the critical section at the same time.

14.5 A Message Delivery System

We now wish to specify a simple message delivery system, which models the sending of a message by a SENDER process to a RECEIVER process. The SENDER and RECEIVER are not directly connected; rather, the message is routed through some MEDIUM. For example, the SENDER and RECEIVER may be two devices on a local area network connected by an Ethernet; or they may be computers on opposite sides of the globe connected by a complex mesh of links between routers. In our simple system, we ignore the actual content of the message being sent, as well as the address of the RECEIVER, as there will only be a single RECEIVER process.

When the SENDER accepts a message to send to the RECEIVER (modelled by an "in" action), it sends the message to the MEDIUM (modelled by a "snd" action), and awaits an acknowledgement that the message has been successfully delivered to the RECEIVER (modelled by an "ack" action); when an acknowledgement is received, the SENDER will be ready to accept the next message to send. It may be the case that the message is lost or corrupted by the MEDIUM, in which case the MEDIUM signals to the SENDER that a fault has occurred (modelled by an "err" action); this typically occurs in practice through a time-out mechanism. The SENDER responds to this fault by re-transmitting the message to the MEDIUM. The behaviour of the SENDER is thus modelled by the process Sender defined as follows:

Sender $\overset{\text{def}}{=}$ in.snd.S

$\quad$ S $\overset{\text{def}}{=}$ ack.Sender + err.snd.S

Sort(Sender) = {snd, ack, err}

Its transition system is depicted thus:

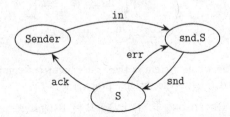

The MEDIUM accepts a message from the SENDER(via the snd action), and either delivers it to the RECEIVER (modelled by a rcv action) and returns to its original state to await the next message to be sent, or it signals to the SENDER that a fault has occurred (via the err action) and again returns to its original state to await the retransmission of the previous message. The behaviour of the MEDIUM is thus modelled by the process Medium defined as follows:

Medium $\overset{\text{def}}{=}$ snd.(rcv.Medium + err.Medium)

Sort(Medium) = {snd, rcv, err}

Its transition system is depicted thus:

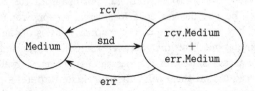

Finally, the RECEIVER awaits the delivery of a message (via the rcv action), and outputs the message (modelled by an out action) before issuing an acknowledgement (via the ack action) that the message has been successfully received and delivered. Its behaviour is modelled by the process Receiver defined as follows:

Receiver $\overset{\text{def}}{=}$ rcv.out.ack.Receiver

Sort(Receiver) = {rcv, ack}

The transition system is depicted thus:

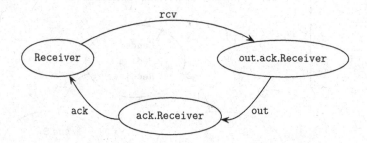

The complete system is defined to be the composition of these three components:

System $\overset{\text{def}}{=}$ Sender || Medium || Receiver

and has the following configuration:

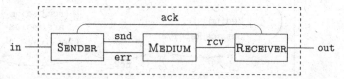

Note that in this simple model, the acknowledgement is communicated directly from the RECEIVER to the SENDER, which is unrealistic given the purpose of the System to relay messages between them. In reality, the acknowledgement would be routed through the MEDIUM resulting in a second phase which is identical to the first but with the roles of the SENDER and RECEIVER reversed.

The behaviour of the complete system is thus depicted by the transition system depicted in Figure 14.3.

Exercise 14.7   (Solution on page 485)

Enhance the simple message passing system so that acknowledgements are routed through the MEDIUM from the RECEIVER to the SENDER. Don't neglect the possibility of acknowledgements being lost.

14.6 Alternating Bit Protocol

The message passing system in the previous section is an example of a very important concept in communication networks: that of *communication protocols*.

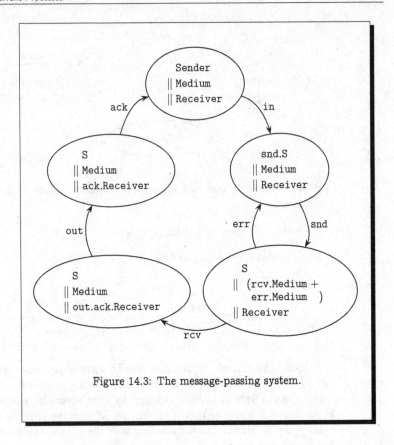

Figure 14.3: The message-passing system.

When you click on a link in your favourite browser to a Web site on the opposite side of the globe, or send an email to your friend who is perhaps thousands of miles away, a complicated procedure is carried out between dozens of computers in relaying your message to its destination (either the computer hosting the Web site you are wanting to access, or the computer on which your friend reads email). Your message gets relayed, bit-by-bit, through a long chain of intermediate computers as it gets routed towards its destination. At any point in this chain, a bit of your message can get lost in transmission, and the particular computer which sent the bit that got lost needs to know that the bit was lost so that it can retransmit it.

Of course, one computer cannot tell another computer that it didn't get a message, as it wouldn't know that it was supposed to get one; and in fact the most common cause of a message being lost in transmission is due to a receiving computer being broken, and thus unable to receive the message or send an acknowledgement. Thus, when messages are passed from one computer to another, the sending computer will wait for an acknowledgement from the receiving computer; if this doesn't come within a reasonable amount of time, the sending computer will assume that the message got lost and retransmit it. Of course, it may be the acknowledgement that got lost: the receiving computer may receive a message and send an acknowledgement and subsequently receive the same message again. In this case, the receiver will assume that its acknowledgement was lost, leading the sender to retransmit the message, in which case the receiver will retransmit the acknowledgement.

There are very many different communication protocols implemented on computers carrying out the above task. In this section we consider a common simple protocol: the alternating bit protocol. This protocol again involves a SENDER and a RECEIVER communicating through a MEDIUM, and works as follows.

- The SENDER accepts a message to be sent to the RECEIVER (modelled by an "in" action), and sends it into the MEDIUM tagged with a protocol bit 0 or 1 (modelled by the actions "s_0" and "s_1", respectively). It then awaits an acknowledgement from the MEDIUM tagged by the same protocol bit (modelled by the actions "ack_0" and "ack_1", respectively).

  When the SENDER receives an acknowledgement tagged by the correct protocol bit, it flips the protocol bit and repeats the protocol for the next message.

  If the SENDER receives an acknowledgement tagged by the wrong protocol bit, or if it times out waiting for the acknowledgement to arrive (modelled by a "t" action), it retransmits the message (again with the corresponding bit attached).

  The behaviour of the SENDER is thus defined by the following process:

$$\text{SENDER} \overset{\text{def}}{=} S_0 \qquad \text{Sort}(\text{SENDER}) = \{s_0, s_1\}$$

$$S_0 \overset{\text{def}}{=} in.S_0' \qquad S_0' \overset{\text{def}}{=} s_0.(ack_0.S_1 + ack_1.S_0' + t.S_0')$$

$$S_1 \overset{\text{def}}{=} in.S_1' \qquad S_1' \overset{\text{def}}{=} s_1.(ack_1.S_0 + ack_0.S_1' + t.S_1')$$

The synchronisation sort of the SENDER process contains the actions s_0 and s_1, as the only way the MEDIUM could receive a message is through a communication with the SENDER; it can only do these actions if and when the SENDER does them.

- When the RECEIVER receives a message from the MEDIUM tagged by the expected protocol bit (modelled by the actions "r_0" and "r_1", respectively), it outputs the message (modelled by an "out" action) and sends an acknowledgement into the MEDIUM tagged by that protocol bit (modelled by the actions "$rack_0$" and "$rack_1$", respectively).

The RECEIVER then awaits a new message tagged by the opposite protocol bit, with which it will repeat this protocol. In the meantime, it will acknowledge any further messages tagged by the old bit.

The behaviour of the RECEIVER is thus defined by the following process:

$$\text{RECEIVER} \stackrel{\text{def}}{=} R_0 \qquad \text{Sort}(\text{RECEIVER}) = \{rack_0, rack_1\}.$$

$$R_0 \stackrel{\text{def}}{=} r_0.out.rack_0.R_1 + r_1.rack_1.R_0$$

$$R_1 \stackrel{\text{def}}{=} r_1.out.rack_1.R_0 + r_0.rack_0.R_1$$

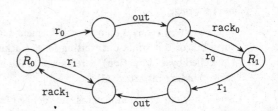

The synchronisation sort of the RECEIVER process contains the actions $rack_0$ and $rack_1$, as the only way the MEDIUM could receive an acknowledgement is through a communication with the RECEIVER; it can only do these actions if and when the RECEIVER does them.

- The MEDIUM merely passes messages from the SENDER to the RECEIVER and acknowledgements from the RECEIVER to the SENDER. Its behaviour is defined by the following process:

$$\text{MEDIUM} \stackrel{\text{def}}{=} M \qquad \text{Sort}(\text{MEDIUM}) = \{r_0, r_1, ack_0, ack_1\}.$$

$$M \stackrel{\text{def}}{=} s_0.r_0.M + s_1.r_1.M + rack_0.ack_0.M + rack_1.ack_1.M$$

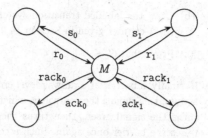

The synchronisation sort of the MEDIUM process does not contain the actions s_0 and s_1; the SENDER may send a message without it being received by the MEDIUM. Nor does it contain the actions $rack_0$ and $rack_1$; the RECEIVER may send an acknowledgement without it being received by the MEDIUM. However, it does contain the actions r_0 and r_1, as the RECEIVER can only receive a message from the MEDIUM; as well as the actions ack_0 and ack_1, as the SENDER can only receiver a message from the MEDIUM.

The complete system is defined to be the composition of these three components

$$\text{System} \overset{\text{def}}{=} \text{Sender} \parallel \text{Medium} \parallel \text{Receiver}$$

and has the following configuration:

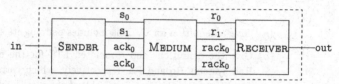

Its complete transition system is large, but by considering it carefully, it can be verified that the in and out actions occur in an alternating fashion and, equally important, that the protocol can never deadlock.

Exercise 14.8   (Solution on page 488)

Argue that the in and out actions occur in alternating fashion in the alternating bit protocol, and that the protocol can never deadlock.

14.7 Additional Exercises

1. (a) Give a definition for a 3-counter C_3, and draw its labelled transition system.

(b) Draw the labelled transition system for $C \parallel C \parallel C$, where C is the 1-counter given in Section 14.2.

(c) Prove that $C_3 \sim C \parallel C \parallel C$.

2. Normally, a barrier at a railway level crossing remains up until a train arrives; this signals the controller, which then lowers the barrier, then turns the signal green, then turns the signal red again, and finally raises the barrier once again. Such a controller C is thus represented by the following LTS:

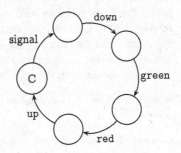

(a) Give a definition for C. This includes defining its sort $\mathcal{L}(C)$.

(b) Give the definitions and associated LTS for the new Road and Rail systems Ro and Ra, respectively, which correspond to the new Controller C. (Keep in mind that the new Road system starts in a state where the barrier is up; and the new Rail system must signal the controller when a train arrives, using the new event "signal" common to their sorts.)

(c) Now consider the liveness properties again.

    i. Is it now the case that, if the barrier is down when a car arrives, then the barrier will eventually go up?

    ii. Is it now the case that, if the signal is red when a train arrives, then the signal will eventually turn green?

3. Argue that the process for Peterson's Algorithm can never deadlock.

4. Model Dekker's Algorithm for mutual exclusion, as outlined as follows.

P_1: while true do	P_2: while true do
$\cdots$ *noncritical section* $\cdots$	$\cdots$ *noncritical section* $\cdots$
b1 := true;	b2 := true;
while b2 do	while b1 do
if k=2 then	if k=1 then
b1 := false	b2 := false
while k=2 do skip	while k=1 do skip
b1 := true	b2 := true
$\cdots$ *critical section* $\cdots$	$\cdots$ *critical section* $\cdots$
k := 2;	k := 1;
b1 := false	b2 := false

5. Argue that the complete alternating bit protocol system can never deadlock, and that the in and out actions alternate as desired.

6. Show that the operator $\parallel$ is commutative, by showing that the transition systems defined by $E \parallel F$ and $F \parallel E$ are isomorphic (ie, identical, disregarding the – irrelevant – labels of the states).

7. Show that the operator $\parallel$ is *not* associative, by showing that $(E \parallel F) \parallel F \nsim E \parallel (F \parallel F)$, where $E \stackrel{\text{def}}{=} a.0$ with $\text{Sort}(E) = \{a\}$ and $F \stackrel{\text{def}}{=} a.b.0$ with $\text{Sort}(F) = \{b\}$.

   What restriction on synchronisation sorts would make this operator associative?

8. Consider a new parallel operator $E \mid F$ defined by the following transition rules:

   (a) If $E \stackrel{a}{\to} E'$ and $a \notin \text{Sort}(F)$ then $E \mid F \stackrel{a}{\to} E' \mid F$.
   (b) If $F \stackrel{a}{\to} F'$ and $a \notin \text{Sort}(E)$ then $E \mid F \stackrel{a}{\to} E \mid F'$.
   (c) If $E \stackrel{a}{\to} E'$ and $F \stackrel{a}{\to} F'$ and $a \in \text{Sort}(E) \cap \text{Sort}(F)$ then $E \mid F \stackrel{a}{\to} E' \mid F'$.

   This is identical to the synchronisation merge except that the transition rule for synchronising processes requires the action on which the processes synchronise to be in the sorts of *both* processes rather than just one of them.

   Show that the operator $\mid$ is both commutative and associative.

Chapter 15

★ # Temporal Properties

> *When I eventually met Mr Right I had no idea that his first name was Always.*
>
> - Rita Rudner.

The modal logic **HML** of Chapter 13, while faithfully characterising properties which are relevant for distinguishing between process behaviours, has a fundamental drawback: a given formula $P \in$ **HML** can only explore the initial behaviour of a process, namely its first k steps where $k = md(P)$ is the modal depth of the formula. We cannot write a single formula that will explore a process to an unbounded length of its execution.

Consider, for example, the property of being *deadlockable*. In various examples of real-world system verifications, a common question is whether or not the system in question might at some point in time deadlock, that is, reach a state from which no action is possible. For example, we might like to verify that a new operating system design can never get in to a deadlocked state, one in which the machine on which it is running simply "hangs" leaving the user to apply the age-old solution of turning it off and on again.

Such properties are referred to as *temporal properties*, as they refer to the long-term behaviour of a system throughout the lifetime of its execution. Note that, since these properties are still based on the behaviour of systems, any such property which is true of a given process will be true of any equivalent process. These properties typically fall under one of the following two categories:

- *Safety properties* assert that *nothing bad ever happens*. Typical examples of safety properties include: the operating system will never deadlock; or a car will never be able to enter a level crossing at the same time as a train.

- *Liveness properties* assert that *something good eventually happens*. Typical examples of liveness properties include: having pressed the elevator button the elevator will eventually arrive; or if a train arrives its signal will eventually turn green.

F. Moller, G. Struth, *Modelling Computing Systems*,
Undergraduate Topics in Computer Science,
DOI 10.1007/978-1-84800-322-4_16, © Springer-Verlag London 2013

In this chapter we will explore various standard temporal properties, as well as the means to define our own temporal properties from recursive equations involving operators of the modal logic.

15.1 Three Standard Temporal Operators

A variety of fundamental temporal operators have been devised for expressing properties. Some of these are described as follows.

15.1.1 Always: $\Box P$

The most basic safety property, that *nothing bad ever happens*, is catered for by the temporal operator $\Box P$ (pronounced "box P") which asserts that the property P is true in *every* state into which the process may evolve. Formally,

$E \models \Box P$ if, and only if,

$\quad F \models P$ for all F such that $E \xrightarrow{w} F$ for some $w \in A^*$.

This is similar to the action "box" operator $[a]P$ except that the transitions involve arbitrary strings $w \in A^*$ rather than a single action $a \in A$.

We can view this property as an infinite conjunction; the property asserts that P is true after *any* number of transitions:

$$ P \ \land \ [-]P \ \land \ [-][-]P \ \land \ [-][-][-]P \ \land \ \cdots $$

That is to say, P is true at the start; and after any single transition; and after any two transitions; and after any three transitions; and .... Another way to view this is as a recursive property: the above infinite conjunction expresses a property X which satisfies the recursive equation

$$ X \ = \ P \ \land \ [-]X $$

which describes a property expressing the fact that P is true, and no matter what transition happens the property defined by X must hold (that is, P is true, and no matter what transition happens the property defined by X must hold (that is, ...)). Note that *every* one of an infinite number of properties must be true in order to satisfy $\Box P$.

Deadlock-freedom is an example of a property which can be defined with this operator: being free of deadlocks means that the property of being able to perform *some* action, *ie* $\langle - \rangle$true, is true in *every* state into which the process may evolve:

$$ \text{Deadlock-free} \ = \ \Box \langle - \rangle \text{true}. $$

Example 15.1

Consider the two clock processes Clock and Clock$_*$ from Example 11.11 pictured in Figure 11.7 (page 298). Recall that these two processes could not be distinguished by any formula of **HML**, since they were n-game equivalent for every $n \in \mathbb{N}$, despite the fact that they are not bisimilar (that is, they are not ∞-game equivalent). What distinguishes Clock$_*$ from Clock is the possibility that it may evolve into a deadlock-free state:

$$\text{Clock}_* \models \langle tick \rangle \Box \langle tick \rangle \text{true} \quad \text{but} \quad \text{Clock} \not\models \langle tick \rangle \Box \langle tick \rangle \text{true}.$$

Example 15.2

The safety property for our railway level crossing example of Section 14.3 is that at no time can a car cross at the same time as a train. This is expressed as

$$\Box \big([\text{ccross}]\text{false} \lor [\text{tcross}]\text{false} \big).$$

That is to say, it is always the case that either I cannot do a ccross action (a car cannot cross) or I cannot do a tcross action (a train cannot cross).

15.1.2   Possibly: $\Diamond P$

If we wish to express the *possibility* that *something bad may happen*, we can use another standard temporal operator, $\Diamond P$ (pronounced "diamond P"), which asserts that the property P is true in *some* state into which the process may evolve. Formally,

$$E \models \Diamond P \quad \text{if, and only if,}$$

$$F \models P \text{ for some } F \text{ such that } E \overset{w}{\to} F \text{ for some } w \in A^*.$$

This is similar to the action "diamond" operator $\langle a \rangle P$ except that the transitions involve arbitrary strings $w \in A^*$ rather than a single action $a \in A$.

We can view this property as an infinite disjunction; the property asserts that P is true after *some* number of transitions:

$$P \quad \lor \quad \langle - \rangle P \quad \lor \quad \langle - \rangle \langle - \rangle P \quad \lor \quad \langle - \rangle \langle - \rangle \langle - \rangle P \quad \lor \quad \cdots$$

That is to say, P is true: either at the start; or after some single transition; or after some two transitions; or after some three transitions; or . . .. Another way to view this is as a recursive property: the above infinite disjunction expresses a property X which satisfies the recursive equation

$$X \ = \ P \ \lor \ \langle - \rangle X$$

which describes a property expressing the fact that either P is true, or there is some transition which may happen after which the property defined by X will hold (that is, either P is true, or there is some transition which may happen after which the property defined by X will hold (that is, . . .)). Note that *some* one of an infinite number of properties must be true in order to satisfy this $\Diamond P$.

The property of being deadlockable, *ie* the opposite of Deadlock-freedom, is an example of a property which can be defined with this operator: being deadlockable means that the property of *not* being able to perform *any* action, *ie* $[-]$false, is true in *some* state into which the process may evolve:

$$\text{Deadlockable} = \Diamond[-]\text{false}.$$

The properties $\Diamond P$ and $\Box P$ are related in the same way that $\langle a \rangle P$ and $[a]P$ are related, in that each is used to express the negation of the other:

$$\neg \Diamond P = \Box \neg P \quad \text{and} \quad \neg \Box P = \Diamond \neg P$$

These operations are thus inter-definable:

- $\Diamond P = \neg \Box \neg P$: P is true in *some* reachable state if, and only if, it is *not* true that P is *false* in *every* reachable state;
- $\Box P = \neg \Diamond \neg P$: P is true in *every* reachable state if, and only if, it is *not* true that P is *false* in *some* reachable state.

(Exercise 15.2)   (Solution on page 488)

Use the above relationships between $\Box$ and $\Diamond$ to show that

$$\text{Deadlock-free} = \neg \text{Deadlockable}$$

where Deadlock-free $= \Box \langle - \rangle$true and Deadlockable $= \Diamond[-]$false.

15.1.3   Until: $P \cup Q$

It is often desirable to express that some property remains true until some other property becomes true, and that this latter property eventually does at some time become true. For example, we might wish to assert that when we send a document to a printer, the document will remain on the printer queue until it is scheduled to be printed, and it will eventually be printed. This type of property is expressed by the temporal operator $P \cup Q$ which asserts two things: that a particular property Q will eventually be true; and that until that time the property P will remain true. Formally:

$$E \models P \cup Q \quad \text{if, and only if,}$$

$$\text{if } E = E_0 \xrightarrow{a_1} E_1 \xrightarrow{a_2} E_2 \xrightarrow{a_3} \cdots \xrightarrow{a_n} E_n \not\rightarrow$$

$$\text{or } E = E_0 \xrightarrow{a_1} E_1 \xrightarrow{a_2} E_2 \xrightarrow{a_3} E_3 \xrightarrow{a_4} \cdots$$

$$\text{then } \exists k \text{ such that } E_k \models Q \text{ and } E_i \models P \text{ for all } i < k.$$

Note that $P \cup Q$ is true if Q initially holds; and that P can remain true when Q eventually holds but doesn't have to.

We can view the property $P \cup Q$ as a property X which satisfies the recursive equation

$$X \;=\; Q \vee (P \wedge \langle - \rangle\mathrm{true} \wedge [-]X)$$

which describes a property expressing the fact that: either Q is true; or P is true, and it is possible to do something, and no matter what you do the property defined by X must hold (that is: either Q is true; or P is true, and it is possible to do something, and no matter what you do the property defined by X must hold (that is: ...)). Again note that *some* one of an infinite number of properties must be true in order to satisfy $P \cup Q$.

Exercise 15.3   (Solution on page 488)

The generic liveness property asserts that *something good eventually happens*. Show how to express the temporal operator $\mathrm{Ev}\,P$ (pronounced "eventually P") using the above standard temporal operators.

15.2 Recursive Properties

The temporal operators considered in the previous section could all be viewed as solutions to recursive equations over the language **HML** of modal logic. For example, we noted above that $\Box P$ expresses a property X which satisfies the recursive equation

$$X \;=\; P \wedge [-]X.$$

Thus we would want $E \models \Box P$ to hold if, and only if, the following is true:

$$E \models X \text{ if, and only if, } E \models P \wedge [-]X.$$

To incorporate this idea into the logic **HML**, we need to introduce variables such as X into the language of properties. However, the semantic definition from Section 13.2 gives us no means by which we can determine if $E \models X$. It is not enough to assume that each variable X is defined by some equation $X = P$ and declare that $E \models X$ if, and only if, $E \models P$. For example, if $E \stackrel{\mathrm{def}}{=} a.E$ and X is defined by $X = \langle a \rangle X$, we would only be able to infer that $E \models X$ if, and only if, $E \models X$; either answer – $E \models X$ or $E \not\models X$ – is consistent with this observation.

To get around this deficiency, we need to introduce some mechanism to determine which states satisfy a variable property like X. This is provided by a *valuation function*

$$\mathsf{V} : \text{Variables} \rightarrow \mathcal{P}\,(\text{States})$$

where Variables is a set of variables (such as X above), and States is the set of states of the labelled transition system in which we are interested. Modal formulæ involving variables are then interpreted with respect to a valuation function as follows:

$$E \models_V \text{true} \qquad \text{for } all \ E.$$

$$E \models_V \text{false} \qquad \text{for } no \ E.$$

$$E \models_V X \qquad \text{if, and only if, } E \in \mathsf{V}(X).$$

$$E \models_V \neg P \qquad \text{if, and only if, } E \not\models_V P.$$

$$E \models_V P \wedge Q \quad \text{if, and only if, } E \models_V P \ and \ E \models_V Q.$$

$$E \models_V P \vee Q \quad \text{if, and only if, } E \models_V P \ or \ E \models_V Q.$$

$$E \models_V \langle a \rangle P \quad \text{if, and only if, } F \models_V P \text{ for } some \ F \text{ such that } E \xrightarrow{a} F.$$

$$E \models_V [a]P \qquad \text{if, and only if, } F \models_V P \text{ for } all \ F \text{ such that } E \xrightarrow{a} F.$$

This is identical to the original definition for $E \models P$ but for the extra clause which determines when a state E satisfies a variable property X; this case is catered for by the valuation function V which is now attached to the satisfaction relation $\models_V$.

In a similar fashion we can extend the global semantic definition $\|P\|$ to incorporate the valuation function as follows.

$$\|\text{true}\|_V = \text{States}$$

$$\|\text{false}\|_V = \emptyset$$

$$\|X\|_V = \mathsf{V}(X)$$

$$\|\neg P\|_V = \overline{\|P\|_V}$$

$$\|P \wedge Q\|_V = \|P\|_V \cap \|Q\|_V$$

$$\|P \vee Q\|_V = \|P\|_V \cup \|Q\|_V$$

$$\|\langle a \rangle P\|_V = \{\, E \in \text{States} : E \xrightarrow{a} E' \text{ for some } E' \in \|P\|_V \,\}$$

$$\|[a]P\|_V = \{\, E \in \text{States} : E \xrightarrow{a} E' \text{ implies } E' \in \|P\|_V \,\}$$

Theorem 13.12 then extends directly to properties with variables as follows.

Theorem 15.3

$E \models_V P$ if, and only if, $E \in \|P\|_V$.

Exercise 15.4   (Solution on page 489)

Prove Theorem 15.3

15.2.1 Solving Recursive Equations

In order to determine if a state E satisfies the property being expressed by a recursive equation $X = P$, where P is an **HML** formula possibly involving the variable X, we need to somehow "solve" the equation $X = P$. This equation simply declares that the set of states which satisfy the property X being defined is precisely the set of states which satisfy the property P; in other words, we need to equate the following two sets:

$$\|X\|_V = \|P\|_V.$$

To solve this equation we need to find a valuation V which makes this a valid set equation. Since $\|X\|_V = V(X)$, the answer we seek is the set S which such a valuation V assigns to the variable X. That is, the solution is a set $S \subseteq$ States satisfying

$$S = \|P\|_{V[X \mapsto S]}$$

where $V[X \mapsto S]$ denotes the valuation V updated by assigning the set S to the variable X:

$$\big(V[X \mapsto S]\big)(Y) = \begin{cases} S & \text{if } Y = X \\ V(Y) & \text{if } Y \neq X. \end{cases}$$

Example 15.4

Consider a property X which satisfies the equation

$$X = \langle a \rangle X.$$

Informally, this equation suggests that an infinite sequence of consecutive a actions can be performed:

$$E \models X \;\Leftrightarrow\; E \xrightarrow{a} E' \text{ for some } E' \text{ such that } E' \models X$$
$$\Leftrightarrow\; E \xrightarrow{a} E' \xrightarrow{a} E'' \text{ for some } E' \text{ and } E'' \text{ such that } E'' \models X$$
$$\Leftrightarrow\; \cdots$$
$$\Leftrightarrow\; E \xrightarrow{a} E' \xrightarrow{a} E'' \xrightarrow{a} E''' \xrightarrow{a} \cdots \text{ for some } E', E'', E''', \ldots$$

Let $S \subseteq$ States be the set of such states:

$$S = \{ E \in \text{States} \;:\; E \xrightarrow{a} \cdot \xrightarrow{a} \cdot \xrightarrow{a} \cdots \}.$$

Then

$$\|\langle a \rangle X\|_{V[X \mapsto S]} = \{ E \in \text{States} \;:\; E \xrightarrow{a} E' \text{ for some } E' \in S \} = S.$$

Thus, as intended, S satisfies the equation $S = \|\langle a \rangle X\|_{V[X \mapsto S]}$.

One problem that we have is that an arbitrary recursive equation needn't necessarily have a solution. For example, if we take the recursive equation

$$X = \neg X$$

then given any valuation V,

$$\|X\|_V = V(X)$$
$$\neq \overline{V(X)} = \overline{\|X\|_V} = \|\neg X\|_V.$$

Another problem is that an equation may be satisfied by many different solutions, as illustrated in the following.

Exercise 15.5   (Solution on page 489)

Show that the set $S = \emptyset$ satisfies the equation $S = \|\langle a \rangle X\|_{V[X \mapsto S]}$ from Example 15.4.

However, we will show here that any recursive equations which does not involve negation has a solution, and moreover we will show how to solve it to obtain the intended solution.

15.2.2 Fixed Point Solutions

Let $f : \mathcal{P}(\text{States}) \to \mathcal{P}(\text{States})$ be defined by

$$f(S) = \|P\|_{V[X \mapsto S]}.$$

Then a solution to the recursive equation $X = P$ is merely a *fixed point* of this function: a set $S \subseteq$ States such that $S = f(S)$. By the Knaster-Tarski Theorem (Theorem 6.18, page 174), this function is guaranteed to have a fixed point – in fact both greatest and least fixed points – so long as the function is monotonic. That this function is monotonic is an immediate corollary of the following result.

Theorem 15.5

Let P be an **HML** formula which does not involve negation, and let V and W be valuations such that $V(X) \subseteq W(X)$ for all X. Then $\|P\|_V \subseteq \|P\|_W$.

Proof: By induction – and arguing by cases – on the structure of P.

$\underline{P = \textbf{true}}$: $\|\text{true}\|_V = \text{States} = \|\text{true}\|_W$.

$\underline{P = \textbf{false}}$: $\|\text{false}\|_V = \emptyset = \|\text{false}\|_W$.

$\underline{P = X}$: $\|X\|_V = V(X) \subseteq W(X) = \|X\|_W$.

$\underline{P = Q_1 \wedge Q_2}$: $\begin{aligned}[t] \|Q_1 \wedge Q_2\|_V &= \|Q_1\|_V \cap \|Q_2\|_V \\ &\subseteq \|Q_1\|_W \cap \|Q_2\|_W \\ &= \|Q_1 \wedge Q_2\|_W \end{aligned}$

$\underline{P = Q_1 \vee Q_2}$: $\|Q_1 \vee Q_2\|_V = \|Q_1\|_V \cup \|Q_2\|_V$

$\subseteq \|Q_1\|_W \cup \|Q_2\|_W$

$= \|Q_1 \vee Q_2\|_W$

$\underline{P = \langle a \rangle Q}$: $\|\langle a \rangle Q\|_V = \{ E \in \text{States} : E \xrightarrow{a} E' \text{ such that } E' \in \|Q\|_V \}$

$\subseteq \{ E \in \text{States} : E \xrightarrow{a} E' \text{ such that } E' \in \|Q\|_W \}$

$= \|\langle a \rangle Q\|_W$

$\underline{P = [a]Q}$: $\|[a]Q\|_V = \{ E \in \text{States} : E \xrightarrow{a} E' \text{ implies } E' \in \|Q\|_V \}$

$\subseteq \{ E \in \text{States} : E \xrightarrow{a} E' \text{ implies } E' \in \|Q\|_W \}$

$= \|[a]Q\|_W$ $\square$

The Knaster-Tarski Theorem thus tells us that recursive equations which do not involve negation have two identifiable solutions, corresponding to their greatest and least fixed point solutions. This begs the question, when considering a recursively-defined property, as to *which* solution – if indeed either of them – represents the *intended* solution. That is, if we express a property as a recursive equation $X = P$, the set of states which satisfy the property we have in mind is a fixed point of the function $f(S) = \|P\|_{V[X \mapsto S]}$; but is it the greatest fixed point, or the least fixed point, or some fixed point in between?

This question will be explored in Section 15.4, where we will show that the answer is roughly:

- least fixed points express liveness properties; and

- greatest fixed points express safety properties.

Before we do this, though, we first look more carefully at adding these two fixed points to the modal logic **HML** without negation. The resulting logic with fixed points is called the *modal mu-calculus*, and is one of the most fundamental logics used in the specification of computer systems.

Exercise 15.6   (Solution on page 490)

What are the least and greatest fixed points of the function

$$f(S) = \|\langle a \rangle X\|_{V[X \mapsto S]}$$

corresponding to the property considered in Example 15.4?

Can you find a fixed point which is neither least nor greatest?

$$\|\text{true}\|_V = \text{States}$$

$$\|\text{false}\|_V = \emptyset$$

$$\|P \wedge Q\|_V = \|P\|_V \cap \|Q\|_V$$

$$\|P \vee Q\|_V = \|P\|_V \cup \|Q\|_V$$

$$\|\langle a \rangle P\|_V = \{ E \in \text{States} : E \xrightarrow{a} E' \text{ for some } E' \in \|P\|_V \}$$

$$\|[a]P\|_V = \{ E \in \text{States} : E \xrightarrow{a} E' \text{ implies } E' \in \|P\|_V \}$$

$$\|X\|_V = V(X)$$

$$\|\mu X.P\|_V = \bigcap \{ S \subseteq \text{States} : S \supseteq \|P\|_{V[X \mapsto S]} \}$$

$$\|\nu X.P\|_V = \bigcup \{ S \subseteq \text{States} : S \subseteq \|P\|_{V[X \mapsto S]} \}$$

Figure 15.1: Global semantics of the modal mu-calculus.

15.3 The Modal Mu-Calculus

The syntax of the modal mu-calculus consists of the logic **HML** – minus negation – extended with variables and constructs for defining both greatest and least fixed points. Formally, it is presented by the following BNF equation:

$$P, Q ::= \text{true} \mid \text{false} \mid P \wedge Q \mid P \vee Q \mid \langle a \rangle P \mid [a]P$$
$$\mid X \mid \mu X.P \mid \nu X.P$$

The symbols μ and ν are the characters "mu" and "nu" from the Greek alphabet (from which the name "mu-calculus" derives). The formula $\mu X.P$ is used to represent the least fixed point of the equation $X = P$ (or more correctly, of the function $f(S) = \|P\|_{V[X \mapsto S]}$) whereas $\nu X.P$ is used to represent its greatest fixed point.

An inductive definition of the semantic function $\|P\|_V$ – defining which states satisfy the property P with respect to the valuation V – is given in Figure 15.1. The clauses are identical to those presented for the basic modal logic **HML** in Figure 13.2 with the inclusion of the obvious clause for variables, and clauses for the fixed points as given by the Knaster-Tarski Theorem (Theorem 6.18, page 174), That the Knaster-Tarski Theorem applies follows from the fact that the function $f(S) = \|P\|_{V[X \mapsto S]}$ is monotonic. We demonstrated this for **HML** in Theorem 15.5, but we need to extend this result for the larger logic.

Theorem 15.6

Let P be a modal mu-calculus formula which does not involve negation, and let V and W be valuations such that $V(X) \subseteq W(X)$ for all X. Then $\|P\|_V \subseteq \|P\|_W$.

Proof: By induction – and arguing by cases – on the structure of P. All of the cases have been catered for in the proof of Theorem 15.5 – and carry over directly to the present setting – apart from the cases of variables and fixed point formulæ, which we handle here.

$\underline{P = X}$: $\quad \|X\|_V = V(X) \subseteq W(X) = \|X\|_W$

$\underline{P = \mu X.Q}$: $\quad E \in \|\mu X.Q\|_V \Leftrightarrow E \in S$ whenever $\|Q\|_{V[X \mapsto S]} \subseteq S$

$\qquad\qquad\qquad\qquad \Rightarrow E \in S$ whenever $\|Q\|_{W[X \mapsto S]} \subseteq S$

$\qquad\qquad\qquad\qquad \Leftrightarrow E \in \|\mu X.Q\|_W$

$\underline{P = \nu X.Q}$: $\quad E \in \|\nu X.Q\|_V \Leftrightarrow E \in S$ for some S where $S \subseteq \|Q\|_{V[X \mapsto S]}$

$\qquad\qquad\qquad\qquad \Rightarrow E \in S$ for some S where $S \subseteq \|Q\|_{W[X \mapsto S]}$

$\qquad\qquad\qquad\qquad \Leftrightarrow E \in \|\nu X.Q\|_W \qquad\qquad\qquad\qquad \square$

Definition 15.6

A direct definition of when a state E satisfies a property P of the modal mu-calculus with respect to a valuation V for interpreting free variables which appear in P is as follows:

$E \models_V true \quad$ for all E.

$E \models_V false \quad$ for no E.

$E \models_V P \wedge Q \quad$ if, and only if, $E \models_V P$ and $E \models_V Q$.

$E \models_V P \vee Q \quad$ if, and only if, $E \models_V P$ or $E \models_V Q$.

$E \models_V \langle a \rangle P \quad$ if, and only if, $F \models_V P$ for some F such that $E \xrightarrow{a} F$.

$E \models_V [a]P \quad$ if, and only if, $F \models_V P$ for all F such that $E \xrightarrow{a} F$.

$E \models_V X \quad$ if, and only if, $E \in V(X)$.

$E \models_V \mu X.P \quad$ if, and only if, $\forall S \subseteq$ States : if $E \notin S$ then $\qquad\qquad\qquad\qquad\qquad\qquad \exists F \notin S$ such that $F \models_{V[X \mapsto S]} P$

$E \models_V \nu X.P \quad$ if, and only if, $\exists S \subseteq$ States : $E \in S$ and $\qquad\qquad\qquad\qquad\qquad\qquad \forall F \in S : F \models_{V[X \mapsto S]} P$

We leave it as an exercise (Exercise 5, page 402) to prove (by induction on the structure of P) that $E \models_V P$ if, and only if, $E \in \|P\|_V$.

Leaving negation out of the logic is not a real restriction, as the result from Section 13.3 that negation is definable in the modal logic **HML** extends to the whole of the modal mu-calculus. This is justified by the following.

Exercise 15.7   (Solution on page 490)

The negation of a modal mu-calculus formula can be inductively defined as follows:

$$\text{neg(true)} = \text{false} \qquad\qquad \text{neg}(\langle a \rangle P) = [a]\text{neg}(P)$$
$$\text{neg(false)} = \text{true} \qquad\qquad \text{neg}([a]P) = \langle a \rangle \text{neg}(P)$$

$$\text{neg}(P \wedge Q) = \text{neg}(P) \vee \text{neg}(Q) \qquad \text{neg}(\mu X.P) = \nu X.\text{neg}(P)$$
$$\text{neg}(P \vee Q) = \text{neg}(P) \wedge \text{neg}(Q) \qquad \text{neg}(\nu X.P) = \mu X.\text{neg}(P)$$

$$\text{neg}(X) = X$$

Prove that $E \models_{\overline{V}} \text{neg}(P)$ if, and only if, $E \not\models_V P$, where $\overline{V}(X) = \overline{V(X)}$.

15.4 Least versus Greatest Fixed Points

We now understand how to interpret recursive logical properties as fixed points of particular functions between sets of states. However, we are left with the problem of understanding why the property we intend is expressed by either the greatest or the least fixed point of this function, as well as the problem of knowing which. To solve this, we shall explore how such a recursive property can be understood by "unrolling" it.

Given a property X defined by a recursive equation $X = P$, we can unroll the equation by replacing each occurrence of X on the right-hand-side by P itself. Clearly this will not change the meaning of the property being defined by the recursive equation.

Example 15.7

Suppose we wish to express the property that an infinite sequence of consecutive a actions can happen. That is, denoting this property by X, we would like the following to be the case:

$$E \models X \text{ if, and only if, } E \xrightarrow{a} \cdot \xrightarrow{a} \cdot \xrightarrow{a} \cdots .$$

This property is expressed by the recursive equation

$$X = \langle a \rangle X$$

which we can repeatedly unroll as follows:

$$\begin{aligned}
X &= \langle a \rangle X \\
&= \langle a \rangle \langle a \rangle X \\
&= \langle a \rangle \langle a \rangle \langle a \rangle X \\
&= \langle a \rangle \langle a \rangle \langle a \rangle \langle a \rangle X \\
&= \langle a \rangle \langle a \rangle \langle a \rangle \langle a \rangle \langle a \rangle \cdots
\end{aligned}$$

Example 15.8

Suppose we wish to express the property that an a action must eventually occur. This property is expressed by the recursive equation

$$X = \langle - \rangle \text{true} \wedge [-a]X.$$

That is: some action is possible; and if anything other than an a action occurs, then an a action must eventually occur in the resulting state. We can repeatedly unroll this recursive equation as follows:

$$\begin{aligned}
X &= \langle - \rangle \text{true} \wedge [-a]X \\
&= \langle - \rangle \text{true} \wedge [-a]\big(\langle - \rangle \text{true} \wedge [-a]X\big) \\
&= \langle - \rangle \text{true} \wedge [-a]\big(\langle - \rangle \text{true} \wedge [-a](\langle - \rangle \text{true} \wedge [-a]X)\big) \\
&= \langle - \rangle \text{true} \wedge [-a]\big(\langle - \rangle \text{true} \wedge [-a](\langle - \rangle \text{true} \wedge [-a](\cdots))\big)
\end{aligned}$$

15.4.1 Approximating Fixed Points

By repeatedly unrolling a recursive equation, we seem to eliminate the variable from the formula. Of course we would have to unroll the equation infinitely often in order to get rid of the variable altogether. However, we don't have any means for determining whether or not an infinitely-long property is satisfied. We can, however, define better and better approximations for such properties, by replacing the variable in the rolled-out formula by either false or true. To this end, we can define the nth mu- and nu-approximants as follows.

Definition 15.8

Given a recursive equation $X = P$, the nth *mu-approximant* $\mu^n X.P$ and the nth *nu-approximant* $\nu^n X.P$ are defined inductively as follows:

$$\mu^0 X.P = \text{false} \qquad\qquad \nu^0 X.P = \text{true}$$
$$\mu^{n+1} X.P = P[X \mapsto \mu^n X.P] \qquad\qquad \nu^{n+1} X.P = P[X \mapsto \nu^n X.P]$$

These definitions extend to all ordinal numbers (see Section 12.5.1), with the following definitions for the approximants corresponding to a limit ordinal λ:

$$\mu^\lambda X.P = \bigvee_{\alpha < \lambda} \mu^\alpha X.P \qquad\qquad \nu^\lambda X.P = \bigwedge_{\alpha < \lambda} \nu^\alpha X.P$$

Example 15.9

Recall the property from Example 15.7 that an infinite sequence of consecutive a actions can happen:

$$X = \langle a \rangle X.$$

Its mu-approximants Φ_n and nu-approximants Ψ_n are as follows:

Φ_0	=	false	Ψ_0 = true
Φ_1	=	$\langle a \rangle$false	Ψ_1 = $\langle a \rangle$true
Φ_2	=	$\langle a \rangle \langle a \rangle$false	Ψ_2 = $\langle a \rangle \langle a \rangle$true
Φ_3	=	$\langle a \rangle \langle a \rangle \langle a \rangle$false	Ψ_3 = $\langle a \rangle \langle a \rangle \langle a \rangle$true

$\vdots$ $\vdots$

Clearly none of the mu-approximants Φ_n can be satisfied by any state. However, every one of the nu-approximants Ψ_n must be satisfied in order for our intended property to be satisfied.

This is suggestive of a *safety property*: checking that *something bad never happens* (in this case, that an a action is impossible) amounts to checking the validity of *every* unrolling of the formula, starting from true.

Example 15.10

Recall the property from Example 15.7 that an a action must eventually occur:

$$X = \langle - \rangle true \wedge [-a] X.$$

Its mu-approximants Φ_n and nu-approximants Ψ_n are as follows:

$$\Phi_0 = \text{false}$$

$$\Phi_1 = \langle - \rangle \text{true} \wedge [-a] \text{false}$$

$$\Phi_2 = \langle - \rangle \text{true} \wedge [-a] \big(\langle - \rangle \text{true} \wedge [-a] \text{false} \big)$$

$$\Phi_3 = \langle - \rangle \text{true} \wedge [-a] \big(\langle - \rangle \text{true} \wedge [-a] (\langle - \rangle \text{true} \wedge [-a] \text{false}) \big)$$

$$\vdots$$

$$\Psi_0 = \text{true}$$

$$\Psi_1 = \langle - \rangle \text{true} \wedge [-a] \text{true}$$

$$\Psi_2 = \langle - \rangle \text{true} \wedge [-a] \big(\langle - \rangle \text{true} \wedge [-a] \text{true} \big)$$

$$\Psi_3 = \langle - \rangle \text{true} \wedge [-a] \big(\langle - \rangle \text{true} \wedge [-a] (\langle - \rangle \text{true} \wedge [-a] \text{true}) \big)$$

$$\vdots$$

With a little thought, it is apparent that one of the mu-approximants Φ_n must be satisfied in order for our intended property to be satisfied. However, every one of the nu-approximants is satisfied, for example, by a process which runs forever without ever doing an a action.

This is indicative of a *liveness property*: checking that *something good eventually happens* (in this case, that an a action occurs) amounts to checking the validity of *some* unrolling of the formula, starting from false.

In the first of the above two examples, the property which we wished to express was interpreted as the conjunction of all of the nu-approximants (the unrollings starting from true); while in the second of the two examples, the property of interest was interpreted as the disjunction of all of the mu-approximants (the unrollings starting from false). In what follows, we shall see that the first corresponds to the greatest fixed point interpretation of the recursive property, while the second corresponds to the least fixed point interpretation.

In Section 6.5 we described how the least and greatest fixed points of a monotonic function f defined on the powerset of a given set S could be constructed, by repeatedly applying the function f to either the empty set $\emptyset$ (for the least fixed point) or to the whole set S (for the greatest fixed point); this result was given in Theorem 6.19. This is just the result we are looking for here, as the nth mu- and nu-approximants correspond, respectively, to applying the relevant function n times either to the empty set $\emptyset$ or to the whole set States. These facts are recorded in the following.

Theorem 15.10

$$f^n(\emptyset) = \|\mu^n X.P\|_V \text{ and } f^n(\text{States}) = \|\nu^n X.P\|_V, \text{ where } f(S) = \|P\|_{V[X \mapsto S]}.$$

Proof: We prove only the first result, by induction on n, and leave the proof of the second as an exercise (Exercise 6, page 402).

For the base case $n = 0$, we have

$$f^0(\emptyset) \;=\; \emptyset \;=\; \|\text{false}\|_V \;=\; \|\mu^0 X.P\|_V$$

For the induction step, we have that

$$
\begin{aligned}
f^{n+1}(\emptyset) &= f(f^n(\emptyset)) \\
&= f(\|\mu^n X.P\|_V) \qquad \textit{(by induction)} \\
&= \|P\|_{V[X \mapsto \|\mu^n X.P\|_V]} \\
&= \|P[X \mapsto \mu^n X.P]\|_V \\
&= \|\mu^{n+1} X.P\|_V \qquad\qquad \square
\end{aligned}
$$

Example 15.11

Consider the recursive equation for $\Box P$, the property that P holds in every reachable state:

$$X \;=\; P \wedge [-]X.$$

This gives rise to the function

$$f(S) \;=\; \{\, E \in \|P\|_V \;:\; E \to E' \text{ implies } E' \in S \,\}.$$

Using the construction from Theorem 6.19 (page 175) starting from the empty set $\emptyset$, we discover that

$$f(\emptyset) \;=\; \emptyset$$

demonstrating that the least fixed point is the empty set. This certainly does not correspond to the property $\Box P$. However, starting from the universal set States, we get that

$$
\begin{aligned}
f^0(\text{States}) &= \text{States} \\
f^1(\text{States}) &= \|P\|_V \\
f^2(\text{States}) &= \{\, E \in \|P\|_V \;:\; E \to E' \text{ implies } E' \in \|P\|_V \,\} \\
f^3(\text{States}) &= \{\, E \in \|P\|_V \;:\; E \to E' \text{ or } E \to\to E' \\
&\qquad\qquad\qquad\qquad \text{implies } E' \in \|P\|_V \,\}
\end{aligned}
$$

$$\vdots$$

$$
\begin{aligned}
f^n(\text{States}) &= \text{states in which } P \text{ is true throughout} \\
&\qquad \text{the duration of the first } n \text{ transitions.}
\end{aligned}
$$

This sequence is approaching the set S of sets in which P is true in every reachable state, which is the desired solution to our recursive equation and easily seen to be a fixed point of the function f.

Example 15.12

Consider the recursive equation expressing that a process is deadlockable:

$$X = [-]\text{false} \vee \langle-\rangle X$$

This gives rise to the function

$$f(S) = \{ E \in \text{States} : E \nrightarrow \text{ or } E \rightarrow E' \text{ with } E' \in S \}.$$

Using the construction from Theorem 6.19 starting from the universal set States, we discover that

$$f(\text{States}) = \text{States}$$

demonstrating that the greatest fixed point is the set of all states. This certainly does not correspond to the property that a process is deadlockable. However, starting from the empty set $\emptyset$, we get that

$$f^0(\emptyset) = \emptyset$$
$$f^1(\emptyset) = \{ E \in \text{States} : E \nrightarrow \}$$
$$f^2(\emptyset) = \{ E \in \text{States} : E \nrightarrow \text{ or } E \rightarrow E' \nrightarrow \}$$
$$f^3(\emptyset) = \{ E \in \text{States} : E \nrightarrow \text{ or } E \rightarrow E' \nrightarrow$$
$$\text{or } E \rightarrow E' \rightarrow E'' \nrightarrow \}$$
$$\vdots$$
$$f^n(\emptyset) = \text{states which can deadlock within the first } n \text{ transitions.}$$

This sequence is approaching the set S of states that can deadlock, which is the desired solution to our recursive equation and easily seen to be a fixed point of the function f.

15.5 Expressing Standard Temporal Operators

The intuition which you should have drawn from above is the following.

- Greatest fixed point properties are those for which you need to unroll the underlying recursive equation top-down (from true, or the full set of states) an *infinite* number of times in order to verify that the property is always true; if the property fails for some finite unrolling, then the fixed point property itself fails.

In this sense, greatest fixed point properties are representative of *safety properties* which assert that *nothing bad ever happens.*

- Least fixed point properties are those for which you need to unroll the underlying recursive equation bottom-up (from false, or the empty set of states) a *finite* number of times in order to verify that the property is eventually true; if the property fails for every finite unrolling, then the fixed point property itself fails.

In this sense, least fixed point properties are representative of *liveness properties* which assert that *something good eventually happens.*

We now consider how to express each of the standard temporal operators introduced in Section 15.1 in the mu-calculus.

15.5.1 Always: $\Box P$

The temporal operator $\Box P$, expressing that the property P is true in every state into which the process may evolve, is defined by the recursive equation

$$X = P \wedge [-]X.$$

In order to establish the truth of $\Box P$, the recursive equation would need to be unrolled forever, to make sure nothing goes wrong. As such, this property is expressed by the *greatest* fixed point formula:

$$\Box P = \nu X.P \wedge [-]X.$$

15.5.2 Possibly: $\Diamond P$

The temporal operator $\Diamond P$, expressing that the property P is true in some state into which the process may evolve, is defined by the recursive equation

$$X = P \vee \langle - \rangle X.$$

In order to establish the truth of $\Diamond P$, the recursive equation would need to be unrolled only until the property can be verified – that is, only until the property P becomes true. As such, this property is expressed by the *least* fixed point formula:

$$\Diamond P = \mu X.P \vee \langle - \rangle X.$$

15.5.3 Until: $P \,U\, Q$

The temporal operator $P\,U\,Q$, expressing that the property P remains true until the property Q becomes true, which it must eventually do, is defined by the recursive equation

$$X = Q \lor (P \land \langle - \rangle \text{true} \land [-]X)$$

In order to establish the truth of $P \cup Q$, the recursive equation would need to be unrolled only until the property can be verified – that is, only until the property Q becomes true (verifying along the way that P remains true). As such, this property is expressed by the *least* fixed point formula:

$$P \cup Q = \mu X.Q \lor (P \land \langle - \rangle \text{true} \land [-]X).$$

Exercise 15.12   (Solution on page 491)

1. $\Box P = \nu Z.P \land [-]Z$ means P holds in *every* state.

   What does $\mu Z.P \land [-]Z$ mean?

2. $\Diamond P = \mu Z.P \lor \langle - \rangle Z$ means P holds in *some* (reachable) state.

   What does $\nu Z.P \lor \langle - \rangle Z$ mean?

3. $P \cup Q = \mu Z.Q \lor \big(P \land \langle - \rangle \text{true} \land [-]Z \big)$ means Q *will* become true, and until then P will remain true.

   What does $\nu Z.Q \lor \big(P \land \langle - \rangle \text{true} \land [-]Z \big)$ mean?

15.6 Further Fixed Point Properties

In this section we look at a collection of properties that can be expressed in the mu-calculus.

There is an a^ω path.

By this, we mean that we can do an infinite number of consecutive a transitions starting from the state in question.

If we let X represent this property, then X satisfies the recursive equation

$$X = \langle a \rangle X \quad \textit{(It is possible to do an a transition and} \\ \textit{go to a state in which the property holds.)}$$

As we are clearly wanting to unroll this fixed point equation infinitely often, to verify that the property holds forever, we are in this case interested in the greatest fixed point solution:

$$\nu X.\langle a \rangle X.$$

There is no a^ω path.

This is the negation of the previous property, and the most straightforward way to find a mu-calculus formula which expresses it is to use the construction from Exercise 15.7:

$$\text{neg}(\nu X.\langle a\rangle X) = \mu X.[a]X.$$

Unrolling the underlying recursive equation

$$X = [a]X \quad \textit{(If I do an a transition, I must end up} \\ \textit{in a state in which the property holds.)}$$

suggests an exploration of each a^ω path; as this is a least fixed point property, this search must terminate, namely at a state in which an a transition is not possible.

P holds at every state along some a^ω path.

This is a simple adaptation of the first property above:

$$\nu X.P \wedge \langle a\rangle X.$$

P holds somewhere along some a^ω path.

We note first that we want to get to a state at which the property P holds by only doing a transitions. This property is expressed by the recursive equation

$$X = P \vee \langle a\rangle X \quad \textit{(Either P is true, or it is possible to do an a tran-} \\ \textit{sition} \\ \textit{and end up in a state in which the property} \\ \textit{holds.)}$$

As we need P to be true at some point, this is a least fixed point property:

$$\mu X.P \vee \langle a\rangle X.$$

This is not the end of the story, as we require that this path of a transitions leading up to the state in which P is true be the start of an a^ω path. In other words, the point at which P is true must be the start of an a^ω path – which is the first property we considered above: $\nu X.\langle a\rangle X$. Hence, the property we seek is as follows:

$$\mu X.(P \wedge \nu X.\langle a\rangle X) \vee \langle a\rangle X.$$

This formula is fine; however, to avoid confusion it is best to use different variables for the two fixed point constructions:

$$\mu X.(P \wedge \nu Y.\langle a\rangle Y) \vee \langle a\rangle X.$$

P holds at every state along every a^ω path.

We first express the property that P holds along every (finite or infinite) path of a transitions:

$$\nu X.P \wedge [a]X.$$

As a greatest fixed point property, the only way this property can fail is if we reach a state after some number of a transitions in which P does not hold. If this is the case, then we want to ensure that we are *not* on an a^ω path; that is, that from this state in which P fails to hold we cannot continue along an a^ω path – which is the second property we considered above: $\mu X.[a]X$. Hence the property we seek is as follows:

$$\nu X.(P \vee \mu X.[a]X) \wedge [a]X.$$

Again it is sensible to use different variables for the two fixed point constructions:

$$\nu X.(P \vee \mu Y.[a]Y) \wedge [a]X.$$

Exercise 15.13   (Solution on page 491)

Give mu-calculus formulæ for the following properties. In each case give an intuitive explanation of your solution.

1. P almost always holds along some a^ω path.

   Note: To say that something holds *almost always* means *always apart from a finite number of times*. Thus, this property says that P holds everywhere along some a^ω path after some point along this path.

2. P holds infinitely often along some a^ω path.

15.7 Additional Exercises

1. Give a semantic definition for the *weak until* temporal operator $P \, W \, Q$ which asserts that the property P remains true until the property Q becomes true, but allows that the property Q may never become true (in which case the property P remains true for as long as the process evolves).

2. We noted in Section 13.2 that we can express the property that the action a *must* happen as:

$$P = \langle a \rangle \text{true} \wedge \bigwedge_{b \neq a} [b]\text{false}$$

which says that a may happen and nothing other than a may happen.

What is the difference between the property *"eventually a must happen"* as expressed by the temporal formula Ev P (where the Eventually temporal operator was defined in Exercise 15.3) and the property *"a must eventually happen"*?

(Hint: exactly one of these properties holds for the process $b.0 + a.a.0$.)

3. Prove Theorem 15.3.

4. Prove that if X is not a free variable of P then for any $S \subseteq$ States, $\|P\|_{V[X \mapsto S]} = \|P\|_V$.

5. Prove that $E \models_V P$ if, and only if, $E \in \|P\|_V$, where E ranges over formulæ of the modal mu-calculus.

6. Prove the second part of Theorem 15.10, that $f^n(\text{States}) = \|\nu^n X.P\|_V$ where $f(S) = \|P\|_{V[X \mapsto S]}$.

7. Express the following properties in the modal mu-calculus.

   (a) P holds at some state along every a^ω path.

   (b) P almost always holds along every a^ω paths

   (c) P holds infinitely often along every a^ω path.

8. Express the property of mutual exclusion in the modal mu-calculus; that is, the property that whenever an entry action occurs (signifying that a process has entered the critical region), then a further entry action cannot occur until an exit action occurs.

9. The extended modality $\langle a \rangle^* P$ expresses that the property P holds after some number of a transitions, while the extended modality $[a]^* P$ expresses that the property P holds after any number of a transitions.

   Express these extended modalities as mu-calculus formulæ.

10. Express the following properties in the modal mu-calculus.

    (a) In some run the action a does not happen.

    (b) The actions a and b happen alternately forever (starting with the action a), with any number of occurrences of other actions before and between the a and b actions.

    (c) In any run, a and b happen infinitely often.

    (d) If a and b happen infinitely often, then P is true infinitely often.

    (e) In any run, P is true at least twice.

    (f) In any run, P is true exactly twice.

11. Let $E \overset{\text{def}}{=} a.E + a.F$, $F \overset{\text{def}}{=} b.G$, and $G \overset{\text{def}}{=} a.G$, and consider the following two properties:

$$\Phi_1 = \mu Y. \big(\nu X. \langle a \rangle \text{true} \wedge [-]X \big) \vee [-]Y$$

$$\Phi_2 = \mu Y.\nu X.\big(\langle a\rangle \text{true} \ \wedge \ [-]X\big) \ \vee \ [-]Y$$

The process state E satisfies Φ_2 but not Φ_1.

Can you understand and explain why this is the case?

Can you work out what these two properties are expressing?

Solutions to Exercises

In the book of life, the answers aren't in the back

- Charlie Brown.

Chapter 1

Exercise 1.1 (page 20)

1, 2, 4, 6, 7 and 8 are statements, while 3 and 5 are not.

Note that 7 refers to some unspecified utterance by Felix, upon which the truth of this statement depends. Statement 8 however is more complicated: if the sentence it refers to is itself, then there is no consistent way to determine its truth value: if what it says is true, then it must be false; and if what it says is false, then it must be true.

Exercise 1.2 (page 20)

1. This is *not* a valid deduction. There may be some other reason that everyone is leaving the building, e.g., it may be closing for the night.

2. Ideally this would be true, but it is not a valid deduction. It would be valid to deduce that everyone *must* leave the building; however, saying someone (or something) *must* behave in a particular fashion does not make it so; for example, some people may ignore fire alarms, considering fire alarm testing to be a nuisance.

3. This is *not* a valid deduction. The conclusion is no doubt true, as there is surely a rule that states that a train must wait at a red signal; but this rule is not provided in the argument. It might be that the rules for the railway in question do not state that trains must wait at a red light.

4. This is *not* a valid deduction. The conclusion is true, but not for the reasons provided in the two premises.

5. This is *not* a valid deduction. The rook that has already moved might not be the one involved in the castling.

F. Moller, G. Struth, *Modelling Computing Systems*,
Undergraduate Topics in Computer Science,
DOI 10.1007/978-1-84800-322-4, © Springer-Verlag London 2013

6. A judgement as to the validity of this deduction cannot be made on purely logical grounds, due to the ambiguity of the language of the city by-law. Specifically, what is the status of the conjunction *"and"* in the by-law? As Charles does not keep any cats, and certainly no more than three, it could be argued that it is within his rights to keep five (or even fifty) dogs on his property. Even worse, the "more than" in "more than three dogs and three cats" might only apply to dogs and not cats, thereby making the keeping of, say, five dogs and any number of cats allowed except when there are exactly three cats.

Exercise 1.3 (page 21)

1. This is a valid deduction.

2. This is a valid deduction. The conclusion that Epimenides is a liar follows from the premises, as a truth-telling Cretan cannot say that all Cretans (including himself) are liars.

3. This is *not* a valid deduction. It *may* be that all Cretans are liars; or it may be that Epimenides is the only liar. Also, from the previous deduction we already know that Epimenides is a liar based on the given premises, so the conclusion – being precisely what Epimenides claims, must be false.

4. This is a valid deduction. We know that the premises imply that Epimenides is a liar, so his claim that all Cretans are liars must be false.

5. This is *not* a valid deduction. Aristotle may be a liar.

Exercise 1.4 (page 23)

1. "The earth does not revolve around the sun."

2. "I have at least one daughter."

3. $2 + 2 > 4$.

Exercise 1.5 (page 24)

1. This is true, as the second disjunct is true (although the first disjunct is false).

2. This is false, as neither disjunct is true.

3. This is true, as both disjuncts are true.

Exercise 1.6 (page 24)

1. Inclusive. This statement implies that Joel could *not* have come in last place if he beat both Felix and Oskar, so he must have lost to one

of them; but he may well have lost to both of them.

2. Exclusive. It is impossible for a light to be both on and off at the same time.

3. Exclusive. The server no doubt intends to offer the guest only one of the beverages. However, if the guest is so odd as to ask for a cup of both (either one cup with a mix of coffee and tea; or two cups, one with coffee and the other with tea), the server will no doubt reluctantly oblige.

Exercise 1.7 (page 25)

1. This is false, as only the second conjunct is true (the first conjunct is false).

2. This is false, as neither conjunct is true.

3. This is true, as both conjuncts are true.

Exercise 1.8 (page 26)

1. AmandaHappy $\Rightarrow$ JoelHappy.

2. JoelHappy $\Rightarrow$ AmandaHappy.

3. AmandaHappy $\Rightarrow$ JoelHappy.

Exercise 1.9 (page 26)

It may well be true that barking dogs don't bite (i.e., Bark $\Rightarrow$ ¬Bite), but this says nothing about the habits of dogs that *don't* bark; they may bite, or they might not.

Exercise 1.10 (page 29)

1. $p \mid q = \neg(p \wedge q)$.

2. $p \downarrow q = \neg(p \vee q)$.

3. $q \lhd p \rhd r = (p \wedge q) \vee (\neg p \wedge r)$.

Exercise 1.12 (page 31)

1. $P \Rightarrow Q \Leftrightarrow Q \Rightarrow P$ has the following syntax tree:

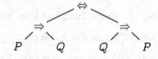

It would be sensible in this example to include redundant parentheses for readability, and to write the formula as $(P \Rightarrow Q) \Leftrightarrow (Q \Rightarrow P)$

2. This is not a well-formed formula.

3. $(P \vee Q) \wedge P$ has the following syntax tree:

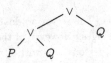

Due to the precedence rules, the parentheses are not redundant; $P \vee Q \wedge P$ would be interpreted as $P \vee (Q \wedge P)$.

4. This is not a well-formed formula.

5. $P \vee Q \wedge R \Leftrightarrow P \vee Q \wedge (P \vee R)$ has the following syntax tree:

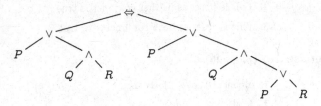

In this case, only one pair of parentheses is redundant; however, it would be sensible to avoid confusion by including all of the redundant parentheses.

Exercise 1.14 (page 33)

This example hints at the many complicated ways that English (or any natural language) can be used to express simple facts. We can draw the conclusion that Lewis Carroll is after by making clear what each of the above assumptions is saying.

Firstly, we introduce propositional variables to represent the different atomic propositions that appear in the argument.

Love: "Amos Judd loves cold mutton."

Police: "Amos Judd is a policeman on this beat."

Sup: "Amos Judd sups with our cook."

Long: "Amos Judd has long hair."

Poet: "Amos Judd is a poet."

Prison: "Amos Judd has been to prison."

Cousin: "Amos Judd is our cook's cousin."

We wish to deduce, formally and logically, the truth of the atomic proposition Love, which asserts that "Amos Judd loves cold mutton." Notice that we modelled the problem by instantiating the properties of all men to apply only to Amos Judd, as he is the only man in whom we have any interest.

The seven assumptions above then translate into the following propositional formulæ:

1. Police ⇒ Sup.

2. Long ⇒ Poet.

3. ¬Prison.

4. Cousin ⇒ Love.

5. Poet ⇒ Police.

6. Sup ⇒ Cousin.

7. ¬Prison ⇒ Long.

You should think carefully about each of these translations, and make sure that you understand why they are correct. Assumptions 5 and 6 are particularly tricky. For example, when 5 says that "None but policemen on this beat are poets," it is asserting that in order to be a poet you must be a policeman on this beat. Thus, if Amos Judd is a poet (Poet), then Amos Judd must be a policeman on this beat (Police): Poet ⇒ Police. Also, when 7 says that "Men with short hair have all been in prison," it is asserting that anyone who has *not* been to prison must have long hair; thus if Amos Judd has not been to prison (¬Prison), then Amos Judd must have long hair (Long): ¬Prison ⇒ Long.

We can finally work out the logic, step-by-step, behind the claim that "Amos Judd loves cold mutton" (Love):

	¬Prison	(by 3).
Thus	Long	(by 7, ¬Prison ⇒ Long).
Thus	Poet	(by 2, Long ⇒ Poet).
Thus	Police	(by 5, Poet ⇒ Police).
Thus	Sup	(by 1, Police ⇒ Sup).
Thus	Cousin	(by 6, Sup ⇒ Cousin).
Thus	Love	(by 4, Cousin ⇒ Love).

The last line is the conclusion that we sought. (Along the way, we also deduced that Amos Judd has long hair; he is a poet; he is a policeman on this beat; he sups with our cook; and he is a cousin of the cook.)

Exercise 1.15 (page 34)

The first clause states that the right to castle with a particular rook (either the left rook or the right rook) has been lost if either the king or the rook in question has already moved:

KingMoved ∨ LeftRookMoved

⇒ ¬RightToCastleLeft.

KingMoved ∨ RightRookMoved

⇒ ¬RightToCastleRight.

The second clause states that the player may not castle with a particular rook if the right to do so has been lost, or if there is a piece between the king and the rook in question, or if the square on which the king stands, or the square which it must cross, or the square which it is to occupy is under attack:

¬RightToCastleLeft	¬RightToCastleRight
∨ PieceBetweenLeft	∨ PieceBetweenRight
∨ KingAttack	∨ KingAttack
∨ LeftSquareAttack	∨ RightSquareAttack
∨ KingMoveLeftAttack	∨ KingMoveRightAttack
⇒ ¬MayCastleLeft	⇒ ¬MayCastleRight

Exercise 1.16 (page 35)

1. We need to express the property that the piece of paper held by each boy has exactly one of the other's names on it, *and* that each name is written on a piece of paper held by exactly one other boy.

    The following proposition p expresses that the piece of paper held by each boy has exactly one of the other's names on it:

    $$p = (\text{FonJ} \lor \text{OonJ}) \land (\neg\text{FonJ} \lor \neg\text{OonJ})$$
    $$\land\ (\text{JonF} \lor \text{OonF}) \land (\neg\text{JonF} \lor \neg\text{OonF})$$
    $$\land\ (\text{JonO} \lor \text{FonO}) \land (\neg\text{JonO} \lor \neg\text{FonO}).$$

    For succinctness, we could have used the exclusive-or connective:

    $$p = (\text{FonJ} \oplus \text{OonJ}) \land (\text{JonF} \oplus \text{OonF}) \land (\text{JonO} \oplus \text{FonO}).$$

    The following proposition q expresses that each name is written on a piece of paper held by exactly one other boy:

$$q = (\text{JonF} \lor \text{JonO}) \land (\neg\text{JonF} \lor \neg\text{JonO})$$
$$\land (\text{FonJ} \lor \text{FonO}) \land (\neg\text{FonJ} \lor \neg\text{FonO})$$
$$\land (\text{OonJ} \lor \text{OonF}) \land (\neg\text{OonJ} \lor \neg\text{OonF}).$$

Again this could be expressed more succinctly:

$$p = (\text{JonF} \oplus \text{JonO}) \land (\text{FonJ} \oplus \text{FonO}) \land (\text{OonJ} \oplus \text{OonF}).$$

The formula we seek is then $p \land q$.

2. From OonJ we can deduce ¬FonJ from p, from which we can deduce FonO from q, from which we can deduce ¬JonO from p, from which we can deduce JonF from q.

In summary, we have "Oskar" on Joel's piece of paper, "Joel" on Felix's piece of paper, and "Felix" on Oskar's piece of paper.

Exercise 1.19 (page 40)

If you answer this question quickly, you might conclude that I would reject the white circle. However, this would be wrong if, for example, I had the white square in mind.

In fact, you cannot conclude that I will reject any particular symbol (though you can conclude that I will reject one of them, you just cannot determine which).

Exercise 1.20 (page 40)

Nine.

The point of this old joke is that four and five are nine *irrespective of the premise of the conditional statement*.

Exercise 1.21 (page 42)

Define the following atomic propositions.

$U = $ You understand implication.
$P = $ You pass the exam.

The statement translates to $U \Rightarrow P$ which has the following truth table:

U	P	$U \Rightarrow P$
F	F	T
F	T	T
T	F	F
T	T	T

The *only* scenario in which the above statement can be considered false is if U is true and P is false – that is, if you do not pass the exam despite understanding induction.

Exercise 1.22 (page 44)

Each new variable *doubles* the number of combinations of truth values. Thus, a truth table involving four propositional variables will have 16 rows, and one involving five variables will have 32 rows. In general, a truth table involving n propositional variables will have 2^n rows.

Truth tables grow very quickly with the number of propositional variables. Building truth tables for propositions with many variables, such as in the Amos Judd example (Exercise 1.14), can therefore be frustrating or even infeasible.

Exercise 1.23 (page 44)

1.

P	Q	$\neg$	$(P$	$\Leftrightarrow$	$\neg$	$Q)$
F	F	T	F	F	T	F
F	T	F	F	T	F	T
T	F	F	T	T	T	F
T	T	T	T	F	F	T

2.

P	Q	$(P$	$\wedge$	$Q)$	$\vee$	$(\neg$	P	$\wedge$	$\neg$	$Q)$
F	F	F	F	F	T	T	F	T	T	F
F	T	F	F	T	F	T	F	F	F	T
T	F	T	F	F	F	F	T	F	T	F
T	T	T	T	T	T	F	T	F	F	T

3.

P	Q	R	S	(P	∧	Q)	⇒	(¬	R	∨	S)
F	F	F	F	F	F	F	**T**	T	F	T	F
F	F	F	T	F	F	F	**T**	T	F	T	T
F	F	T	F	F	F	F	**T**	F	T	F	F
F	F	T	T	F	F	F	**T**	F	T	T	T
F	T	F	F	F	F	T	**T**	T	F	T	F
F	T	F	T	F	F	T	**T**	T	F	T	T
F	T	T	F	F	F	T	**T**	F	T	F	F
F	T	T	T	F	F	T	**T**	F	T	T	T
T	F	F	F	T	F	F	**T**	T	F	T	F
T	F	F	T	T	F	F	**T**	T	F	T	T
T	F	T	F	T	F	F	**T**	F	T	F	F
T	F	T	T	T	F	F	**T**	F	T	T	T
T	T	F	F	T	T	T	**T**	T	F	T	F
T	T	F	T	T	T	T	**T**	T	F	T	T
T	T	T	F	T	T	T	**F**	F	T	F	F
T	T	T	T	T	T	T	**T**	F	T	T	T

Exercise 1.24 (page 44)

p	q	p	⇔	¬	q
F	F	F	**F**	T	F
F	T	F	**T**	F	T
T	F	T	**T**	T	F
T	T	T	**F**	F	T

Exercise 1.25 (page 46)

1. $p \vee (\neg p \wedge q)$ is neither a tautology nor a contradiction:

p	q	p	∨	(¬	p	∧	q)
F	F	F	**F**	T	F	F	F
F	T	F	**T**	T	F	T	T
T	F	T	**T**	F	T	F	F
T	T	T	**T**	F	T	F	T

2. $(p \wedge q) \wedge \neg(p \vee q)$ is a contradiction:

p	q	(p	∧	q)	∧	¬	(p	∨	q)
F	F	F	F	F	**F**	T	F	F	F
F	T	F	F	T	**F**	F	F	T	T
T	F	T	F	F	**F**	F	T	T	F
T	T	T	T	T	**F**	F	T	T	T

3. $(p \Rightarrow \neg p) \Leftrightarrow \neg p$ is a tautology:

p	$(p$	$\Rightarrow$	$\neg$	$p)$	$\Leftrightarrow$	$\neg$	p
F	F	T	T	F	**T**	T	F
T	T	F	F	T	**T**	F	T

4. $(p \Rightarrow q) \Rightarrow p$ is neither a tautology nor a contradiction:

p	q	$(p$	$\Rightarrow$	$q)$	$\Rightarrow$	p
F	F	F	T	F	**F**	F
F	T	F	T	T	**F**	F
T	F	T	F	F	**T**	T
T	T	T	T	T	**T**	T

5. $p \Rightarrow (q \Rightarrow p)$ is a tautology:

p	q	p	$\Rightarrow$	$(q$	$\Rightarrow$	$p)$
F	F	F	**T**	F	T	F
F	T	F	**T**	T	F	F
T	F	T	**T**	F	T	T
T	T	T	**T**	T	T	T

Exercise 1.26 (page 47)

The following is a truth table for these three propositions:

Pressure	Height	Land	p	q	r
F	F	F	T	T	T
F	F	T	T	T	T
F	T	F	F	*T*	F
F	T	T	T	T	T
T	F	F	F	*T*	F
T	F	T	T	T	T
T	T	F	F	F	F
T	T	T	T	T	T

The formula p representing the original program code is not equivalent to the formula q representing the first optimisation, as there are interpretations of the atomic propositions which give rise to different truth values for p and q, highlighted in the third and fifth rows of the above truth table.

However, the formulæ p and q are equivalent, as the truth values of these formulæ are the same under all interpretations, and hence the second optimisation is valid.

Exercise 1.27 (page 50)

1. $p \wedge (\neg p \vee q)$

$\Leftrightarrow \ (p \wedge \neg p) \vee (p \wedge q)$ *(Distributivity)*

$\Leftrightarrow \ \text{false} \vee (p \wedge q)$ *(Excluded Middle)*

$\Leftrightarrow \ (p \wedge q) \wedge \text{false}$ *(Commutativity)*

$\Leftrightarrow \ p \wedge q$ *(Tautology)*

2. $\neg(p \Rightarrow q)$

$\Leftrightarrow \ \neg(\neg p \vee q)$ *(Implication)*

$\Leftrightarrow \ \neg\neg p \wedge \neg q$ *(De Morgan)*

$\Leftrightarrow \ p \wedge \neg q$ *(Double Negation)*

3. $p \Rightarrow (q \vee r)$

$\Leftrightarrow \ \neg p \vee (q \vee r)$ *(Implication)*

$\Leftrightarrow \ (\neg p \vee \neg p) \vee (q \vee r)$ *(Idempotence)*

$\Leftrightarrow \ (\neg p \vee q) \vee (\neg p \vee r)$ *(Associativity, Commutativity)*

$\Leftrightarrow \ (p \Rightarrow q) \vee (p \Rightarrow r)$ *(Implication)*

4. $p \Rightarrow (q \wedge r)$

$\Leftrightarrow \ \neg p \vee (q \wedge r)$ *(Implication)*

$\Leftrightarrow \ (\neg p \vee q) \wedge (\neg p \vee r)$ *(Distributivity)*

$\Leftrightarrow \ (p \Rightarrow q) \wedge (p \Rightarrow r)$ *(Implication)*

5. $(p \wedge q) \Rightarrow r$

$\Leftrightarrow \ \neg(p \wedge q) \vee r$ *(Implication)*

$\Leftrightarrow \ (\neg p \vee \neg q) \vee r$ *(De Morgan)*

$\Leftrightarrow \ (\neg p \vee \neg q) \vee (r \vee r)$ *(Idempotence)*

$\Leftrightarrow \ (\neg p \vee r) \vee (\neg q \vee r)$ *(Associativity, Commutativity)*

$\Leftrightarrow \ (p \Rightarrow r) \vee (q \Rightarrow r)$ *(Implication)*

6. $(p \vee q) \Rightarrow r$

$\Leftrightarrow \ \neg(p \vee q) \vee r$ *(Implication)*

$\Leftrightarrow \ (\neg p \wedge \neg q) \vee r$ *(De Morgan)*

$\Leftrightarrow \ (\neg p \vee r) \wedge (\neg p \vee r)$ *(Distributivity)*

$\Leftrightarrow \ (p \Rightarrow r) \wedge (q \Rightarrow r)$ *(Implication)*

Chapter 2

Exercise 2.1 (page 59)

1. $\{1, 3, 5, 7\}$.
2. $\{$ Tuesday, Thursday, Friday, Saturday $\}$.
3. $\{$ Catherine of Aragon, Anne Boleyn, Jane Seymour, Anne of Cleves, Catherine Howard, Catherine Parr $\}$.
4. $\{$ Sean Connery, George Lazenby, Roger Moore, Timothy Dalton, Pierce Brosnan, Daniel Craig $\}$.

Exercise 2.2 (page 60)

1. $\{2, 4, 6, 8, 10\}$.
2. $\{1, 2\}$.

Exercise 2.3 (page 60)

1, 3 and 5 are true, while 2 and 4 are false.

Exercise 2.4 (page 60)

$A = E$ and $C = D$.

Exercise 2.5 (page 62)

1 is true, while 2 and 3 are false.

Exercise 2.7 (page 65)

If $R \in R$, then by definition of R we would have $R \notin R$, which cannot be true. Therefore we must have that $R \notin R$.

This is no longer a problem, as $R \notin R$ now means that either $R \notin A$ or $R \in R$; since we know that $R \notin R$, this simply means that $R \notin A$.

Exercise 2.15 (page 69)

The Venn diagram is depicted in Figure 15.2

1. $A \cap C = \{5, 7, 9\}$.
2. $(A \cap B) \cup C = \{3, 5, 6, 7, 8, 9\}$.
3. $A \cap (B \cup C) = \{3, 5, 7, 9\}$.
4. $(A \cup B) \setminus C = \{1, 3, 4\}$.
5. $\overline{(A \cup B)} \cap C = \{6, 8\}$.

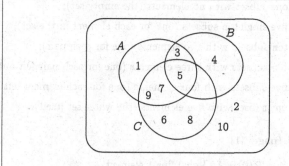

Figure 15.2: Venn diagram for Exercise 2.15.

Exercise 2.16 (page 69)

You can use Venn diagrams to verify these properties.

1. If $A \subseteq B$, then $A \cup B = B$ and $A \cap B = A$.
2. If $A \subseteq B$, then $\overline{B} \subseteq \overline{A}$.
3. $\overline{\overline{A}} = A$.
4. If $C \subseteq A$ and $C \subseteq B$, then $C \subseteq A \cap B$.
5. If $A \subseteq C$ and $B \subseteq C$, then $A \cup B \subseteq C$.

Exercise 2.17 (page 71)

Letting $D = $ Daniel, $E = $ Ella, $M = $ Mia, $R = $ Rhodri and $Z = $ Zoe, we get

$$\mathcal{P}(\{D, E, M, R, Z\})$$
$$= \{\emptyset,$$

$\{D\}, \{E\}, \{M\}, \{R\}, \{Z\},$

$\{D, E\}, \{D, M\}, \{D, R\}, \{D, Z\},$
$\{E, M\}, \{E, R\}, \{E, Z\},$
$\{M, R\}, \{M, Z\}, \{R, Z\},$

$\{D, E, M\}, \{D, E, R\}, \{D, E, Z\},$
$\{D, M, R\}, \{D, M, Z\}, \{D, R, Z\},$
$\{E, M, R\}, \{E, M, Z\}, \{E, R, Z\}, \{M, R, Z\},$

$\{D, E, M, R\}, \{D, E, M, Z\}, \{D, E, R, Z\},$
$\{D, M, R, Z\}, \{E, M, R, Z\},$

$\{D, E, M, R, Z\}\}.$

More specifically, there are the following subsets:

- one subset with no elements (the empty set);
- five singleton subsets (one for each element in the set);
- ten subsets with two elements (one for each pair);
- ten subsets with three elements (one for each pair left out);
- five subsets with four elements (one for each element left out); and
- one subset with five elements (the whole set itself).

Exercise 2.18 (page 71)

1. $A = \mathcal{P}(\emptyset) = \{\emptyset\}$ contains 1 element.
2. $B = \mathcal{P}(A) = \{\emptyset, \{\emptyset\}\}$ contains 2 elements.
3. $C = \mathcal{P}(B) = \{\emptyset, \{\emptyset\}, \{\{\emptyset\}\}, \{\emptyset, \{\emptyset\}\}\}$ contains 4 elements.

Exercise 2.19 (page 71)

$\mathcal{P}(A) \cap \emptyset = \emptyset$ and $\mathcal{P}(A) \cap \{\emptyset\} = \{\emptyset\}$.

Exercise 2.20 (page 73)

$\bigcap \mathcal{P}_{\text{fin}}(A) = \emptyset$ and $\bigcup \mathcal{P}_{\text{fin}}(A) = A$.

Note that the union of infinitely-many finite sets may well be infinite, although the union of finitely-many finite sets will of course be finite.

Exercise 2.23 (page 75)

$(p, q) + (r, s) = (ps + qr, qs)$ and $(p, q) \times (r, s) = (pr, qs)$.

Exercise 2.24 (page 76)

Consider the following sets of people:

$\quad$ Love $=$ the set of people who love cold mutton.
$\quad$ Police $=$ the set of policemen on this beat.
$\quad$ Sup $=$ the set of people who sup with our cook.
$\quad$ Long $=$ the set of long-haired people.
$\quad$ Poet $=$ the set of poets.
$\quad$ NoPrison $=$ the set of people who have never been to prison.
$\quad$ Cousin $=$ the set of cousins of our cook.

The above seven assumptions then translate to the following set inclusions:

1. Police $\subseteq$ Sup.

2. Long $\subseteq$ Poet.

3. Amos $\in$ NoPrison.

4. Cousin $\subseteq$ Love.

5. Poet $\subseteq$ Police.

6. Sup $\subseteq$ Cousin.

7. NoPrison $\subseteq$ Long.

We can then conclude that Amos $\in$ LOVE, that is, that Amos Judd loves cold mutton, as follows:

	Amos $\in$ NoPrison	(by 3).
Thus	Amos $\in$ Long	(by 7, NoPrison $\subseteq$ Long).
Thus	Amos $\in$ Poet	(by 2, Long $\subseteq$ Poet).
Thus	Amos $\in$ Police	(by 5, Poet $\subseteq$ Police).
Thus	Amos $\in$ Sup	(by 1, Police $\subseteq$ Sup).
Thus	Amos $\in$ Cousin	(by 6, Sup $\subseteq$ Cousin).
Thus	Amos $\in$ Love	(by 4, Cousin $\subseteq$ Love).

The last line is the conclusion that we sought. (Again, along the way, we also deduced that Amos Judd has long hair; he is a poet; he is a policeman on this beat; he sups with our cook; and he is a cousin of the cook.)

Exercise 2.25 (page 77)

Let B stand for the set of all babies, I for the set of all illogical persons, D for the set of despised persons and C for the set of those persons who can manage a crocodile. Then the premises become :

$$B \subseteq I, \ C \cap D = \emptyset, \text{ and } I \subseteq D$$

which are reflected in the following Venn diagram:

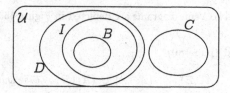

It is clear from this that no baby can manage a crocodile, as a baby would be illogical ($B \subseteq I$) and hence despised ($I \subseteq D$); and no despised person, such as this baby, could manage a crocodile.

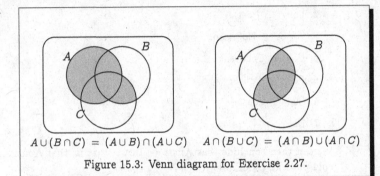

$$A \cup (B \cap C) = (A \cup B) \cap (A \cup C) \qquad A \cap (B \cup C) = (A \cap B) \cup (A \cap C)$$

Figure 15.3: Venn diagram for Exercise 2.27.

Exercise 2.26 (page 77)

Consider the following Venn diagram:

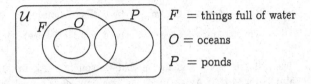

F = things full of water

O = oceans

P = ponds

The first premise in the argument says that $O \subseteq F$; and the second premise in the argument says that $P \cap O = \emptyset$. These premises are satisfied by the above Venn diagram. However, the conclusion of the argument says that $P \cap F = \emptyset$, which is not (necessarily) satisfied by the above Venn diagram.

The argument is thus not valid, as the above Venn diagram suggests a counter-example to the argument: there may well be ponds which are not oceans yet are nonetheless full of water.

Exercise 2.27 (page 80)

The two Venn diagrams are depicted in Figure 15.3.

Exercise 2.28 (page 80)

$$
\begin{aligned}
A \cap (\overline{A} \cup B) &= (A \cap \overline{A}) \cup (A \cap B) && \textit{(Distributive Law)} \\
&= \emptyset \cup (A \cap B) && \textit{(Complement Law)} \\
&= (A \cap B) \cap \emptyset && \textit{(Commutative Law)} \\
&= A \cap B && \textit{(Empty Set Law)}
\end{aligned}
$$

Exercise 2.29 (page 81)

- By Associativity, Commutativity and Idempotence, $(A \cap B) \cap A = A \cap B$.

- Letting $X = A \cap B$ and $Y = A$, this says that $X \cap Y = X$.

- This means that $X \subseteq Y$; that is, that $A \cap B \subseteq A$.

Exercise 2.30 (page 83)

1. $A \subseteq B$ if, and only if, $\overline{B} \subseteq \overline{A}$.
2. $A = B$ if, and only if, $(A \subseteq B) \wedge (B \subseteq A)$.

Exercise 2.31 (page 83)

We might naïvely translate the law

$$\neg(P \Rightarrow Q) \quad \Leftrightarrow \quad P \wedge \neg Q$$

into $A \not\subseteq B$ if, and only if, $A \cap \overline{B} = \mathcal{U}$. This law for sets is blatantly false: $A \cap \overline{B} = \mathcal{U}$ can only be true if $A = \mathcal{U}$ and $B = \emptyset$; and this is certainly not the only situation in which we can have $A \not\subseteq B$.

The problem arises from attempting to translate the negation of an implication. To get a correct law for sets corresponding to the given law for propositions, we first simplify the law by negating both sides:

$$P \Rightarrow Q \quad \Leftrightarrow \quad \neg(P \wedge \neg Q)$$

Translating $P \Rightarrow Q$ into $A \subseteq B$, and expressing $\neg(P \wedge \neg Q)$ as $P \wedge \neg Q \Leftrightarrow \mathbf{F}$ gives rise to the following valid law for sets:

$$A \subseteq B \text{ if, and only if, } A \cap \overline{B} = \emptyset.$$

Chapter 3

Exercise 3.3 (page 89)

It is straightforward, if a bit tedious, verifying that each of these laws holds for every combination of values of x, y and z. For example, to verify that the first Distributivity Law

$$x + (y \times z) = (x+y) \times (x+z)$$

is true, we need only use the tables defining $+$ and $\times$ to check the following eight equations are true (one for each of the eight combinations of values for x, y and z):

$$0 + (0 \times 0) = (0+0) \times (0+0) \qquad 1 + (0 \times 0) = (1+0) \times (1+0)$$
$$0 + (0 \times 1) = (0+0) \times (0+1) \qquad 1 + (0 \times 1) = (1+0) \times (1+1)$$
$$0 + (1 \times 0) = (0+1) \times (0+0) \qquad 1 + (1 \times 0) = (1+1) \times (1+0)$$
$$0 + (1 \times 1) = (0+1) \times (0+1) \qquad 1 + (1 \times 1) = (1+1) \times (1+1)$$

The details are omitted.

Exercise 3.9 (page 92)

Since $0+1 = 1$ (by Ident1) and $0 \times 1 = 0$ (by Ident2), the Uniqueness of Complement Theorem 3.8 says that $0' = 1$.

But then $1' = (0')' = 0$ by the Involution Law (Theorem 3.9).

An alternative proof which avoid the use of the Uniqueness of Complement Theorem is as follows:

$$\begin{aligned} 0' &= 0' + 0 \quad &(Ident1) \\ &= 0 + 0' \quad &(Comm1) \\ &= 1 \quad &(Compl1) \end{aligned} \qquad \begin{aligned} 1' &= 1' \cdot 1 \quad &(Ident2) \\ &= 1 \cdot 1' \quad &(Comm2) \\ &= 0 \quad &(Compl2) \end{aligned}$$

Exercise 3.10 (page 93)

1.
$$\begin{aligned} (xy + x'y')' &= (x' + y')(x + y) &&(De\ Morgan,\ Involution) \\ &= xx' + xy' + x'y + yy' &&(Distr,\ Comm,\ Assoc) \\ &= xy' + x'y &&(Compl,\ Ident, \\ & &&\quad Comm,\ Assoc) \end{aligned}$$

2. Assume that $x + y = x + z$ and $x' + y = x' + z$. Then

$$\begin{aligned} xy &= xx' + xy &&(Compl2,\ Ident1,\ Comm1) \\ &= x(x' + y) &&(Distr2) \\ &= x(x' + z) &&(Assumption\ 2) \\ &= xx' + xz &&(Distr2) \\ &= xz &&(Compl2,\ Ident1,\ Comm1) \end{aligned}$$

Thus, with Assumption 1, we have from Theorem 3.7 that $y = z$.

3. If $x + y = 0$ then $x' = x + y + x' = (x+x') + y = 1 + y = 1$, so $x = 0$.

By similar reasoning, if $x + y = 0$ then $y = 0$.

4. If $x = 0$ then $x' = 1$ and thus $y = 0y' + 1y = xy' + x'y$.

Conversely, if $y = xy' + x'y$ for all y, then taking $y = 0$, and thus $y' = 1$, we get that $0 = xy' + x'y = x1 + x'0 = x$.

Exercise 3.11 (page 94)

1. $\big((x+y)(x'+y')\big)' = (x+y')(x'+y)$.

2. If $xy = xz$ and $x'y = x'z$ then $y = z$.

3. If $xy = 1$ then $x = y = 1$.

4. $x = 1$ if, and only if, $y = (x+y')(x'+y)$ for all y.

Exercise 3.14 (page 99)

We start by expressing $x \oplus y$ in terms of the three basic operations:

$$x \oplus y \;=\; (x + y)(xy)'$$

The circuit for this is then as follows:

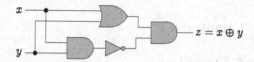

Exercise 3.15 (page 99)

We start by annotating the diagram with variables for all of the intermediate values which are computed:

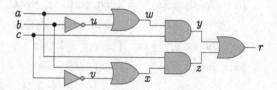

We can then calculate the intermediate and final values by considering their Boolean expressions:

		a	b	c	u	v	w	x	y	z	r
u	$= b'$	0	0	0	1	1	1	1	0	0	0
v	$= c'$	0	0	1	1	0	1	0	1	0	1
w	$= a + u$	0	1	0	0	1	0	1	0	0	0
x	$= b + v$	0	1	1	0	0	0	1	0	0	0
y	$= wc$	1	0	0	1	1	1	1	0	1	1
z	$= ax$	1	0	1	1	0	1	0	1	0	1
r	$= y + z$	1	1	0	0	1	1	1	0	1	1
		1	1	1	0	0	1	1	1	1	1

Exercise 3.16 (page 99)

The output R to be computed is given by the formula $R = M(B' + C')$, which by De Morgan's Law can be rewritten as $R = M(BC)'$. Thus either of the following two Boolean circuits will give a valid implementation.

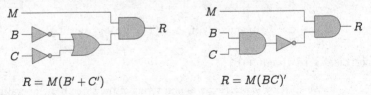

$$R = M(B' + C') \qquad\qquad R = M(BC)'$$

Exercise 3.17 (page 102)

$$29 + 22 = 51.$$

Chapter 4

Exercise 4.2 (page 111)

1. $\{x : Even(x)\} = \{x \in \mathbb{Z} : x \text{ is even}\}$
$$= \{\ldots, -6, -4, -2, 0, 2, 4, 6, \ldots\}.$$

2. $\{x : EvenPrime(x)\} = \{2\}.$

3. $\{x : DeadlySin(x)\} = \{lust, \, gluttony, \, greed,$
$$sloth, \, wrath, \, envy, \, pride\}.$$

4. $\{x : Sum(x, y, z)\} = \{(x, y, z) \in \mathbb{Z}^3 : x + y = z\}$

5. $\{x : Sum(u, 5, v)\} = \{(u, v) \in \mathbb{Z}^2 : u + 5 = v\}$
$$= \{\ldots, (-3, 2), (-2, 3), (-1, 4),$$
$$(0, 5), (1, 6), (2, 7), (3, 8), \ldots\}.$$

Exercise 4.5 (page 115)

1. $\forall x \, \forall y \, \big(B(x) \wedge F(y) \Rightarrow L(x, y) \big).$

2. $\forall x \, \forall y \, \big(B(x) \wedge L(x, y) \Rightarrow F(y) \big).$

3. $\forall x \, \forall y \, \big(F(y) \wedge L(x, y) \Rightarrow B(x) \big).$

Exercise 4.7 (page 117)

1. $\forall x \, \big(Male(x) \oplus Female(x) \big).$

2. $\forall x \, \big(\exists y \, Mother(x, y) \Rightarrow Parent(x, y) \wedge Female(x) \big).$

3. $\forall x\, \exists m\, \exists f\, \forall y\, \big((Mother(y, x) \Leftrightarrow y{=}m) \land (Father(y, x) \Leftrightarrow y{=}f) \big)$.

4. $\forall x\, \forall y\, \big(Sibling(x, y) \Rightarrow \forall z\, (Parent(z, x) \Leftrightarrow Parent(z, y)) \big)$.

5. $\forall x\, \forall y\, \big(Cousin(x, y) \Rightarrow$
$$\exists u\, \exists v\, (Parent(u, x) \land Parent(v, y) \land Sibling(u, v)) \big).$$

Exercise 4.8 (page 117)

The premise of the argument translates into

$$\forall h\big(Horse(h) \Rightarrow Animal(h) \big)$$

which says that any thing h which is a horse is an animal.

The conclusion of the argument translates into

$$\forall x\Big(\exists h(Horse(h) \land Head(x, h)) \Rightarrow$$
$$\exists a(Animal(a) \land Head(x, a)) \Big)$$

which says that any thing x which is the head of some horse h is the head of some animal a.

This argument is valid: for suppose x is the head of some horse h (Black Beauty, say). Since the premise says that all horses are animals, this particular horse h (Black Beauty) is an animal; and hence this thing x is the head of some animal a, namely h (Black Beauty).

Exercise 4.9 (page 120)

1. $\exists! c\, (T(Alice, c) \land T(Bob, c))$

2. $\exists c_1\, \big(T(Alice, c_1) \land T(Bob, c_1)$
$$\land\ \exists! c_2\, (T(Alice, c_2) \land T(Bob, c_2) \land c_1 \neq c_2) \big)$$

Exercise 4.10 (page 121)

1. $\exists x\, LikesMaths(x)$, where $LikesMaths(x) = $ "x likes maths".

   Its negation is (b).

   (a) $\exists x\, \neg LikesMaths(x)$.

   (b) $\forall x\, \neg LikesMaths(x)$.

   (c) $\forall x\, LikesMaths(x)$.

2. $\forall x\, (Fur(x) \land Tail(x))$, where $Fur(x) = $ "x has fur" and $Tail(x) = $ "x has a tail".

   Its negation is (c).

   (a) $\neg\exists x\, (Fur(x) \land Tail(x))$.

   (b) $\exists x\,(\,\neg Fur(x)\ \wedge\ \neg Tail(x)\,)$.

   (c) $\exists x\,(\,\neg Fur(x)\ \vee\ \neg Tail(x)\,)$.

3. $\forall x\,(\,\neg Vaccinated(x)\ \Rightarrow\ Sick(x)\,)$, where $Vaccinated(x) =$ "x has been vaccinated" and $Sick(x) =$ "x got Sick".

Its negation is (c).

   (a) $\forall x\,(\,Vaccinated(x)\ \Rightarrow\ \neg Sick(x)\,)$.

   (b) $\exists x\,(\,Vaccinated(x)\ \wedge\ Sick(x)\,)$.

   (c) $\exists x\,(\,\neg Vaccinated(x)\ \wedge\ \neg Sick(x)\,)$.

Exercise 4.12 (page 125)

Let $Loves(x, y) =$ "x loves y", where the universe of discourse is the set of people.

1. *Everybody loves somebody:* $\forall x\,\exists y\,Loves(x, y)$.
   Somebody is loved by everybody: $\exists x\,\forall y\,Loves(y, x)$.

   These English statements are ambiguous, as each may be interpreted as saying precisely what the other is saying. However, the likely interpretation for each is as formalised in predicate logic above.

   This argument is *not* valid. For example, perhaps Alice only loves herself, but everyone else loves Bob (including Bob himself); in this scenario, the premise is true, but the conclusion is false.

2. *Somebody loves everybody:* $\exists x\,\forall y\,Loves(x, y)$.
   Everybody is loved by somebody: $\forall x\,\exists y\,Loves(y, x)$.

   This argument is valid. The premise of the argument says that there is some person – Theresa say – who loves everybody. This means that the conclusion of the argument must be true as well: everybody *is* loved by someone, in particular by Theresa.

Exercise 4.13 (page 127)

2	9	8	1	3	5	4	6	7
4	1	7	8	9	6	2	3	5
3	6	5	2	7	4	9	8	1
7	4	9	5	2	3	6	1	8
8	2	3	9	6	1	7	5	4
6	5	1	7	4	8	3	9	2
1	3	4	6	8	7	5	2	9
5	7	2	3	1	9	8	4	6
9	8	6	4	5	2	1	7	3

Chapter 5

Exercise 5.2 (page 134)

(Fact 15.14)

$A \cup B \subseteq C \;\Rightarrow\; A \subseteq C \wedge B \subseteq C.$

Proof: Assume that $A \cup B \subseteq C$; we must show that $A \subseteq C \wedge B \subseteq C$. This means that we must show both $A \subseteq C$ and $B \subseteq C$.

We consider $A \subseteq C$ first. By the definition of the set inclusion $A \subseteq C$, we choose an arbitrary element $x \in A$ and we show that $x \in C$. Since $x \in A$, it is thus also the case that $x \in A \cup B$. Hence, by our assumption, $x \in C$. We have thus shown that $A \subseteq C$.

The proof that $B \subseteq C$ is very similar. □

Exercise 5.3 (page 136)

Fact: If a and b are both odd integers, then ab is an odd integer.

Proof: Assume that a and b are odd integers.

An odd integer is one more than twice an integer.

Thus $a = 2p+1$ and $b = 2q+1$ for some integers p and q.

Hence $ab = (2p+1)(2q+1) = 4pq + 2p + 2q + 1$

$= 2(2pq + p + q) + 1$

$= 2k+1$ for the integer $k = 2pq + p + q$.

Therefore, ab is an odd integer. □

Exercise 5.5 (page 138)

If the sum of the digits of a number is divisible by 3, then that number itself is divisible by 3.

- The sum of the digits of 45 is $4+5 = 9$, which is divisible by 3; so by modus ponens, 45 itself is divisible by 3.
- The sum of the digits of 9 839 853 is $9+8+3+9+8+5+3 = 45$, which is divisible by 3; so by modus ponens, 9 839 853 itself is divisible by 3.

Exercise 5.9 (page 141)

Fact: There is no smallest positive rational number.

Proof: Assume to the contrary that $a > 0$ is the smallest rational number.

Then $b = a/2$ is a positive rational number which is smaller than a, contradicting our assumption that a is the smallest such number.

Hence there cannot be a smallest positive rational number.   □

Exercise 5.10 (page 142)

Fact: Every integer greater than 1 can be written as a product of prime numbers.

Proof: Assume to the contrary that not all integers greater than 1 can be written as a product of prime numbers,

Let n be the smallest such integer; thus, every smaller integer greater than 1 can be written as a product of primes.

By assumption, n cannot be prime, so $n = pq$ where p and q are two smaller integers greater than 1.

Since p and q are smaller than n, they must themselves each be a product of primes.

But then n must be a product of primes as well. namely the product of those primes making up p and q, contradicting the definition of n.

Hence every integer greater than 1 can be written as a product of prime numbers.   □

Exercise 5.13 (page 145)

Fact: If a and b are integers and ab is even, then either a is even or b is even.

Proof: Assume that a and b are integers and that ab is even. That is, $ab = 2p$ for some integer p.

Suppose that a is odd; that is, suppose that $a = 2q+1$ for some integer q.

Then $ab = (2q+1)b = 2qb + b$; and since $ab = 2p$, this means that $2p = 2qb + b$, and thus that $b = 2p - 2qb = 2(p - qb)$.

Since $p - qb$ is an integer, this means that b must be even.

Thus, if a is *not* even, then b must be even; that is, either a or b is even.

Exercise 5.14 (page 146)

Fact: If $A \subseteq B$ then either $x \notin A$ or $x \in B$.

Proof: Assume that $A \subseteq B$.

Suppose that $x \in A$; that is, that it is *not* the case that $x \notin A$.

Then since $A \subseteq B$, we must have that $x \in B$.

Thus, either $x \notin A$, or $x \in B$.

Exercise 5.15 (page 147)

Fact: For real numbers a and b, $|a + b| \leq |a| + |b|$.

Proof: Since $|a + b| = |b + a|$, we can assume without any loss of generality that $|a| \geq |b|$.

- Either a and b have the same sign – that is, they are both nonnegative (i.e., greater than or equal to 0) or they are both negative;
- or a and b have opposite signs – that is, one is nonnegative and the other is negative.

We shall consider these two cases in turn.

- If a and b have the same sign, then $|a + b| = |a| + |b| \leq |a| + |b|$.
- If a and b have opposite signs, then $|a + b| = |a| - |b| \leq |a| + |b|$.

In either case, the result is true. □

Exercise 5.16 (page 147)

Fact: If n is an integer, then the final digit of n^2 is 0, 1, 4, 5, 6 or 9.

Proof: We can prove this by breaking down the problem into cases depending on the final digit of n:

- If the final digit of n is 0, then the final digit of n^2 will be 0.
- If the final digit of n is 1 or 9, then the final digit of n^2 will be 1.
- If the final digit of n is 2 or 8, then the final digit of n^2 will be 4.
- If the final digit of n is 3 or 7, then the final digit of n^2 will be 9.
- If the final digit of n is 4 or 6, then the final digit of n^2 will be 6.
- If the final digit of n is 5, then the final digit of n^2 will be 5.

This exhausts all possibilities for the final digit of n, and hence the result must be true. □

Exercise 5.17 (page 147)

Saying that it is *not* the case that $x \neq 7$ and $y \neq 8$ means that either $x = 7$ or $y = 8$, *not* that both of these equalities holds.

Exercise 5.18 (page 149)

Fact: If A and $B \setminus C$ are disjoint then $A \cap B \subseteq C$.

Proof: Assume that A and $B \setminus C$ are disjoint. From this assumption, we need to prove that $A \cap B \subseteq C$; that is, that for any x, if $x \in A \cap B$ then $x \in C$:

$$\forall x \, (x \in A \cap B \ \Rightarrow \ x \in C).$$

To this end, let a be an arbitrary value.

To show that $a \in A \cap B \Rightarrow a \in C$, we assume that $a \in A \cap B$ and prove from this assumption that $a \in C$.

Assume then that $a \in A \cap B$; that is, that $a \in A$ and $a \in B$.

Since A and $B \setminus C$ are disjoint (from a premise of the proposition) and $a \in A$, we must have that $a \notin B \setminus C$.

But since $a \in B$, $a \notin B \setminus C$ means we must have that $a \in C$.   □

Exercise 5.19 (page 151)

Fact: $\forall x {>} 0 \, \exists y \, (y(y{+}1) = x)$.

Proof: Let $x > 0$ be arbitrary, and let $y = \frac{1}{2}\left(-1 + \sqrt{1+4x}\right)$.

$$\begin{aligned}
\text{Then } y(y{+}1) &= \tfrac{1}{2}\left(-1 + \sqrt{1+4x}\right)\left(\tfrac{1}{2}\left(-1 + \sqrt{1+4x}\right) + 1\right) \\
&= \tfrac{1}{4}\left(\sqrt{1+4x} - 1\right)\left(\sqrt{1+4x} + 1\right) \\
&= \tfrac{1}{4}\left((1{+}4x) - 1\right) \ = \ \tfrac{1}{4}(4x) \ = \ x \qquad \square
\end{aligned}$$

Where did this value of y come from? Given $x{>}0$, we want a value y satisfying $y(y{+}1) = x$, or in other words, by expanding and rewriting this equation, a solution y to the quadratic equation

$$y^2 + y - x = 0.$$

The quadratic formula tells us that the two values for y which solve this equation are

$$y = \frac{-1 \pm \sqrt{1 + 4x}}{2}.$$

Only one of these two solutions is positive as required, namely

$$y = \tfrac{1}{2}\left(-1+\sqrt{1+4x}\right).$$

Exercise 5.20 (page 152)

Fact: $\exists x\left(P(x)\vee Q(x)\right) \Leftrightarrow \exists x\,P(x)\ \vee\ \exists x\,Q(x)$.

Proof: ($\Rightarrow$) Suppose $\exists x\left(P(x)\vee Q(x)\right)$.

Then $P(a)\vee Q(a)$ holds for some a.

For this value a, either $P(a)$ holds or $Q(a)$ holds.

- If $P(a)$ holds, then $\exists x\,P(x)$, and thus $\exists x\,P(x)\ \vee\ \exists x\,Q(x)$.
- If $Q(a)$ holds, then $\exists x\,Q(x)$, and thus $\exists x\,P(x)\ \vee\ \exists x\,Q(x)$.

In either case, $\exists x\,P(x)\ \vee\ \exists x\,Q(x)$.

($\Leftarrow$) Suppose $\exists x\,P(x)\ \vee\ \exists x\,Q(x)$.

Then either $\exists x\,P(x)$ holds, or $\exists x\,Q(x)$ holds.

- If $\exists x\,P(x)$ holds, then $P(a)$ holds for some value a.
  For this a, $P(a)\vee Q(a)$ holds, and so $\exists x\left(P(x)\vee Q(x)\right)$.
- If $\exists x\,Q(x)$ holds, then $Q(a)$ holds for some value a.
  For this a, $P(a)\vee Q(a)$ holds, and so $\exists x\left(P(x)\vee Q(x)\right)$.

Thus, in either case, $\exists x\left(P(x)\vee Q(x)\right)$. $\square$

Exercise 5.22 (page 153)

Fact: There is a unique set A such that, for every set B, $A\cup B = B$.

Proof: To show existence of such a set, we simply note that the empty set $\emptyset$ clearly has the desired property, as $\emptyset\cup B = B$ for every set B.

To show that $\emptyset$ is the only set with this property, assume that some set A satisfies this property; in particular, taking $B = \emptyset$, this means that $A\cup\emptyset = \emptyset$. But then $A = A\cup\emptyset = \emptyset$. $\square$

Chapter 6

Exercise 6.2 (page 158)

1. range(score) $= \{\,46, 54, 59, 64, 68, 75, 78, 88, 92, 100\,\}$.
2. score$^{-1}(\{n\in\mathbb{N}\ :\ n\geq 70\})$.

Exercise 6.3 (page 158)

1. *Mother* is a function as every person has one and exactly one (biological) mother.

2. *Parent* is not a function as people have two parents not one.

3. *Child* is not a function as a person may have any number of children.

4. *FirstBornChild* is not a function as a person may have no children.

Exercise 6.4 (page 160)

graph(f) = { $(1,c), (2,a), (3,c)$ }.

Exercise 6.5 (page 161)

1. The function score is *not* one-to-one as, for example, score(Collins) = score(Parker). Also, score(Evans) = score(Williams).

2. The function $f : \mathbb{R} \to \mathbb{R}$ defined by $f(x) = x^2$ is *not* one-to-one as, for example, $f(-1) = f(1)$. (In fact, $f(x) = f(-x)$ for any value $x \in \mathbb{R}$.)

3. The function $f : \mathbb{N} \to \mathbb{N}$ defined by $f(x) = x^2$ *is* one-to-one.

Exercise 6.6 (page 161)

1. The function score is *not* onto as, for example, no one has scored 0.

2. The function $f : \mathbb{R} \to \mathbb{R}$ defined by $f(x) = x^2$ is *not* onto as $f(x) \geq 0$ for all $x \in \mathbb{R}$ so, for example, for no $x \in \mathbb{R}$ do we have $x^2 = -1$.

3. The function $f : \mathbb{N} \to \mathbb{N}$ defined by $f(x) = x^2$ is *not* onto as, for example, for no $x \in \mathbb{N}$ do we have $x^2 = 3$.

Exercise 6.7 (page 161)

- f_1 is one-to-one but not onto, as there is an element of the codomain (the third element from the top) which is not in the range of the function.

- f_2 is onto but not one-to-one, as f_2 maps two elements of the domain (the top and bottom elements) to the same element of the codomain (the middle of the three elements).

- f_3 is not one-to-one, as it maps two elements of the domain (the first two elements) to the same element of the codomain (the third element from the top); nor is f_3 onto, as there is an element of the codomain (the second element from the top) which is not in the range of the function.

- f_4 is both one-to-one and onto,

Exercise 6.8 (page 163)

$$
\begin{array}{cccccccccccc}
a & b & c & d & e & f & g & h & i & j & k & l & m \\
\downarrow & \downarrow & \downarrow & \downarrow & \downarrow & \downarrow & \downarrow & \downarrow & \downarrow & \downarrow & \downarrow & \downarrow & \downarrow \\
x & i & e & p & d & t & o & v & z & s & b & m & j
\end{array}
$$

$$
\begin{array}{cccccccccccc}
n & o & p & q & r & s & t & u & v & w & x & y & z \\
\downarrow & \downarrow & \downarrow & \downarrow & \downarrow & \downarrow & \downarrow & \downarrow & \downarrow & \downarrow & \downarrow & \downarrow & \downarrow \\
w & r & q & n & u & f & c & h & l & g & y & a & k
\end{array}
$$

Exercise 6.9 (page 164)

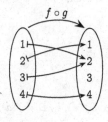 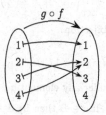

Exercise 6.10 (page 165)

If $f : A \to B$ and $g : B \to C$ are both bijections, then they are both one-to-one and onto. Therefore, $g \circ f : A \to C$ is both one-to-one (by Theorem 6.9) and onto (by Theorem 6.10), and thus it is a bijection.

Exercise 6.11 (page 165)

Let $a \in A$ be arbitrary. By the definition of the inverse of a bijection, Definition 6.7 (page 162), if $f^{-1}(f(a)) = x$ then $f(x) = f(a)$. Since f is one-to-one, this means that $x = a$. Hence $f^{-1}(f(a)) = a$ for any $a \in A$; that is, $f^{-1} \circ f = \mathrm{id}_A$.

Let $b \in B$ be arbitrary. Again by Definition 6.7, if $f(f^{-1}(b)) = y$ then $y = b$. Hence $f(f^{-1}(b)) = b$ for any $b \in B$; that is, $f \circ f^{-1} = \mathrm{id}_B$.

Exercise 6.12 (page 166)

$$
\begin{aligned}
\big(h \circ (g \circ f)\big)(x) = h\big((g \circ f)(x)\big) &= h\big(g(f(x))\big) \\
&= (h \circ g)(f(x)) = \big((h \circ g) \circ f\big)(x).
\end{aligned}
$$

Exercise 6.14 (page 170)

$$
f^{-1}(n) = \begin{cases} 2n-1, & \text{if } n > 0; \\ -2n, & \text{if } n \le 0 \end{cases}
$$

Exercise 6.15 (page 170)

Take $h = g \circ f^{-1}$ which, by Exercise 6.10, is guaranteed to be a bijection.

Exercise 6.16 (page 172)

Given the bijection $f : \mathbb{N} \to \mathbb{Q}^+$ from Example 6.15, the function $g : \mathbb{N} \to \mathbb{Q}$ defined by

$$g(n) = \begin{cases} 0, & \text{if } n = 0, \\ f\left(\frac{n-1}{2}\right), & \text{if } n > 0 \text{ is odd}, \\ -f\left(\frac{n}{2}\right), & \text{if } n > 0 \text{ is even}, \end{cases}$$

is a bijection.

Exercise 6.17 (page 173)

Consider any element $a \in A$.

- If $a \in B$ then by definition of B, $a \notin f(a)$, so $B \neq f(a)$.
- If $a \notin B$ then by definition of B, $a \in f(a)$, so again $B \neq f(a)$.

We thus have that $B \neq f(a)$ for every $a \in A$, that is, f cannot be onto.

Exercise 6.18 (page 174)

Let $S = \{1, 2\}$, and let $f : \mathcal{P}(S) \to \mathcal{P}(S)$ be defined by:

$$f(\emptyset) = f(\{1\}) = \{1\} \quad \text{and} \quad f(\{2\}) = f(S) = \{2\}.$$

The subsets $\{1\}$ and $\{2\}$ are clearly fixed points of f, and are the only fixed points of f. As $\{1\} \not\subseteq \{2\}$ and $\{2\} \not\subseteq \{1\}$, these are neither greatest nor least fixed points.

Exercise 6.20 (page 176)

1. If $S \subseteq T$, then

$$f(S) = \{0\} \cup \{n+2 : n \in S\}$$
$$\subseteq \{0\} \cup \{n+2 : n \in T\} = f(T).$$

2.
$$f(\emptyset) = \{0\} \qquad\qquad f(\mathbb{N}) = \mathbb{N} \setminus \{1\}$$
$$f^2(\emptyset) = \{0, 2\} \qquad\qquad f^2(\mathbb{N}) = \mathbb{N} \setminus \{1, 3\}$$
$$f^3(\emptyset) = \{0, 2, 4\} \qquad\qquad f^3(\mathbb{N}) = \mathbb{N} \setminus \{1, 3, 5\}$$
$$\vdots \qquad\qquad\qquad\qquad \vdots$$
$$f^n(\emptyset) = \{0, 2, \ldots, 2n-2\} \qquad f^n(\mathbb{N}) = \mathbb{N} \setminus \{1, 3, \ldots, 2n-1\}$$

3. $L = G = \{0, 2, 4, 6, \ldots\}$.

Chapter 7

Exercise 7.1 (page 180)

1. $Q = \{r \in BondFilms : r$ was directed by Lewis Gilbert$\}$
   $= \{r03, r06, r07\}$.

2. $Q = \{r \in BondFilms : r$ was released in the 1970s$\}$
   $= \{r05, r06, r07\}$.

Exercise 7.3 (page 183)

$StarsIn$ = { (Sean Connery, Dr. No),
(Sean Connery, Thunderball),
(Sean Connery, You Only Live Twice),
(George Lazenby, On Her Majesty's Secret Service),
(Sean Connery, Diamonds Are Forever),
(Roger Moore, The Spy Who Loved Me),
(Roger Moore, Moonraker),
(Roger Moore, For Your Eyes Only),
(Sean Connery, Never Say Never Again),
(Roger Moore, Octopussy),
(Roger Moore, A View to a Kill),
(Timothy Dalton, The Living Daylights),
(Timothy Dalton, Licence to Kill),
(Pierce Brosnan, Golden Eye),
(Pierce Brosnan, Tomorrow Never Dies),
(Pierce Brosnan, The World Is Not Enough),
(Pierce Brosnan, Die Another Day),
(Daniel Craig, Casino Royale),
(Daniel Craig, Quantum of Solace),
(Daniel Craig, Skyfall) }.

Exercise 7.5 (page 184)

Letting SC, GL, TD, PB, and DC stand for Sean Connery, George Lazenby, Roger Moore, Timothy Dalton, Pierce Brosnan and Daniel Craig, respectively, the binary relation *Before* consists of the following pairs:

$Before$ = { (SC, SC),   (SC, GL),   (SC, RM), (SC, TD),

$$(SC, PB), \quad (SC, DC), \quad (GL, SC), \quad (GL, RM),$$
$$(GL, TD), \quad (GL, PB), \quad (GL, DC), \quad (RM, SC),$$
$$(RM, RM), \quad (RM, TD), \quad (RM, PB), \quad (RM, DC),$$
$$(TD, TD), \quad (TD, PB), \quad (TD, DC), \quad (PB, PB),$$
$$(PB, DC), \quad (DC, DC) \ \}.$$

This binary relation can be visualised as follows:

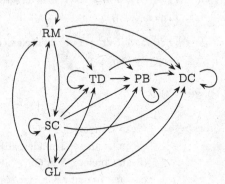

The binary relation *FirstBefore* consists of the following pairs:

$$FirstBefore \ = \ \{ \ (SC, GL), \ (SC, RM), \ (SC, TD),$$
$$(SC, PB), \ (SC, DC), \ (GL, RM),$$
$$(GL, TD), \ (GL, PB), \ (GL, DC),$$
$$(RM, TD), \ (RM, PB), \ (RM, DC),$$
$$(TD, PB), \ (TD, DC), \ (PB, DC) \ \}.$$

This binary relation can be visualised as follows:

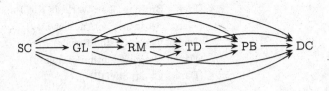

Exercise 7.6 (page 185)

$$Child \ = \ \{ \ (Donald, Quackmore), \ (Donald, Hortense),$$
$$(Della, Quackmore), \ (Della, Hortense),$$
$$(Huey, Della), \ (Louis, Della), \ (Dewey, Della) \ \}.$$

$$Brother = \big\{ \text{(Scrooge, Hortense), (Donald, Della),}$$
$$\text{(Huey, Louis), (Huey, Dewey), (Louis, Huey),}$$
$$\text{(Louis, Dewey), (Dewey, Huey), (Dewey, Louis)} \big\}.$$

$$Sister = \big\{ \text{(Hortense, Scrooge), (Della, Donald)} \big\}.$$

$$Sibling = \big\{ \text{(Scrooge, Hortense), (Hortense, Scrooge),}$$
$$\text{(Donald, Della), (Della, Donald),}$$
$$\text{(Huey, Louis), (Louis, Huey),}$$
$$\text{(Huey, Dewey), (Dewey, Huey),}$$
$$\text{(Louis, Dewey), (Dewey, Louis)} \big\}.$$

The *Child* relation can be visualised as follows.

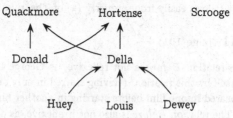

Exercise 7.7 (page 187)

1. $R_1 \cup R_2 = R_3$.
2. $R_3 \cap \overline{R_2} = R_1$.
3. $R_3 \setminus R_1 = R_2$.

Exercise 7.8 (page 188)

$Sibling^{-1} = Sibling$.

Exercise 7.9 (page 189)

- $Uncle = Parent \circ Brother$ (an uncle is a brother of a parent).

  In the case of the Duck family, we have:

  $$Uncle = \{ \text{(Scrooge, Donald), (Scrooge, Della),}$$
  $$\text{(Donald, Huey), (Donald, Louis), (Donald, Dewey)} \}.$$

The first two pairs arise from the fact that Scrooge is a brother of Hortense, who is a parent of Donald and Della.

The final three pairs arise from the fact that Donald is a brother of Della, who is a parent of Huey, Louis and Dewey.

- $Nephew = Sibling \circ Son$ (a nephew is a son of a sibling).

In the case of the Duck family, we have:

$$Nephew = \{\,(\text{Donald}, \text{Scrooge}), (\text{Huey}, \text{Donald}),$$
$$(\text{Louis}, \text{Donald}), (\text{Dewey}, \text{Donald})\,\}.$$

The first pair arises from the fact that Donald is a son of Hortense, who is a sibling of Scrooge.

The final three pairs arise from the fact that Huey, Louis and Dewey are sons of Della, who is a sibling of Donald.

Exercise 7.10 (page 190)

This follows easily from property $(\star)$ of Theorem 7.6.

Exercise 7.11 (page 191)

The relation *Before* is not reflexive, as George Lazenby is not related to himself by this relation. (Having starred in only one film, he could not have appeared in one film before starring in another film.)

The relation *Before* is also not irreflexive, as all of the other actors who have played James Bond have done so on more than one occasion, so each of them is related to himself by the *Before* relation.

The relation *FirstBefore* is irreflexive (and thus it is not reflexive), as an actor could not have starred as James Bond before starring as James Bond.

Exercise 7.12 (page 191)

The relation *Before* is not symmetric; for example, it contains the pair (SC,TD) but not the pair (TD,SC). Nor is it antisymmetric; for example, it contains the pairs (SC,GL) and (GL,SC), and SC≠GL.

The relation *FirstBefore* is not symmetric; for example, it contains the pair (SC,GL) but not the pair (GL,SC). However, it is antisymmetric: given two James Bond actors, one of the two will not have starred as James Bond before the other.

Exercise 7.13 (page 192)

The relation *Before* is not transitive; for example, it contains the pairs (RM,SC) and (SC,GL), but not the pair (RM,GL).

The relation *FirstBefore* is transitive: if one actor starred as James Bond before a second actor, who in turned starred as James Bond before a third actor, then the first actor will naturally have starred as James Bond before the third actor.

Exercise 7.14 (page 192)

The *is-an-ancestor-of* relation is

- not reflexive, but in fact irreflexive, as a person cannot be their own ancestor;
- not symmetric, but in fact antisymmetric, as a person cannot be an ancestor of their own ancestor; and
- transitive, as an ancestor of an ancestor is again an ancestor.

The *is-married-to* relation is

- not reflexive, but in fact irreflexive, as a person cannot be married to themselves;
- symmetric, and not antisymmetric, as the person you are married to is of course married to you; and
- not transitive, as otherwise a married person, by symmetry, would then have to be married to themselves.

Exercise 7.18 (page 194)

1. This is a partial order but not a total order; and it is an equivalence relation.
2. This is not a partial order (it is not antisymmetric), and hence not a total order; but it is an equivalence relation.
3. This is not a partial order (it is not antisymmetric), and hence not a total order; but it is an equivalence relation.

Exercise 7.19 (page 194)

R_1 is an equivalence relation, as it is clearly reflexive (a student takes all the same courses as themselves), symmetric (if x takes all the same courses as y then y takes all the same courses as x) and transitive (if x takes all the same courses as y and y takes all the same courses as z then x takes all the same courses as z).

R_2 is not an equivalence relation, as it is not transitive (though it is reflexive and symmetric). For example, Alice and Bob might take the same Mathematics course, and Bob and Carol might take the same Computing course, while Alice and Carol do not take any of the same courses.

Exercise 7.21 (page 195)

The finest partition of a set A consists of singletons: $\{\{a\} : a \in A\}$.

The coarsest partition of a set A consists of one set: $\{A\}$.

Exercise 7.23 (page 196)

The equivalence relation defined by the finest partition of a set A is the identity relation: $I_A = \{(a, a) \;:\; a \in A\}$.

The equivalence relation defined by the coarsest partition of a set A is the universal relation: $U_A = \{(a, b) \;:\; a, b \in A\}$.

Exercise 7.24 (page 196)

The relation R partitions the set A into the following 18 equivalence classes:

$$[1] = \{1\} \qquad [2] = \{2, 4, 8, 16\} \qquad [3] = \{3, 9, 27\}$$
$$[5] = \{5, 25\} \qquad [6] = \{6, 12, 18, 24\} \qquad [7] = \{7\}$$
$$[10] = \{10, 20\} \qquad [11] = \{11\} \qquad [13] = \{13\}$$
$$[14] = \{14, 28\} \qquad [15] = \{15\} \qquad [17] = \{17\}$$
$$[19] = \{19\} \qquad [21] = \{21\} \qquad [22] = \{22\}$$
$$[23] = \{23\} \qquad [26] = \{26\} \qquad [29] = \{29\}$$

Chapter 8

Exercise 8.1 (page 203)

$4 \in \mathbb{N}$: By clause (1), $0 \in \mathbb{N}$, so by clause (2), $1 \in \mathbb{N}$; so by clause (2), $2 \in \mathbb{N}$; so by clause (2), $3 \in \mathbb{N}$; and so finally by clause (2), $4 \in \mathbb{N}$.

$4.5 \notin \mathbb{N}$: Since $4.5 \neq 0$, clause 1 does not apply, so we could only infer that $4.5 \in \mathbb{N}$ from clause (2), and thus from first inferring that $3.5 \in \mathbb{N}$; but by a similar reasoning we could only infer this by first inferring that $2.5 \in \mathbb{N}$; which we could only infer by first inferring that $1.5 \in \mathbb{N}$; which we could only infer by first inferring that $0.5 \in \mathbb{N}$; which we could only infer by first inferring that $-0.5 \in \mathbb{N}$; which we could only infer by first inferring that $-1.5 \in \mathbb{N}$; *et cetera ad infinitum*. This process would never "bottom out", so we could never infer that any of these were in $\mathbb{N}$.

Alternatively, we can easily see that the set $\{0, 1, 2, 3, 4, \ldots\}$ satisfies clauses (1) and (2) of the definition; and since $\mathbb{N}$ is being defined to be the *smallest* set satisfying these clauses, $\mathbb{N}$ must be a subset of this; since this set does not contain 4.5, $4.5 \notin \mathbb{N}$.

Exercise 8.2 (page 204)

ODD is defined to be the *smallest* set satisfying the two clauses. The fact that N satisfies these two clauses only implies that ODD $\subseteq$ N; that is, N is not necessarily (and in fact is not) the smallest such set.

Exercise 8.3 (page 204)

POWERS-OF-2 is the smallest set satisfying the following:

1. $1 \in$ POWERS-OF-2.

2. If $n \in$ POWERS-OF-2 then $2n \in$ POWERS-OF-2.

Exercise 8.4 (page 205)

Given a set A, the smallest set $P(A)$ satisfying:

1. $\emptyset \in P(A)$;   and

2. if $X \in P$ and $a \in A$ then $X \cup \{a\} \in P(A)$

is the set of all *finite* subsets of A. This is the same as the powerset $\mathcal{P}(A)$ of A only in the case when A is a finite set.

Exercise 8.5 (page 207)

POSDECIMALNUMBERS is inductively defined as the smallest set satisfying the following:

1. $1, 2, 3, 4, 5, 6, 7, 8, 9 \in$ POSDECIMALNUMBERS;

2. If $w \in$ POSDECIMALNUMBERS and $x \in$ DECIMALDIGITS
   then $wx \in$ POSDECIMALNUMBERS.

Exercise 8.6 (page 208)

The following is a BNF equation for formulæ of predicate logic.

$$p, q ::= \text{true} \mid \text{false} \mid P(x_1, \ldots, x_n)$$
$$\mid \neg p \mid p \vee q \mid p \wedge q \mid p \Rightarrow q \mid p \Leftrightarrow q \mid \forall x\, p \mid \exists x\, p$$

Here, $P(x_1, \ldots, x_n)$ is taken to range over the set of predicates with free variables taken from $x_1, \ldots, x_n$ and x is taken to range over all variables.

Exercise 8.7 (page 212)

The dictionary data structure can be defined using the following BNF equation:

$$d \;=\; \star \;\mid\; N(w, d_1, d_2)$$

where w ranges over words (representing names). That is, a dictionary is either a leaf (if it is empty), or it consists of a name along with two sub-dictionaries d_1 and d_2. (Note that the semantic understanding of a dictionary, i.e., the property that the stored names are ordered lexicographically throughout the dictionary, is not reflected in this data structure definition, only its syntactic structure.)

Exercise 8.8 (page 213)

- $s_1 = s_0 + 2\cdot1 - 1 = 0 + 2 - 1 = 1$
- $s_2 = s_1 + 2\cdot2 - 1 = 1 + 4 - 1 = 4$
- $s_3 = s_2 + 2\cdot3 - 1 = 4 + 6 - 1 = 9$
- $s_4 = s_3 + 2\cdot4 - 1 = 9 + 8 - 1 = 16$
- $s_5 = s_4 + 2\cdot5 - 1 = 16 + 10 - 1 = 25$
- $s_6 = s_5 + 2\cdot6 - 1 = 25 + 12 - 1 = 36$

It would appear (though it is as yet uncertain) that $s_n = n^2$.

Exercise 8.9 (page 213)

We could readily compute

$$H_6 \;=\; \tfrac{1}{1} + \tfrac{1}{2} + \tfrac{1}{3} + \tfrac{1}{4} + \tfrac{1}{5} + \tfrac{1}{6} \;=\; \tfrac{49}{20} \;=\; 2.45.$$

However, by the inductive definition we would proceed as follows:

- $H_1 = H_0 + \tfrac{1}{1} = 0 + 1 = 1$
- $H_2 = H_1 + \tfrac{1}{2} = 1 + \tfrac{1}{2} = \tfrac{3}{2} = 1.5$
- $H_3 = H_2 + \tfrac{1}{3} = \tfrac{3}{2} + \tfrac{1}{3} = \tfrac{11}{6} \approx 1.833$
- $H_4 = H_3 + \tfrac{1}{4} = \tfrac{11}{6} + \tfrac{1}{4} = \tfrac{25}{12} \approx 2.083$
- $H_5 = H_4 + \tfrac{1}{5} = \tfrac{25}{12} + \tfrac{1}{5} = \tfrac{137}{60} \approx 2.283$
- $H_6 = H_5 + \tfrac{1}{6} = \tfrac{137}{60} + \tfrac{1}{6} = \tfrac{49}{20} = 2.45$

Exercise 8.10 (page 214)

At the start of month n you will have f_n pairs of rabbits, where f_n is the nth Fibonacci number.

- For a start, at the start of month 1 you have 1 pair, and at the start of month 2 you still have just the 1 pair. At the start of month 3, though, you will have 2 pairs, and at the start of month 4 you will have 3 pairs.

- In general, at the start of month n you will have $f_n = f_{n-1} + f_{n-1}$ pairs of rabbits, as you will have as many pairs as you had at the start of month $n-1$, namely f_{n-1}, plus a new pair for each pair you had at the start of month $n-2$, namely f_{n-2}.

Exercise 8.11 (page 215)

$$\text{mult}(m, 0) = 0; \quad and$$
$$\text{mult}(m, s(n)) = \text{add}(\text{mult}(m, n), m).$$

Exercise 8.12 (page 215)

$$sum([\,]) = 0$$
$$sum(n : L) = n + sum(L)$$

Thus for example,

$$\begin{aligned}
sum([6, 2, 5]) &= 6 + sum([2, 5]) \\
&= 6 + 2 + sum([5]) \\
&= 6 + 2 + 5 + sum([\,]) \\
&= 6 + 2 + 5 + 0 \\
&= 13.
\end{aligned}$$

Exercise 8.13 (page 215)

$$[\,] \mathbin{+\!\!+} L_2 = L_2$$
$$(h : L) \mathbin{+\!\!+} L_2 = h : (L \mathbin{+\!\!+} L_2).$$

Exercise 8.14 (page 216)

$$fv(\text{true}) = fv(\text{false}) = \emptyset$$
$$fv(P(x_1, \ldots, x_n)) = \{x_1, \ldots, x_n\}$$
$$fv(\neg p) = fv(p)$$
$$fv(p \vee q) = fv(p \wedge q) = fv(p \Rightarrow q) = fv(p \Leftrightarrow q) = fv(p) \cup fv(q)$$
$$fv(\forall x\, p) = fv(\exists x\, p) = fv(p) \setminus \{x\}$$

Exercise 8.15 (page 216)

By definition, $f(n) = n-10$ for each $n > 100$. Thus we need only consider the value of $f(n)$ for each n from 0 to 100 and verify that $f(n) = 91$ in each case. We can do this starting from $n = 100$ and working down, using the values we calculate along the way.

- $f(100) = f(f(111)) = f(101) = 91.$
- $f(99) = f(f(110)) = f(100) = 91.$
- $f(98) = f(f(109)) = f(99) = 91.$
  $\vdots$
- $f(91) = f(f(102)) = f(92) = 91.$
- $f(90) = f(f(101)) = f(91) = 91.$
- $f(89) = f(f(100)) = f(91) = 91.$
- $f(88) = f(f(99)) = f(91) = 91.$
  $\vdots$
- $f(1) = f(f(12)) = f(91) = 91.$
- $f(0) = f(f(11)) = f(91) = 91.$

Exercise 8.16 (page 218)

$(insert\ a)\ [\ ]\quad = [a]$

$(insert\ a)\ (b:L) = if\ a{<}b\ then\ a:(b:L)\ else\ b:((insert\ a)\ L)$

Exercise 8.17 (page 220)

Moving a pyramid of n discs can be done as follows.

1. If $n{=}1$ then simply move the single disc to the new peg. Otherwise do the following.

2. Move the pyramid of $n{-}1$ discs sitting on top of the largest disc to a different peg.

3. Move the largest disc to the other empty peg.

4. Move the pyramid of $n{-}1$ discs onto the disc holding the largest disc.

Note the two recursive calls in steps 2 and 4.

Carried out on a tower of five discs, this would require 31 individual moves.

Chapter 9

Exercise 9.1 (page 226)

It would appear that the number of regions doubles every time a new spot is added, so it is tempting to guess that 32 regions will be created by connecting 6 spots. In general, our intuition is suggesting that 2^{n-1} regions are created by connecting n spots, based on the evidence with $n = 1, 2, 3, 4$ and 5.

Unfortunately for our intuition, this guess is wrong: no matter how hard you try, you can only create 31 regions by connecting 6 spots.

In fact, the formula for the number of regions created by connecting n spots is not 2^{n-1}, but the following rather astonishing formula:

$$\frac{n^4 - 6n^3 + 23n^2 - 18n + 24}{24}.$$

Where does this formula come from? Starting with no lines, the circle has just one region. Each time you draw a new line across the circle, you increase the number of regions by 1 more than the number of existing lines which this new line crosses. Thus the number of regions is one more than the total number of lines added to the total number of intersections.

The number of lines you can draw using n spots is $n(n-1)/2$, which is just the number of pairs of endpoints you can choose for the line; and the number of intersections is $n(n-1)(n-2)(n-3)/24$, which is the number of pairs of endpoints of two intersecting lines you could choose. The number of regions created is thus

$$1 + \frac{n(n-1)}{2} + \frac{n(n-1)(n-2)(n-3)}{24}$$

which simplifies to the formula given above.

Exercise 9.2 (page 228)

She would assume that the 26th child would confirm that the first 26 numbers add up to $\frac{26 \times 27}{2}$, and from this show that the first 27 numbers add up to $\frac{27 \times 28}{2}$ as follows:

$$
\begin{aligned}
1 + 2 + 3 + \cdots + 27 &= \underbrace{1 + 2 + 3 + \cdots + 26} + 27 \\
&= \frac{26 \times 27}{2} + 27 \\
&= 27 \times \left(\frac{26}{2} + 1\right) \\
&= 27 \times \left(\frac{28}{2}\right) \\
&= \frac{27 \times 28}{2}
\end{aligned}
$$

Exercise 9.3 (page 228)

Young Gauss is reputed to have carried out the following calculation, all in his head:

$$
\begin{aligned}
X = \quad & 1 + \quad 2 + \quad 3 + \cdots + \quad 48 + \quad 49 + \quad 50 \\
+ \; & 100 + \; 99 + \; 98 + \cdots + \; 53 + \; 52 + \; 51 \\
\hline
= \quad & 101 + 101 + 101 + \cdots + 101 + 101 + 101 \\
= \quad & 50 \times 101 = 5050
\end{aligned}
$$

That is, he had noted that the sum consists of 50 pairs of numbers, where each pair sums to 101.

There are many stories about the prodigious Young Gauss; however – without taking away from his greatness as a mathematician – his biographers do note that these stories are mostly attributed to Old Gauss.

Exercise 9.4 (page 231)

1. For all $n \geq 0$, $1^2 + 2^2 + 3^2 + \cdots + n^2 = \frac{n(n+1)(2n+1)}{6}$.

Proof: By induction on n.

Base Case: We note that
$$1^2 + 2^2 + 3^2 + \cdots + 0^2 = 0 = \frac{0(0+1)(2(0)+1)}{6}.$$

Induction Step: We assume that, for *some* k,
$$1^2 + 2^2 + 3^2 + \cdots + k^2 = \frac{k(k+1)(2k+1)}{6},$$

and from this inductive hypothesis we prove that
$$1^2 + 2^2 + 3^2 + \cdots + k^2 + (k+1)^2 = \frac{(k+1)(k+2)(2k+3)}{6}$$

That is, we demonstrate that if the statement of the theorem is true when $n = k$, then it must also be true when $n = k+1$.

$1^2 + 2^2 + 3^2 + \cdots + k^2 + (k+1)^2$

$= \dfrac{k(k+1)(2k+1)}{6} + (k+1)^2$ *(by the inductive hypothesis)*

$= \dfrac{(k+1)}{6}\Big(k(2k+1) + 6(k+1) \Big)$

$= \dfrac{(k+1)}{6}\Big(2k^2 + 7k + 6) \Big)$

$= \dfrac{(k+1)}{6}\Big((k+2)(2k+3) \Big)$

$= \dfrac{(k+1)(k+2)(2k+3)}{6}$ □

2. For all $n \geq 0$, $1 + 3 + 5 + \cdots + (2n-1) = n^2$.

Proof: By induction on n.

Base Case: We note that
$$1 + 3 + 5 + \cdots + (2(0)-1) = 0 = 0^2.$$

Induction Step: We assume that, for *some* k,
$$1 + 3 + 5 + \cdots + (2k-1) = k^2.$$

and from this inductive hypothesis we prove that

$$1 + 3 + 5 + \cdots + \Big(2(k+1)-1\Big) = (k+1)^2.$$

That is, we demonstrate that if the statement of the theorem is true when $n = k$, then it must also be true when $n = k+1$.

$$1 + 3 + 5 + \cdots + (2k - 1) + \Big(2(k+1) - 1\Big)$$

$$= k^2 + \Big(2(k+1) - 1\Big) \quad \textit{(by the inductive hypothesis)}$$

$$= k^2 + 2k + 1$$

$$= (k + 1)^2 \qquad\qquad \square$$

3. For all $n \geq 0$,

$$1{\cdot}2 + 2{\cdot}3 + 3{\cdot}4 + \cdots + n(n+1) = \frac{n(n+1)(n+2)}{3}.$$

Proof: By induction on n.

Base Case: We note that

$$1{\cdot}2 + 2{\cdot}3 + 3{\cdot}4 + \cdots + 0(0+1) = 0 = \frac{0(0+1)(0+2)}{3}.$$

Induction Step: We assume that, for *some* k,

$$1{\cdot}2 + 2{\cdot}3 + 3{\cdot}4 + \cdots + k(k+1) = \frac{k(k+1)(k+2)}{3}$$

and from this inductive hypothesis we prove that

$$1{\cdot}2 + 2{\cdot}3 + 3{\cdot}4 + \cdots + (k+1)(k+2) = \frac{(k+1)(k+2)(k+3)}{3}.$$

That is, we demonstrate that if the statement of the theorem is true when $n = k$, then it must also be true when $n = k+1$.

$$1{\cdot}2 + 2{\cdot}3 + 3{\cdot}4 + \cdots + k(k+1) + (k+1)(k+2)$$

$$= \frac{k(k+1)(k+2)}{3} + (k+1)(k+2) \quad \textit{(by the inductive}$$
$$\textit{hypothesis)}$$

$$= \frac{(k+1)(k+2)}{3}(k + 3)$$

$$= \frac{(k+1)(k+2)(k+3)}{3} \qquad\qquad \square$$

Exercise 9.5 (page 232)

For all $n \geq 0$, $F_0 \times F_1 \times \cdots \times F_n = F_{n+1} - 2$, where $F_n = 2^{2^n} + 1$.

Proof: By induction on n.

Base Case: For the base case ($n=0$), we note that

$$F_0 = 3 = 5 - 2 = F_1 - 2.$$

Induction Step: For the induction step, we assume that, for *some k*,

$$F_0 \times F_1 \times \cdots \times F_k = F_{k+1} - 2$$

and from this assumption (the *"inductive hypothesis"*) we prove that

$$F_0 \times F_1 \times \cdots \times F_{k+1} = F_{k+2} - 2.$$

That is, we demonstrate that if the statement of the theorem is true when $n = k$, then it must also be true when $n = k+1$.

$$F_0 \times F_1 \times F_2 \times \cdots \times F_k \times F_{k+1}$$

$$= (F_{k+1} - 2) \times F_{k+1} \qquad \text{(by the inductive hypothesis)}$$

$$= (2^{2^{k+1}} - 1) \times (2^{2^{k+1}} + 1)$$

$$= 2^{2^{k+2}} - 1 = F_{k+2} - 2 \quad \square$$

Exercise 9.6 (page 232)

For any real number $r \neq 1$,

$$1 + r + r^2 + r^3 + \cdots + r^n = \frac{1 - r^{n+1}}{1 - r}$$

for all $n \geq 0$.

Proof: By induction on n.

Base Case: For the base case ($n=0$), we note that

$$1 + r + r^2 + r^3 + \cdots + r^0 = 1 = \frac{1 - r^1}{1 - r}.$$

Induction Step: For the induction step, we assume that, for *some k*,

$$1 + r + r^2 + r^3 + \cdots + r^k = \frac{1 - r^{k+1}}{1 - r},$$

and from this assumption (the *"inductive hypothesis"*) we prove that

$$1 + r + r^2 + r^3 + \cdots + r^{k+1} = \frac{1 - r^{k+2}}{1 - r}$$

That is, we demonstrate that if the statement of the theorem is true when $n = k$, then it must also be true when $n = k+1$.

By the inductive hypothesis we can rewrite the left-hand side of this equation that we want to prove true as

$$\frac{1 - r^{k+1}}{1 - r} + r^{k+1}$$

which we can successively rewrite as

$$\frac{1 - r^{k+1}}{1 - r} + \frac{(1 - r)r^{k+1}}{1 - r} = \frac{1 - r^{k+1} + r^{k+1} - r^{k+2}}{1 - r} = \frac{1 - r^{k+2}}{1 - r}$$

which is the result we seek. □

Exercise 9.7 (page 233)

Drawing $n \geq 1$ circles so that any two intersect at two points but no three intersect at any point divides the plane into $n^2 - n + 2$ regions.

Proof: By induction on n.

Base Case: One circle divides the plane into $1^2 - 1 + 2 = 2$ regions.

Induction Step: For the induction step, we assume that k circles divides the plane into $k^2 - k + 2$ regions, and show that adding a $(k+1)$st circle results in $(k+1)^2 - (k+1) + 2$ regions.

The $(k+1)$st circle must intersect the other k circles at $2k$ points, meaning that $2k$ regions are divided into two. Thus, $2k$ new regions are created, giving a total of $k^2 - k + 2 + 2k = (k+1)^2 - (k+1) + 2$ regions, which is as we needed to demonstrate. □

A Venn diagram depicting 4 sets would have to divide the plane into 16 regions. Therefore, it could not be drawn using circles, as by the above result 4 circles would only divide the plane into $4^2 - 4 + 2 = 14$ regions.

Exercise 9.8 (page 234)

$f(n) = n^2$ for all $n \geq 0$, where $f(n) = \begin{cases} 0, & \text{if } n=0 \\ f(n-1) + 2n - 1, & \text{if } n>0. \end{cases}$

Proof: By induction on n.

Base Case: For the base case ($n=0$), we simply note that $f(0) = 0 = 0^2$.

Induction Step: For the induction step, we assume that, for some k,

$$f(k-1) = (k-1)^2,$$

and from this assumption (the "inductive hypothesis") we prove that

$$f(k) = k^2.$$

That is, we demonstrate that if the statement of the theorem is true when $n = k-1$, then it must also be true when $n = k$.

$$f(k) = f(k-1) + 2k - 1 \qquad \textit{(by definition)}$$

$$= (k-1)^2 + 2k - 1 \quad \textit{(by the inductive hypothesis)}$$

$$= k^2. \qquad\qquad \square$$

Exercise 9.9 (page 235)

Every $n > 1$ is either prime or a product of primes.

Proof: By strong induction on n. Suppose that $n>1$, and that for every integer k with $1<k<n$, k is either prime or the product of primes. If n itself is prime then we have nothing to prove, so suppose that $n = ab$ with $1<a<n$ and $1<b<n$. By the inductive hypothesis, each of a and b is either prime or a product of primes; but then since $n = ab$, n itself is a product of primes.

Exercise 9.10 (page 236)

For all $m \geq 1$ and all $n \geq m$, $H_n - H_m \geq \dfrac{n-m}{n}$.

Proof: We assume that $m \geq 1$ is fixed, and we prove the result by induction on n.

Base Case $(n = m)$: $H_m - H_m = 0 \geq \dfrac{0}{m}$.

Induction Step: $(n > m)$:

$$H_n - H_m = H_{n-1} - H_m + \frac{1}{n}$$

$$\geq \frac{(n-1)-m}{n-1} + \frac{1}{n} \qquad \textit{(by inductive hypothesis)}$$

$$= \frac{(n-1)n - mn + (n-1)}{(n-1)n}$$

$$\geq \frac{(n-1)n - mn + m}{(n-1)n} \qquad \textit{(since } n-1 \geq m\textit{)}$$

$$= \frac{n-m}{n} \qquad\qquad \square$$

Exercise 9.11 (page 236)

Fact: $(f_0)^2 + (f_1)^2 + (f_2)^2 + \cdots (f_n)^2 = f_n f_{n+1}$ for all $n \geq 0$.

Proof: By induction on n.

Base Case $(n = 0)$:

$$(f_0)^2 + (f_1)^2 + (f_2)^2 + \cdots + (f_0)^2 = (f_0)^2 = 0^2 = 0 \times 1 = f_0 f_1.$$

Induction Step $(n > 0)$:

$$(f_0)^2 + (f_1)^2 + (f_2)^2 + \cdots + (f_n)^2 + (f_{n+1})^2$$
$$= f_n f_{n+1} + (f_{n+1})^2 \quad \text{(by the inductive hypothesis)}$$
$$= f_{n+1}(f_n + f_{n+1}) = f_{n+1} f_{n+2} \qquad \square$$

Exercise 9.12 (page 238)

Fact: The quadratic equation $y^2 - xy - x^2 = \pm 1$ is satisfied by the pair $(x, y) = (f_n, f_{n+1})$ for any $n \geq 0$.

Proof: By induction on n.

Base Case $(n = 0)$: With $(x, y) = (f_0, f_1) = (0, 1)$ we have

$$y^2 - xy - x^2 = 1^2 - 0 \cdot 1 - 1^0 = 1.$$

Induction Step $(n > 0)$: Assuming that $(x, y) = (a, b)$ solves this equation, that is,

$$b^2 - ab - a^2 = \pm 1$$

it suffices to show that $(x, y) = (a+b, b)$ also solves this equation; this is because if $(a, b) = (f_n, f_{n+1})$ then $(f_{n+1}, f_{n+2}) = (a+b, b)$.

$$(a+b)^2 - (a+b)b - b^2 = a^2 + 2ab + b^2 - ab - b^2 - b^2$$
$$= a^2 + ab - b^2$$
$$= -(b^2 - ab - a^2) = \mp 1. \qquad \square$$

Exercise 9.13 (page 238)

Fact: The positive integer solutions (x, y) to

$$y^2 - xy - x^2 = \pm 1$$

are of the form (f_n, f_{n+1}) for some $n \geq 0$.

Proof: By induction on $x+y$. We first note that since x and y are positive, we must have that $x \leq y$. If $x = y$ then we would have that $-x^2 = \pm 1$, in which case we must have that $x=y=1$, so $(x, y) = (f_1, f_2)$.

We now assume that $1 \leq x < y$ and that $y^2 - xy - x^2 = \pm 1$, and note that the pair $(a, b) = (y-x, x)$ also satisfies the equation:

$$b^2 - ab - a^2 = x^2 - (y-x)x - (y-x)^2$$
$$= x^2 - xy + x^2 - y^2 + 2xy - x^2$$
$$= -(y^2 - xy - x^2) = \mp 1.$$

By induction, $(a, b) = (f_n, f_{n+1})$ for some n, from which we get

$$x = b = f_{n+1} \quad \text{and}$$
$$y = a + x = f_n + f_{n+1} = f_{n+2},$$

so $(x, y) = (f_{n+1}, f_{n+2})$. □

Exercise 9.14 (page 239)

Fact: $f_{n+1}^2 - f_n f_{n+2} = (-1)^n$ for all $n \geq 0$.

Proof: By induction on n.

Base Case $(n = 0)$: $f_1^2 - f_0 f_2 = 1^2 - 0 \cdot 1 = 1 = (-1)^0$.

Induction Step $(n > 0)$:

$$
\begin{aligned}
f_{n+1}^2 - f_n f_{n+2} &= f_{n+1}(f_n + f_{n-1}) - f_n(f_{n+1} + f_n) \quad \textit{(by definition)} \\
&= f_{n+1} f_n + f_{n+1} f_{n-1} - f_n f_{n+1} - f_n^2 \\
&= -(f_n^2 - f_{n-1} f_{n+1}) \\
&= -(-1)^{n-1} \quad \textit{(by the inductive hypothesis)} \\
&= (-1)^n \qquad\qquad\qquad\qquad\qquad\qquad □
\end{aligned}
$$

Exercise 9.15 (page 239)

The edges that supposedly make up the diagonal of the rectangle do not in fact line up. Drawn more carefully, a gap (or overlap) is discovered in the middle with an area of one unit.

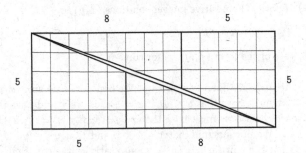

Exercise 9.17 (page 243)

The induction argument cannot be applied when $n=2$: if S' and S'' are overlapping sets which together make up S, then either $S' = S$ or $S'' = S$, in which case you cannot apply induction to this set, as you can only apply induction to sets smaller than S.

Exercise 9.20 (page 246)

Fact: $length(L_1 +\!\!+ L_2) = length(L_1) + length(L_2)$ for all lists L_1 and L_2.

Proof: By induction on the structure of L_1.

Base Case $(L_1 = [\,])$:

$$length([\,] +\!\!+ L_2) = length(L_2) = length([\,]) + length(L_2).$$

Induction Step $(L_1 = h : L)$:

$length((h : L) +\!\!+ L_2)$

$\quad = length(h : (L +\!\!+ L_2))$ *(by definition)*

$\quad = 1 + length(L +\!\!+ L_2)$

$\quad = 1 + length(L) + length(L_2)$ *(by the inductive hypothesis)*

$\quad = length(h : L) + length(L_2).$ $\square$

Chapter 10

Exercise 10.1 (page 257)

1. By brute force reasoning, we get the following table:

n	1	2	3	4	5	6	7	8	9	10
$f(n)$	1	2	3	$\perp$	1	2	3	$\perp$	1	2

$f(n)$ represents the number of coins the first player should take when there are n coins in the pile; we write $f(n) = \perp$ (meaning $f(n)$ is undefined) in the cases in which the second player has the winning strategy.

2. If the first player takes x coins, the second player can respond by taking $(4-x)$ coins, leaving the first player a pile of $(n-4)$ coins when starting from a pile of n coins. This gives a winning strategy for the second player in a game starting with $n=4k$ coins, that is, a number of coins divisible by 4.

In all other cases, starting with $n = 4k+x$ coins (where x is 1, 2, or 3), the first player puts the second player in a losing position by taking x coins and leaving $4k$ coins.

3. The first player is in a losing position if the number of coins n is divisible by $(k+1)$; the second player wins by responding to every move of the first player by taking $(k+1)-x$ coins, where x is the number of coins that the first player takes.

   If the number of coins n is not divisible by $(k+1)$, then the first player wins by taking $n \bmod (k+1)$ coins, leaving the second player in a losing position.

4. The goal in the Misère game is to leave your opponent with one coin. Thus, the second player has a winning strategy when there are $4k+1$ coins, using the same strategy as the normal game.

Exercise 10.2 (page 258)

1. The second player can always place the first two noughts on adjacent sides. (The first nought can be placed on a side which has both adjacent sides empty, and one of these will still be empty after the first player places the second cross.)

   The third nought can then always be placed so that it is aligned with at most one of the first two noughts (i.e., not in the centre square nor in the corner between the two noughts). This is true because there are five such squares, and only three of them can be occupied by crosses.

   The fourth and final nought can then be placed safely, as there can only be at most one square which could create a line of three noughts, yet there will be two empty squares available to chose from.

2. Suppose the first player places the first cross in the centre and then places all subsequent crosses directly opposite the squares on which the second player places noughts. If a line of three crosses should arise, it clearly could not include the centre square, and in fact would imply that there is a line of three noughts already in place directly opposite the line of three crosses.

Exercise 10.3 (page 258)

Following on from the reasoning started in the question:

- 9 o'clock is a winning position (by moving 3 hours ahead to 12 o'clock), and
  10 o'clock is a winning position (by moving 2 hours ahead to 12 o'clock);

- 7 o'clock is a losing position (as you can only move to a winning position: either 2 hours ahead to 9 o'clock or 3 hours ahead to 10 o'clock);

- 4 o'clock is a winning position (by moving 3 hours ahead to 7 o'clock), and
  5 o'clock is a winning position (by moving 2 hours ahead to 7 o'clock);

- 2 o'clock is a losing position (as you can only move to a winning position: either 2 hours ahead to 4 o'clock or 3 hours ahead to 5 o'clock);

- 11 o'clock is a winning position (by moving 3 hours ahead to 2 o'clock), and
12 o'clock is a winning position (by moving 2 hours ahead to 2 o'clock);

- 8 o'clock is a losing position (as you can only move to a winning position: either 2 hours ahead to 10 o'clock or 3 hours ahead to 11 o'clock);

- 6 o'clock is a winning position (by moving 2 hours ahead to 8 o'clock); and

- 3 o'clock is a losing position (as you can only move to a winning position: either 2 hours ahead to 5 o'clock or 3 hours ahead to 6 o'clock).

This is summarised as follows, where the hours on the clock are annotated with the winning move if one is available.

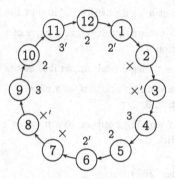

The symbol × indicates that the position is a losing position; and a prime means that the token will pass through the 12 o'clock position once before landing on it the second time around (assuming the losing player uses a particular strategy).

To see that this annotation is correct, it suffices to note that

- every valid move from an hour labelled × (i.e., forward by either two or three hours) leads to a position labelled 2 or 3, neither with a prime, without passing through 12 o'clock;

- every valid move from an hour labelled ×′ leads to a position labelled 2 or 3, (at least) one of which is primed, without passing through 12 o'clock;

- an hour labelled 2 (respectively 3) – by moving forward by 2 (respectively 3) hours – leads either to 12 o'clock, or without passing through 12 o'clock to an hour labelled by ×;

- an hour labelled 2′ (respectively 3′) – by moving forward by 2 (respectively 3) hours – leads either to an hour labelled by ×′ without passing through 12 o'clock, or by first passing through 12 o'clock to an hour labelled by ×.

Exercise 10.4 (page 259)

In this game:

- 9 is a losing position, as the other player will have won by having moved the counter there.

- 8 is a winning position, as a move of 1 takes the counter to the losing position 9.

- 7 is not a legal position, as it is at the head of a snake.

- 6 is a losing position, as moves of 1 and 2 take the counter to the winning positions 4 and 8, respectively.

- 5 is not a legal position, as it is at the foot of a ladder.

- 4 is a winning position, as a move of 1 takes the counter to the losing position 9.

- 3 is not a legal position, as it is at the foot of a ladder.

- 2 is a winning position, as a move of 1 takes the counter to the losing position 6.

- 1 is a winning position, as a move of 2 takes the counter to the losing position 6.

Exercise 10.5 (page 262)

If we ignore one of the piles and consider the column parity of the remaining $n-1$ piles, this indicates what size the final pile would have to be in order to balance the position. (For example, if the $n-1$ piles are balanced, then the final pile would have to be empty; and if all n piles are balanced, then the column parity of any $n-1$ piles would equal the size of the omitted pile.)

Thus, for each pile, there is at most one winning move involving that pile, which consists of leaving the number of coins equal to the column parity of the remaining $n-1$ piles.

Therefore, there are at most n different winning moves possible from a NIM position with n piles.

Exercise 10.6 (page 262)

If the position is initially unbalanced, then the first player need not use this new move in order to win; the first player already has the winning strategy in the original game.

If, on the other hand, the position is initially balanced, then this new move will still not help, as it will produce an unbalanced position, as does any normal move.

Hence this new rule gives the first player no new advantage.

There is another way to see that this new move is obviously of no help to the first player: the second player can respond to this new move by removing all of the coins in the new pile, thus putting the first player back into the same position as before the new move was made.

Exercise 10.7 (page 264)

For $n = f_{k_1} + f_{k_2} + \cdots + f_{k_r} > 0$ with $0 \ll k_1 \ll k_2 \ll \cdots \ll k_r$, let $\mu(n) = f_{k_1}$; that is, $\mu(n)$ is the smallest Fibonacci number appearing in the representation of n in the Fibonacci number system. Also, for convenience, define $\mu(0) = \infty$.

Consider the following two lemmas.

Lemma 1. If $n > 0$ then $\mu(n - \mu(n)) > 2\mu(n)$.

This says that if you take $\mu(n)$ coins on your turn from a pile of n coins – which, in particular, the first person *may* do on their first move if, and only if, n is *not* a Fibonacci number, that is, $\mu(n) \neq n$ – then your opponent will be *unable* to do this in response; that is, they will be faced with some number $m = n - \mu(n)$ of coins and the most coins they can take, namely $2\mu(n)$, will be less than $\mu(m) = \mu(n - \mu(n))$.

Lemma 2. If $0 < m < \mu(n)$ then $\mu(n - m) \leq 2m$.

This says that if you take fewer than $\mu(n)$ coins on your turn from a pile of n coins – which, in particular, the first person *must* do on their first move if, and only if, n *is* a Fibonacci number – then your opponent *will* be able to take $\mu(n - m)$ coins from the pile of $n - m$ coins they are faced with, as $\mu(n - m)$ will be no more than than twice the number m of coins that you have taken.

Theorem 10.7 follows directly from these two lemmas. Given a pile of n_1 coins, if you take $\mu(n_1)$ coins then either you will have taken all coins and won the game, or you will leave some number n_2 of coins from which, by Lemma 1, your opponent cannot take $\mu(n_2)$ coins, and in particular cannot take all of the remaining coins; and by Lemma 2 your opponent will be forced to leave you with some number n_3 of coins from which you can once again use the strategy of taking $\mu(n_3)$ coins; the play will continue in this fashion until you succeed in taking all remaining coins.

It remains only to prove these two lemmas.

Proof of Lemma 1. Let $n = f_{k_1} + f_{k_2} + \cdots + f_{k_r}$ be the Fibonacci number system representation of n.

The result is immediate if n is a Fibonacci number (that is, if $r=1$), as then $n = \mu(n)$, so $\mu(n-\mu(n)) = \mu(0) = \infty > n = \mu(n)$.

Assume, then, that n is *not* a Fibonacci number (that is, $r \geq 2$). Then

$$
\begin{aligned}
\mu(n-\mu(n)) &= f_{k_2} && \text{(as } n-\mu(n) = f_{k_2} + \cdots + f_{k_r}) \\
&\geq f_{k_1+2} && \text{(as } k_1 \ll k_2) \\
&= f_{k_1} + f_{k_1+1} \\
&> f_{k_1} + f_{k_1} \\
&= 2f_{k_1} \\
&= 2\mu(n). && \qquad\qquad \square
\end{aligned}
$$

Proof of Lemma 2. Assume that $0 < m < \mu(n)$. Let the Fibonacci number system representations of n and m be

$$n = f_{k_1} + f_{k_2} + \cdots + f_{k_r} \quad \text{and}$$
$$m = f_{\ell_1} + f_{\ell_2} + \cdots + f_{\ell_s}.$$

By assumption, $m < \mu(n) = f_{k_1}$, so $f_{k_1}-m > 0$. Let the Fibonacci number system representations of $f_{k_1}-m$ be

$$f_{k_1}-m = f_{u_1} + f_{u_2} + \cdots + f_{u_t}.$$

In particular, $f_{u_t} < f_{k_1}$, so $u_t < k_1 \ll k_2$, and hence $u_t \ll k_2$.
Then

$$
\begin{aligned}
n-m &= (f_{k_1}-m) + f_{k_2} + f_{k_3} + \cdots + f_{k_r} \\
&= f_{u_1} + f_{u_2} + \cdots + f_{u_t} + f_{k_2} + f_{k_3} + \cdots + f_{k_r}
\end{aligned}
$$

and since $u_1 \ll u_2 \ll \cdots \ll u_t \ll k_2 \ll k_3 \ll \cdots \ll k_r$, this is the Fibonacci number system representation of $n-m$.

Thus $\mu(n-m) = f_{u_1}$.

Note that

$$
\begin{aligned}
f_{k_1} &= m + f_{u_1} + f_{u_2} + \cdots + f_{u_t} \\
&= f_{\ell_1} + f_{\ell_2} + \cdots + f_{\ell_s} + f_{u_1} + f_{u_2} + \cdots + f_{u_t}.
\end{aligned}
$$

Therefore we must have that $\ell_s \not\ll u_1$, that is, $u_1 \leq \ell_s+1$, as otherwise we would have two different Fibonacci number system representations for f_{k_1}. Thus $f_{u_1} \leq f_{\ell_s+1}$, and hence

$$\mu(n-m) = f_{u_1} \leq f_{\ell_s+1} = f_{\ell_s-1} + f_{\ell_s} \leq 2f_{\ell_s} \leq 2m. \qquad \square$$

Exercise 10.12 (page 270)

Assume to the contrary that the second player has successfully created a path connecting the top border to the bottom border. At some point this path must cross the main diagonal, either horizontally or vertically.

As this path approaches the main diagonal from above, a downwards move cannot turn towards the left border, and a rightwards move cannot turn upwards. Hence this path must continuously travel downwards and to the right. Therefore, it must reach the main diagonally vertically and thus cross it horizontally; it cannot cross the diagonal vertically.

On the other hand, as this path approaches the main diagonal from below, an upwards move cannot turn towards the right border, and a leftwards move cannot turn downwards. Hence this path must continuously travel upwards and to the left. Therefore, it must reach the main diagonally vertically and thus cross it vertically; that is, it cannot cross it horizontally.

Chapter 11

Exercise 11.3 (page 286)

At any given moment, there must be some number i of missionaries and some number j of cannibals on the left bank of the river, and $3-i$ missionaries and $3-j$ cannibals on the right bank. If $i \neq j$ then the cannibals outnumber the missionaries on one of the banks; this would only be safe if the number of missionaries on that bank is in fact zero. Hence the only safe states are those in which $i=3$ (all missionaries together on the left bank) or $i=0$ (all missionaries together on the right bank) or $i=j$ (an equal number of missionaries and cannibals on both banks). There are 10 such pairs of numbers (i, j).

Apart from this, the only information needed to completely describe the state of the system is where the boat is; it may be on the left bank or the right bank. Combined with the 10 possible placements of the missionaries and cannibals, this gives the system a total of 20 possible safe states. However, four of these are not feasible. For a start, we clearly cannot have all the people on one bank ($i=j=3$ or $i=j=0$) and the boat on the other. Furthermore, if the missionaries are all on one bank and the cannibals are all on the other ($\{i, j\} = \{0, 3\}$) then the boat must be with the cannibals; if it were with the missionaries, then one or two of them must have just ferried it across the river from the bank on which all three cannibals are, which would have been an unsafe position.

The remaining 16 states are depicted in Figure 15.4, along with the possible transitions between states drawn in. (To avoid clutter, the transitions are drawn bi-directionally, as they all represent reversible actions.) The groups

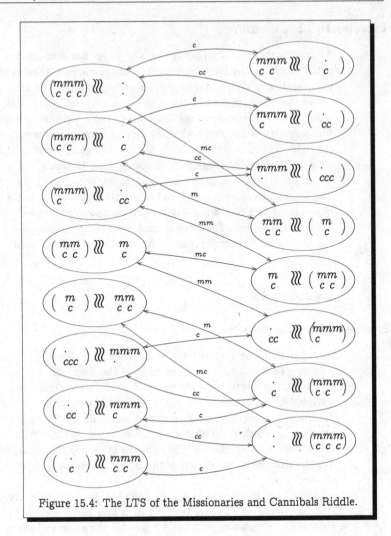

Figure 15.4: The LTS of the Missionaries and Cannibals Riddle.

on the two banks are depicted side-by-side divided by wiggly lines representing the river, with the group holding the boat enclosed in parentheses. There are five possible actions: m (a missionary crosses alone); mm (two missionaries cross together); c (a cannibal crosses alone); cc (two cannibals cross together); and mc (a missionary and a cannibal cross together). Notice that all of the transitions are drawn bi-directionally, as every transition can clearly be reversed.

The group start in the top-left state in which the whole group is on the left bank, and they wish to get to the bottom-right state in which they are all on the right bank. It is not hard to find a such path through the LTS which involves 11 crossings.

Exercise 11.4 (page 287)

A state of the system underlying this riddle consists of a pair of integers (i, j) with $0 \leq i \leq 5$ and $0 \leq j \leq 3$, representing the volume of water in the 5-gallon and 3-gallon jugs A and B, respectively. The initial state is $(0, 0)$ and the final state you wish to reach is $(4, 0)$.

There are six types of moves possible from any given state (i, j):

$$(i,j) \xrightarrow{fillA} (5, j) \qquad\qquad (if\ i{=}0)$$

$$(i,j) \xrightarrow{fillB} (i, 3) \qquad\qquad (if\ j{=}0)$$

$$(i,j) \xrightarrow{emptyA} (0, j) \qquad\qquad (if\ i{>}0)$$

$$(i,j) \xrightarrow{emptyB} (i, 0) \qquad\qquad (if\ j{>}0)$$

$$(i,j) \xrightarrow{AtoB} (\max(0, i{+}j{-}3), \min(3, i{+}j)) \quad (if\ i{>}0\ and\ j{<}3)$$

$$(i,j) \xrightarrow{BtoA} (\min(5, i{+}j), \max(0, i{+}j{-}5)) \quad (if\ i{<}5\ and\ j{>}0)$$

Drawing out the LTS, we identify the following 7-step solution:

$$(0, 0) \xrightarrow{fillA} (5, 0) \xrightarrow{AtoB} (2, 3) \xrightarrow{emptyB} (2, 0) \xrightarrow{AtoB} (0, 2)$$

$$\xrightarrow{fillA} (5, 2) \xrightarrow{AtoB} (4, 3) \xrightarrow{emptyB} (4, 0).$$

Exercise 11.5 (page 287)

The beer mats must start in one of the following three non-winning configurations:

X: 3 one way, the 4th the other way.

Y: 2 face-up and 2 face-down, with diagonally-opposite corners different.

Z: 2 face-up and 2 face-down, with diagonally-opposite corners the same.

Furthermore, there are only three different moves which you may apply to the beer mats:

a: flip one beer mat.

b: flip two adjacent beer mats.

c: flip two diagonally-opposite beer mats.

(Flipping 3 beer mats has the same effect on the possible configurations as flipping 1 beer mat; and flipping all four beer mats has no effect whatsoever on the configuration.)

In the following table, we indicate which non-winning configurations we may go to from each non-winning configuration.

	a	b	c
X	Y/Z	X	X
Y	X	Z	Y
Z	X	Y	—

For example, from an X-configuration, an a-move (flipping one beer mat) could lead to a winning configuration, or to either a Y-configuration or a Z-configuration; and from a Z-configuration, a c-move (flipping diagonally-opposite beer mats) is guaranteed to lead to a winning configuration.

We can use a labelled transition system to keep track of which configurations we may be in at any given time *assuming* that we have never passed through a winning configuration. The LTS looks as follows.

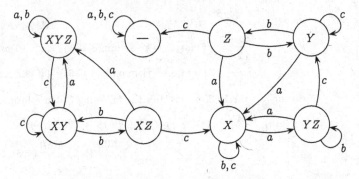

Here, we start in the state labelled "XYZ" signifying that we don't know which state X, Y or Z we are in. If we do an a move or a b move, then we may still be in any of these states; however, if we do a c move, then we will know that we *cannot* be in a Z state.

From this we can see that the shortest sequence of moves which guarantees a win is the sequence "cbcacbc" of seven moves.

Exercise 11.6 (page 288)

1. There are six states. In the graphical presentation of the transition system, these are represented by the following (x, y)-valued pairs:

$$\begin{pmatrix} x = 246 \\ y = 174 \end{pmatrix} \qquad \begin{pmatrix} x = 72 \\ y = 30 \end{pmatrix} \qquad \begin{pmatrix} x = 12 \\ y = 6 \end{pmatrix}$$

$$\begin{pmatrix} x = 72 \\ y = 174 \end{pmatrix} \qquad \begin{pmatrix} x = 12 \\ y = 30 \end{pmatrix} \qquad \begin{pmatrix} x = 0 \\ y = 6 \end{pmatrix}$$

(Of course, how the states are labelled is irrelevant.)

2. There are two actions, namely "x := x mod y" and "y := y mod x".

3. There are five transitions. Labelling the states as above, these transitions are:

$$\begin{pmatrix} x = 246 \\ y = 174 \end{pmatrix} \xrightarrow{\;x\,:=\,x\bmod y\;} \begin{pmatrix} x = 72 \\ y = 174 \end{pmatrix}$$

$$\begin{pmatrix} x = 72 \\ y = 174 \end{pmatrix} \xrightarrow{\;y\,:=\,y\bmod x\;} \begin{pmatrix} x = 72 \\ y = 30 \end{pmatrix}$$

$$\begin{pmatrix} x = 72 \\ y = 30 \end{pmatrix} \xrightarrow{\;x\,:=\,x\bmod y\;} \begin{pmatrix} x = 12 \\ y = 30 \end{pmatrix}$$

$$\begin{pmatrix} x = 12 \\ y = 30 \end{pmatrix} \xrightarrow{\;y\,:=\,y\bmod x\;} \begin{pmatrix} x = 12 \\ y = 6 \end{pmatrix}$$

$$\begin{pmatrix} x = 12 \\ y = 6 \end{pmatrix} \xrightarrow{\;x\,:=\,x\bmod y\;} \begin{pmatrix} x = 0 \\ y = 6 \end{pmatrix}$$

Exercise 11.7 (page 289)

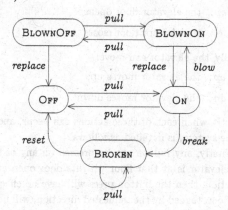

Exercise 11.8 (page 290)

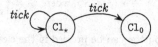

The above process has two states: Cl_* and Cl_0; one action: tick; and two transitions: $Cl_* \xrightarrow{\text{tick}} Cl_*$ and $Cl_* \xrightarrow{\text{tick}} Cl_0$.

Exercise 11.9 (page 291)

As described, a state is a triple $\langle f, d, S \rangle$ where

$$f \in \{1, 2^\uparrow, 2^\downarrow, 3, 1^+, 2^+, 2^-, 3^-\};$$

$$d \in \{open, closed\}; \text{ and}$$

$$S \subseteq \{1, 2^\uparrow, 2^\downarrow, 3\}.$$

There are 11 actions that the system can possibly do. Firstly, any of the call buttons can be pressed on any of the floors:

p_1:   (up) button on floor 1 is pressed;

$p_{2\uparrow}$:   up button on floor 2 is pressed;

$p_{2\downarrow}$:   down button on floor 2 is pressed;

p_3:   (down) button on floor 3 is pressed.

Next, any of the floor buttons can be pressed in the elevator:

e_1:   floor 1 button is pressed in the elevator;

e_2:   floor 2 button is pressed in the elevator;

e_3:   floor 3 button is pressed in the elevator.

Next, the elevator door can open or close:

op: the elevator door opens;

cl: the elevator door closes.

Finally, the elevator can move:

up: the elevator moves up;

dn: the elevator moves down.

Exactly when each of these actions can occur, and their effect on the state of the system, is detailed as follows.

Firstly, any button can be pressed on any of the floors at any time. If the elevator is at that floor with its door open and travelling in the right direction, then the button press will have no effect on the state; otherwise, the floor, tagged by the requested direction, will be added to the destination list:

$$\langle f, d, S \rangle \xrightarrow{p_b} \langle f, d, S' \rangle, \text{ where } S' = \begin{cases} S, & \text{if } f = b \text{ and } d = open \\ S \cup \{b\}, & \text{otherwise} \end{cases}$$

Next, any button can be pressed in the elevator at any time. If the elevator is at the floor being requested with its door open, then the button press will have no effect on the state; otherwise, the floor being requested, tagged by the direction to get to the requested floor (or the current direction being travelled if the elevator is at that floor), will be added to the destination list:

$$\langle f, d, S \rangle \xrightarrow{e_1} \langle f, d, S' \rangle, \text{ where }$$

$$S' = \begin{cases} S, & \text{if } f=1 \text{ and } d=open \\ S \cup \{1\}, & \text{otherwise} \end{cases}$$

$\langle f, d, S \rangle \xrightarrow{e_2} \langle f, d, S' \rangle$, where

$$S' = \begin{cases} S, & \text{if } f \in \{2^\uparrow, 2^\downarrow\} \text{ and } d=open \\ S \cup \{2^\uparrow\}, & \text{if } f \in \{1, 1^+, 2^-\} \\ & \text{or } f=2^\uparrow \text{ and } d=closed \\ S \cup \{2^\downarrow\}, & \text{if } f \in \{3, 3^-, 2^+\} \\ & \text{or } f=2^\downarrow \text{ and } d=closed \end{cases}$$

$\langle f, d, S \rangle \xrightarrow{e_3} \langle f, d, S' \rangle$, where

$$S' = \begin{cases} S, & \text{if } f=3 \text{ and } d=open \\ S \cup \{3\}, & \text{otherwise} \end{cases}$$

Next, the door can either open or close at appropriate times:

$\langle f, open, S \rangle \xrightarrow{cl} \langle f, closed, S \rangle$;

$\langle f, closed, S \rangle \xrightarrow{op} \langle f, open, S \setminus \{f\} \rangle$, if $f \in S$;

$\langle 2^\uparrow, closed, S \rangle \xrightarrow{op} \langle 2^\downarrow, open, S \setminus \{2^\downarrow\} \rangle$, if $2^\downarrow \in S$ and $2^\uparrow, 3 \notin S$;

$\langle 2^\downarrow, closed, S \rangle \xrightarrow{op} \langle 2^\uparrow, open, S \setminus \{2^\uparrow\} \rangle$, if $2^\uparrow \in S$ and $2^\downarrow, 1 \notin S$.

Finally, the elevator can move as and when appropriate:

$\langle 1, closed, S \rangle \xrightarrow{up} \langle 1^+, closed, S \rangle$, if $1 \notin S$ and $S \neq \emptyset$;

$\langle 2^\uparrow, closed, \{3\} \rangle \xrightarrow{up} \langle 2^+, closed, \{3\} \rangle$;

$\langle 2^\downarrow, closed, \{3\} \rangle \xrightarrow{up} \langle 2^+, closed, \{3\} \rangle$;

$\langle 1^+, closed, S \rangle \xrightarrow{up} \langle 1^+, closed, S \rangle$;

$\langle 1^+, closed, S \rangle \xrightarrow{up} \langle 2^\uparrow, closed, S \rangle$;

$\langle 2^+, closed, S \rangle \xrightarrow{up} \langle 2^+, closed, S \rangle$;

$\langle 2^+, closed, S \rangle \xrightarrow{up} \langle 3, closed, S \rangle$;

$\langle 3, closed, S \rangle \xrightarrow{dn} \langle 3^-, closed, S \rangle$, if $3 \notin S$ and $S \neq \emptyset$;

$\langle 2^\downarrow, closed, \{1\} \rangle \xrightarrow{dn} \langle 2^-, closed, \{1\} \rangle$;

$\langle 2^\uparrow, closed, \{1\} \rangle \xrightarrow{dn} \langle 2^-, closed, \{1\} \rangle$;

$\langle 3^-, closed, S \rangle \xrightarrow{dn} \langle 3^-, closed, S \rangle$;

$\langle 3^-, closed, S \rangle \xrightarrow{dn} \langle 2^\downarrow, closed, S \rangle$;

$$\langle 2^-, closed, S \rangle \xrightarrow{\ dn\ } \langle 2^-, closed, S \rangle;$$

$$\langle 2^-, closed, S \rangle \xrightarrow{\ dn\ } \langle 1, closed, S \rangle;$$

Exercise 11.10 (page 291)

The two models for flipping coins are as follows:

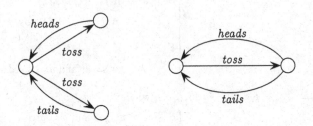

As required, the outcome is determined in the first model already when the coin is tossed: the system will either be in a state in which it can do a *heads* action and not a *tails* action, or it will be in a state in which it can do a *tails* action and not a *heads* action. In contrast to this, when the coin is tossed in the second model, the system will be in a state in which it can do either a *heads* action or a *tails* action.

Which model is more realistic? We might introduce quantum mechanics and allude to the fate of Schrödinger's cat placed in a sealed box with a flask of poison, a radioactive source, and a mechanism which will shatter the flask – releasing the poison and killing the cat – if a Geiger counter detects a radioactive particle; according to quantum mechanics, after a while the cat will be *simultaneously dead and alive* until we open the box; only by observing the cat will its fate be sealed. With this in mind, we might choose the second model to be more realistic.

Barring the complexities of Schrödinger's cat, the first model is more realistic, in that it enforces the principle that the toss itself decides the fate of the coin; having tossed the coin, and with it resting on the back of one hand shielded from view by the palm of the other, no further forces can influence the outcome of the coin flip. The coin is decidedly showing heads or tails.

We can contrast this situation with the model of the simple vending machine from page 282 which accepts a 50p coin and allows the user to decide whether to press a coffee button or a tea button. Having inserted the 50p coin, the user is completely free to choose which button to press, and thus the model for the vending machine closely resembles the second coin-flipping model above. Such a free choice is of course undesirable in a coin flip. (It would be equally undesirable for the vending machine to eliminate the user's free choice of drinks when the 50p coin is inserted, as with the first coin-flipping model above.)

Exercise 11.11 (page 296)

By the rule for action prefix,

$$pull.\textsc{Broken} \xrightarrow{pull} \textsc{Broken} \quad \text{and}$$

$$reset.\textsc{Off} \xrightarrow{reset} \textsc{Off}.$$

Hence, by the rule for choice,

$$pull.\textsc{Broken} + reset.\textsc{Off} \xrightarrow{pull} \textsc{Broken} \quad \text{and}$$

$$pull.\textsc{Broken} + reset.\textsc{Off} \xrightarrow{reset} \textsc{Off}.$$

As $\textsc{Broken} \stackrel{\text{def}}{=} pull.\textsc{Broken} + reset.\textsc{Off}$, the rule for Process Variables gives us to infer our result:

$$\textsc{Broken} \xrightarrow{pull} \textsc{Broken} \quad \text{and}$$

$$\textsc{Broken} \xrightarrow{reset} \textsc{Off}.$$

Exercise 11.12 (page 297)

$$\text{Cl}_* \stackrel{\text{def}}{=} \text{tick.Cl}_* + \text{tick.Cl}_0.$$

Exercise 11.13 (page 298)

1.

$$C_n \stackrel{\text{def}}{=} \begin{cases} i_5.C_5 + i_{10}.C_{10} + i_{20}.C_{20}, & \text{if } n = 0; \\ d_1.C_0, & \text{if } n = 1; \\ d_1.C_{n-1} + d_2.C_{n-2}, & \text{if } 2 \leq n \leq 4; \\ d_1.C_{n-1} + d_2.C_{n-2} + d_5.C_{n-5}, & \text{if } 5 \leq n \leq 20. \end{cases}$$

2. The transition diagram is depicted in Figure 15.5.

Exercise 11.14 (page 300)

The five states of the second vending machine are:

V_2

$10\text{p.coffee.collect}.V_2 + 10\text{p.tea.collect}.V_1$

$\text{coffee.collect}.V_2$

$\text{tea.collect}.V_2$

$\text{collect}.V_2$

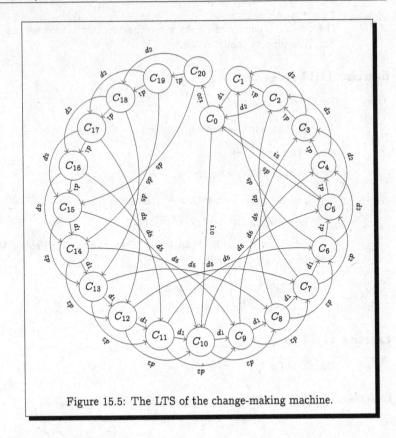

Figure 15.5: The LTS of the change-making machine.

The six states of the third vending machine are:

> V_3
> 10p.coffee.collect.V_3
> 10p.tea.collect.V_3
> coffee.collect.V_3
> tea.collect.V_3
> collect.V_3

Exercise 11.15 (page 301)

No matter how we do a 10p action,
we *must* end up in a state in which
we *may* do a 10p action
and end up in a state in which
we *may* do a tea action.

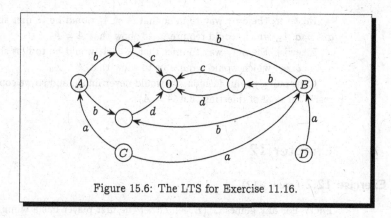

Figure 15.6: The LTS for Exercise 11.16.

Exercise 11.16 (page 301)

1. The transition system is depicted in Figure 15.6.

2. From state C you may do an a action and be in a state in which, no matter how you do a b action you will either not be able to do a c action or you will not be able to do a d action.

    On the other hand, from state D no matter how you do an a action, you will be able to do a b action and end up in a state in which you can both do a c action as well as a d action.

Exercise 11.17 (page 302)

The given definition of equality is:

reflexive: Clearly $E \xrightarrow{a} G$ if, and only if, $E \xrightarrow{a} G$, so $E = E$.

symmetric: Suppose that $E = F$. To show that $F = E$ we need to demonstrate that $F \xrightarrow{a} G$ if, and only if, $E \xrightarrow{a} G$. However, this must be true, as this is exactly the same criterion that makes $E = F$, namely that $E \xrightarrow{a} G$ if, and only if, $F \xrightarrow{a} G$.

transitive: Suppose that $E = F$ and $F = G$. To show that $E = G$ we need to demonstrate that $E \xrightarrow{a} H$ if, and only if, $G \xrightarrow{a} H$. However,

$$E \xrightarrow{a} H \text{ if, and only if, } F \xrightarrow{a} H \quad \text{(since } E = F\text{)}$$

$$\text{if, and only if, } G \xrightarrow{a} H \quad \text{(since } F = G\text{).}$$

Exercise 11.18 (page 303)

The only way to infer that $A = A_0$ would be to first show that $a.A = a.A_1$, which would require us to show that $A = A_1$.

However, the only way to infer that $A = A_1$ would be to first show that $a.A = a.A_2$, which would require us to show that $A = A_2$.

Likewise, the only way to infer that $A = A_2$ would be to first show that $a.A = a.A_3$, which would require us to show that $A = A_3$.

Continuing in this fashion, we would never finish, and so we could never reach our goal of inferring that $A = A_0$.

Chapter 12

Exercise 12.2 (page 312)

Fact: For any games $G_n(E, F)$, either the first player has a winning strategy, or the second player has a winning strategy.

Proof: By induction on n.

For the base case, the second player clearly has a winning strategy for any game $G_0(E, F)$ of length $n=0$.

For the inductive case, assume that for any game $G_n(E', F')$ of length n, either the first player has a winning strategy, or the second player has a winning strategy. Suppose then that the following two properties hold:

- for all actions a and all states E', if $E \xrightarrow{a} E'$ then $F \xrightarrow{a} F'$ for some state F' such that the second player has a winning strategy for the game $G_n(E', F')$; and

- for all actions a and all states F', if $F \xrightarrow{a} F'$ then $E \xrightarrow{a} E'$ for some state E' such that the second player has a winning strategy for the game $G_n(E', F')$.

That is, suppose that no matter what the first player does as her first move in the game $G_{n+1}(E, F)$ – either a move $E \xrightarrow{a} E'$ or a move $F \xrightarrow{a} F'$ – the second player can respond in such a way that he gets into a position in which he has a winning strategy in the game of length n. This clearly defines a winning strategy for the second player in the game $G_{n+1}(E, F)$.

Hence, if the second player does *not* have a winning strategy in the game $G_{n+1}(E, F)$, then one of the above two properties fails to hold. That is, either

- $E \xrightarrow{a} E'$ in such a way that whenever $F \xrightarrow{a} F'$ the second player does not have a winning strategy in the game $G_n(E', F')$; but then by the inductive hypothesis, this implies that the first player has a winning strategy in the game $G_n(E', F')$, which means she can use the $E \xrightarrow{a} E'$ transition as the first move in a winning strategy for the game $G_{n+1}(E, F)$; or

- $F \xrightarrow{a} F'$ in such a way that whenever $E \xrightarrow{a} E'$ the second player does not have a winning strategy in the game $G_n(E', F')$; but then by the inductive hypothesis, this implies that the first player has a winning strategy in the game $G_n(E', F')$, which means she can use the $F \xrightarrow{a} F'$ transition as the first move in a winning strategy for the game $G_{n+1}(E, F)$.   □

Exercise 12.3 (page 313)

$C \sim_2 D$ (and hence $C \sim_0 D$ and $C \sim_1 D$) since from either C or D you can do an a action and nothing else, and regardless of where that takes you, you will be able to do a b action and nothing else. Therefore the second player can obviously copy whatever two moves the first player makes in the Bisimulation Game when the tokens start on the pair of states (C, D).

$C \not\sim_3 D$ (and hence $C \not\sim_n D$ for all $n \geq 3$) since the first player has a strategy which will win her the game within three moves starting from the pair of states (C, D):

- For her first move she can do $C \xrightarrow{a} A$, to which the second player would have to respond with $D \xrightarrow{a} B$; the two tokens will then be on the pair of states (A, B).

- For her second move she could then do $B \xrightarrow{b} c.0 + d.0$, to which the second player would have to respond with either with $A \xrightarrow{b} c.0$ or with $A \xrightarrow{b} d.0$; the two tokens will then be on the pair of states $(c.0, c.0 + d.0)$ or the pair of states $(d.0, c.0 + d.0)$.

- For her third move, if the tokens are on the pair of states $(c.0, c.0+d.0)$ then she should do $c.0 + d.0 \xrightarrow{d} 0$, and if the tokens are on the pair of state $(d.0, c.0 + d.0)$ then she should do $c.0 + d.0 \xrightarrow{c} 0$; in either case the second player will not be able to respond.

Exercise 12.4 (page 315)

Fact: For all $n \in \mathbb{N}$, and for all states E, F, and G, if $E \sim_n F$ and $F \sim_n G$ then $E \sim_n G$.

Proof: By induction on $n \in \mathbb{N}$.

For the base case $n=0$, Theorem 12.3(1) gives us that $E \sim_0 G$.

For the induction step, we assume that $E \sim_{n+1} F$ and $F \sim_{n+1} G$. Referring to the pictorial representations of Theorem 12.3 (page 313), for the induction step the argument will be based on the following picture:

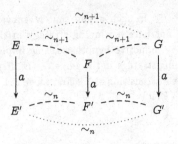

Suppose that the first player makes a transition $E \xrightarrow{a} E'$. By Theorem 12.3(2), since $E \sim_{n+1} F$ we have that $F \xrightarrow{a} F'$ for some F' such that $E' \sim_n F'$; and hence again by Theorem 12.3(2) since $F \sim_{n+1} G$ we have that $G \xrightarrow{a} G'$ for some G' such that $F' \sim_n G'$. Thus, by induction $E' \sim_n G'$. In summary,

- If $E \xrightarrow{a} E'$ then $G \xrightarrow{a} G'$ for some G' such that $E' \sim_n G'$.

Suppose instead that the first player makes a transition $G \xrightarrow{a} G'$. By Theorem 12.3(2), since $F \sim_{n+1} G$ we have that $F \xrightarrow{a} F'$ for some F' such that $F' \sim_n G'$; and hence again by Theorem 12.3(2) since $E \sim_{n+1} F$ we have that $E \xrightarrow{a} E'$ for some E' such that $E' \sim_n F'$. Thus, by induction $E' \sim_n G'$. In summary,

- If $G \xrightarrow{a} G'$ then $E \xrightarrow{a} E'$ for some E' such that $E' \sim_n G'$.

These two bullet points, by Theorem 12.3(2), gives us that $E \sim_{n+1} G$.   □

Exercise 12.5 (page 315)

Fact: For all $n \in \mathbb{N}$, $\mathrm{Cl}_n \sim_n \mathrm{Cl}$, while $\mathrm{Cl}_n \not\sim_{n+1} \mathrm{Cl}$.

Proof: We can show the equivalence by induction on n.

Base Case: $\mathrm{Cl}_0 \sim_0 \mathrm{Cl}$, by Theorem 12.3(1).

Induction Step: Assuming, for some n, that $\mathrm{Cl}_n \sim_n \mathrm{Cl}$, we can conclude from Theorem 12.3(2) that $\mathrm{Cl}_{n+1} \sim_{n+1} \mathrm{Cl}$

The inequivalence follows from noting that in the bisimulation game played with the tokens on Cl_n and Cl, after an exchange of n moves the tokens will necessarily be on Cl_0 and Cl, and the first person will be able to make the move $\mathrm{Cl} \xrightarrow{tick} \mathrm{Cl}$ which the second person cannot match.   □

Fact: For all $n \in \mathbb{N}$, $\mathrm{Clock} \sim_n \mathrm{Clock}_*$, while $\mathrm{Clock} \not\sim_\infty \mathrm{Clock}_*$.

Proof: We can firstly note that if the first player is to have any chance of winning the bisimulation game with the tokens on the states Clock and Clock$_*$, then he must start with the move Clock$_* \xrightarrow{tick}$ Cl; the second player could respond to any other opening move in such a way as to leave the two tokens on the same state, leaving him with the obvious copycat strategy to win.

In response to this opening move Clock$_* \xrightarrow{tick}$ Cl, the second player can move Clock $\xrightarrow{tick}$ Cl$_n$; and since (from above) Cl$_n \sim_n$ Cl for all n, by Theorem 12.3(2) we can deduce that Clock $\sim_n$ Clock$_*$ for all n.

The inequivalence follows from the fact (from above) that Cl$_n \not\sim_{n+1}$ Cl.

$\square$

Exercise 12.6 (page 317)

To prove that $\mathcal{R}$ is a bisimulation relation, we need to demonstrate that the bisimulation property from Definition 12.5 holds of each of the five pairs of states related by $\mathcal{R}$.

- $(P_1, Q_1) \in \mathcal{R}$:
  - $P_1 \xrightarrow{a} P_2$ is matched by $Q_1 \xrightarrow{a} Q_2$, and vice versa, as $(P_2, Q_2) \in \mathcal{R}$.
  - $P_1 \xrightarrow{a} P_3$ is matched by $Q_1 \xrightarrow{a} Q_3$, and vice versa, as $(P_3, Q_3) \in \mathcal{R}$.
- $(P_2, Q_2) \in \mathcal{R}$:
  - $P_2 \xrightarrow{b} P_3$ is matched by $Q_2 \xrightarrow{b} Q_3$, and vice versa, as $(P_3, Q_3) \in \mathcal{R}$.
- $(P_2, Q_4) \in \mathcal{R}$:
  - $P_2 \xrightarrow{b} P_3$ is matched by $Q_4 \xrightarrow{b} Q_5$, and vice versa, as $(P_3, Q_5) \in \mathcal{R}$.
- $(P_3, Q_3) \in \mathcal{R}$:
  - $P_3 \xrightarrow{b} P_1$ is matched by $Q_3 \xrightarrow{b} Q_1$, and vice versa, as $(P_1, Q_1) \in \mathcal{R}$.
  - $P_3 \xrightarrow{b} P_2$ is matched by $Q_3 \xrightarrow{b} Q_4$, and vice versa, as $(P_2, Q_4) \in \mathcal{R}$.
- $(P_3, Q_5) \in \mathcal{R}$:
  - $P_3 \xrightarrow{b} P_1$ is matched by $Q_5 \xrightarrow{b} Q_1$, and vice versa, as $(P_1, Q_1) \in \mathcal{R}$.
  - $P_3 \xrightarrow{b} P_2$ is matched by $Q_5 \xrightarrow{b} Q_2$, and vice versa, as $(P_2, Q_2) \in \mathcal{R}$.

Exercise 12.7 (page 317)

Assume that $\mathcal{R}$ and $\mathcal{S}$ are bisimulation relations over the states of a labelled transition system, and that $(E, G) \in \mathcal{R} \circ \mathcal{S}$. This means that $E \mathcal{S} F$ and $F \mathcal{R} G$ for some state F.

- If $E \xrightarrow{a} E'$, then from $E \mathcal{S} F$ we get that $F \xrightarrow{a} F'$ for some F' such that

$E'SF'$; and thus from FRG we get that $G \overset{a}{\to} G'$ for some G' such that $F'RG'$ and hence such that $(E', G') \in \mathcal{R} \circ \mathcal{S}$.

- If $G \overset{a}{\to} G'$, then from FRG we get that $F \overset{a}{\to} F'$ for some F' such that $F'RG'$; and thus from ERF we get that $E \overset{a}{\to} E'$ for some E' such that $E'RF'$ and hence such that $(E', G') \in \mathcal{R} \circ \mathcal{S}$.

Thus $\mathcal{R} \circ \mathcal{S}$ is a bisimulation relation.

Exercise 12.8 (page 318)

Let $E \prec_n F$ (where n may be ∞) mean that the second player has a winning strategy in the n-round game in which the first player must always move the token which starts on state E. Then clearly $\asymp_n = \prec_n \cap \prec_n^{-1}$.

(a) $\asymp_n$ is an equivalence relation, as it is:

reflexive: If both tokens are on the same node, then the second player has the obvious winning strategy of following the lead of the first player, copying each move.

symmetric: This follows from the fact that $\asymp_n = \prec_n \cap \prec_n^{-1}$.

transitive: We first show, by induction on n, that $E \prec_n G$ whenever $E \prec_n F$ and $F \prec_n G$. If $n = 0$ then we immediately have that $E \prec_n G$, so suppose $n = k+1$, and suppose that the first player makes a transition $E \overset{a}{\to} E'$; then we must have that $F \overset{a}{\to} F'$ with $E' \prec_n F'$, and thus we have that $G \overset{a}{\to} G'$ with $F' \prec_n G'$; hence by induction $E \prec_{n+1} G$.

To demonstrate that $\prec_\infty$ is transitive, we modify Definition 12.5 (page 316) to define a *simulation relation* to be a binary relation $\mathcal{R}$ over states which satisfies the following property: if ERF then

- if $E \overset{a}{\to} E'$ then $F \overset{a}{\to} F'$ for some F' such that $E'RF'$.

We then rephrase Theorem 12.6 (page 316) as

> The second player has a winning strategy in an infinite simulation game with the tokens starting on states E and F if, and only if, ERF for some simulation relation $\mathcal{R}$. Hence in particular, $\mathcal{R} \subseteq \prec_\infty$ for any simulation relation $\mathcal{R}$.

The proof of this result is completely analogous to that for Theorem 12.6. Our result then follows by showing that $\mathcal{R} \circ \mathcal{S}$ is a simulation relation whenever $\mathcal{R}$ and $\mathcal{S}$ are: this is shown as for the solution to Exercise 12.7.

Assume then that $\mathcal{R}$ and $\mathcal{S}$ are simulation relations, and that $(E, G) \in \mathcal{R} \circ \mathcal{S}$. This means that ESF and FRG for some state F.

If $E \xrightarrow{a} E'$, then from ESF we get that $F \xrightarrow{a} F'$ for some F' such that $E'SF'$; and thus from FRG we get that $G \xrightarrow{a} G'$ for some G' such that $F'RG'$ and hence such that $(E', G') \in R \circ S$.

Thus $R \circ S$ is a simulation relation.

(b) If the second player has a winning strategy in the bisimulation game, then he can use this same strategy to win the new game. The new game only restricts the possible moves of the first player

(c) It is easily verified that $a.b.0 \not\sim_2 a.b.0 + a.0$, while $a.b.0 \precsim_2 a.b.0 + a.0$.

Exercise 12.9 (page 321)

Suppose, by way of contradiction, that the first transition system of Figure 12.1 is coloured with a bisimulation colouring which assigns the same colour to the two states X and U. Since $U \xrightarrow{a} V$, there must be an a-labelled transition out of X leading to a state with the same colour as V. This state must be Y, and hence V and Y must have the same colour in this supposed bisimulation colouring. But then since $Y \xrightarrow{c} Z$, there must be a c-labelled transition out of V leading to a state with the same colour as Z. However, there is no such state, of any colour, which provides us with our desired contradiction.

Exercise 12.10 (page 322)

Consider the first transition system of Figure 12.1.

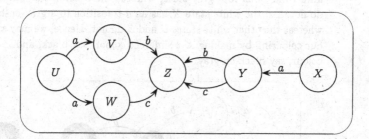

The initial all-white colouring is *not* a bisimulation colouring, as the white states V and Y have b-transition to white states, whereas the other white states U, W, X and Z do not have b-transitions to white states. Hence, by the invariant, states V and Y cannot be equivalent to the other white states; in any bisimulation colouring, states V and Y must each have a different colour from states U, W, X and Z. Hence we may safely refine our colouring by making states V and Y a different colour (gray, say).

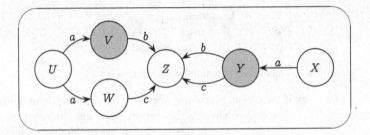

This is still not a bisimulation colouring, as the gray states Y has a c-transitions to a white, whereas the other white state V does not. Hence, by the invariant, states V and Y cannot be equivalent, and so we may safely refine our colouring by making Y a different colour, say gray-on-black. At the same time we can note that the white state W has a c-transitions to a white, whereas the other white states U, X and Z do not, and so we can safely make W a different colour, say gray-on white.

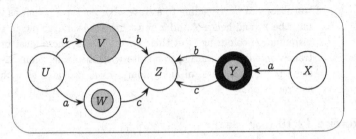

Again this is still not a bisimulation colouring, as the white states U has an a-transitions to a gray state, whereas the other white states X and Z do not; and the white state X has an a-transition to a gray-on-black state, whereas the other white states U and Z do not. Hence, we may safely refine our colouring by making X a different colour, say black, and Z a different colour, say black-on-gray.

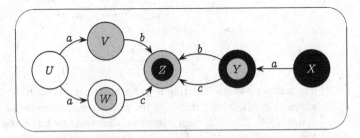

This colouring *is* a bisimulation colouring, which by construction satisfies our invariant. That it is a bisimulation colouring is clear, since there are no two states with the same colour.

Exercise 12.11 (page 326)

Take:

$$E_0 \stackrel{\text{def}}{=} \text{Clock} \qquad\qquad F_0 \stackrel{\text{def}}{=} \text{Clock}_*$$

$$E_{n+1} \stackrel{\text{def}}{=} tick.E_n \qquad F_{n+1} \stackrel{\text{def}}{=} tick.F_n$$

By a straightforward induction argument we can show that, for each $n \in \mathbb{N}$, $E_n \sim_{\omega+n} F_n$ but $E_n \not\sim_{\omega+n+1} F_n$.

Chapter 13

Exercise 13.1 (page 336)

1. $\langle coffee \rangle true$ says:

   we may do a 'coffee' action and end up in a state in which true is true.

   This property will be true if it is possible to do a 'coffee' action.

2. $\langle coffee \rangle false$ says:

   we may do a 'coffee' action and end up in a state in which false is true.

   Such a 'coffee' action cannot therefore be possible, as false could not possibly be true in any subsequent state. Therefore this property can never be satisfied; it is equivalent to the property false.

3. $[coffee]true$ says:

   no matter how we do a 'coffee' action, we must end up in a state in which true is true.

   This property will always be true – regardless of whether or not we can do a 'coffee' action – as true will of course be true in any subsequent state. This property is therefore equivalent to the property true.

4. $[coffee]false$ says:

   no matter how we do a 'coffee' action, we must end up in a state in which false is true.

   A 'coffee' action must therefore not be possible, since false can never be true in any subsequent state. This formula thus says the same thing as $\neg\langle coffee \rangle true$.

Exercise 13.2 (page 336)

$\neg\langle a \rangle\langle a \rangle true$.

In words, this says that it is *not* the case that I can do an 'a' action and get into a state in which I can do another 'a' action.

Exercise 13.3 (page 336)

[tick]⟨tick⟩true.

In words, this says that no matter how I do a 'tick' action, I must end up in a state in which I can do another 'tick' action. This is true of the clock C1 but not of the clock C1$_*$, as the latter clock may stop after just one tick.

Exercise 13.4 (page 340)

1, 4, 5, 6, 7 and 10 are valid, whereas 2, 3, 8, 9, 11 and 12 are not valid.

Exercise 13.5 (page 341)

1. ⟨*pull*⟩⟨*pull*⟩⟨*break*⟩true.

   This is true only of the state ON.

2. ⟨*pull*⟩⟨*pull*⟩⟨*reset*⟩true.

   This is true only of the state BROKEN.

3. ¬⟨*pull*⟩true.

   This is not true of any state; you can do a '*pull*' action from any state.

4. ⟨*pull*⟩true ∧ ¬⟨*break*⟩true ∧ ¬⟨*reset*⟩true.

   This is true only of the state OFF.

   Note that this assumes that the only actions available of the process are '*pull*', '*break*' and '*reset*'. We need to include a conjunct ¬⟨*a*⟩true for every action $a \neq pull$ to explicitly disallow the possibility of such an action '*a*' being possible.

Exercise 13.6 (page 342)

Fact: ¬[a]P ⇔ ⟨a⟩¬P

Proof: $E \models \neg[a]P \Leftrightarrow E \not\models [a]P$

$$\Leftrightarrow \neg \forall F (E \xrightarrow{a} F \Rightarrow F \models P)$$

$$\Leftrightarrow \exists F \neg (E \xrightarrow{a} F \Rightarrow F \models P)$$

$$\Leftrightarrow \exists F (E \xrightarrow{a} F \wedge F \not\models P)$$

$$\Leftrightarrow \exists F (E \xrightarrow{a} F \wedge F \models \neg P)$$

$$\Leftrightarrow E \models \langle a \rangle \neg P \qquad \square$$

Exercise 13.7 (page 343)

Theorem 13.6: For any process E and any property P of **HML**:

1. $E \models \text{pos}(P)$ if, and only if, $E \models P$; and

2. $E \models \text{neg}(P)$ if, and only if, $E \not\models P$.

Proof: By induction on the structure of P. That is, we demonstrate that

1. $E \models \text{pos}(P)$ if, and only if, $E \models P$; and

2. $E \models \text{neg}(P)$ if, and only if, $E \not\models P$

under the assumption that, for any process F and any property Q smaller than P,

1. $F \models \text{pos}(Q)$ if, and only if, $F \models Q$; and

2. $F \models \text{neg}(Q)$ if, and only if, $F \not\models Q$.

We thus argue by cases on the structure of P:

$\underline{P = \text{true}}$:

1. $E \models \text{pos}(\text{true})$

    $\Leftrightarrow$ $E \models \text{true}$ [*by definition of* $\text{pos}(\text{true})$]

2. $E \models \text{neg}(\text{true})$

    $\Leftrightarrow$ $E \models \text{false}$ [*by definition of* $\text{neg}(\text{true})$]

    $\Leftrightarrow$ $E \not\models \text{true}$ [*by semantic definition for* true *and* false]

$\underline{P = \text{false}}$:

1. $E \models \text{pos}(\text{false})$

    $\Leftrightarrow$ $E \models \text{false}$ [*by definition of* $\text{pos}(\text{false})$]

2. $E \models \text{neg}(\text{false})$

    $\Leftrightarrow$ $E \models \text{true}$ [*by definition of* $\text{neg}(\text{false})$]

    $\Leftrightarrow$ $E \not\models \text{false}$ [*by semantic definition for* true *and* false]

$\underline{P = \neg Q}$:

1. $E \models \text{pos}(\neg Q)$

    $\Leftrightarrow$ $E \models \text{neg}(Q)$ [*by definition of* $\text{pos}(\neg Q)$]

    $\Leftrightarrow$ $E \not\models Q$ [*by induction hypothesis 2*]

    $\Leftrightarrow$ $E \models \neg Q$ [*by semantic definition for* $\neg$]

2. $E \models \text{neg}(\neg Q)$

    $\Leftrightarrow$ $E \models \text{pos}(Q)$ [*by definition of* $\text{neg}(\neg Q)$]

    $\Leftrightarrow$ $E \models Q$ [*by induction hypothesis 1*]

    $\Leftrightarrow$ $E \not\models \neg Q$ [*by semantic definition for* $\neg$]

$\underline{P = Q_1 \wedge Q_2}$:

1. $E \models \text{pos}(Q_1 \wedge Q_2)$
   $\Leftrightarrow E \models \text{pos}(Q_1) \wedge \text{pos}(Q_2)$ *[by definition of* $\text{pos}(Q_1 \wedge Q_2)$*]*
   $\Leftrightarrow E \models \text{pos}(Q_1)$ and $E \models \text{pos}(Q_2)$ *[by semantic definition for* $\wedge$*]*
   $\Leftrightarrow E \models Q_1$ and $E \models Q_2$ *[by induction hypothesis 1]*
   $\Leftrightarrow E \models Q_1 \wedge Q_2$ *[by semantic definition for* $\wedge$*]*

2. $E \models \text{neg}(Q_1 \wedge Q_2)$
   $\Leftrightarrow E \models \text{neg}(Q_1) \vee \text{neg}(Q_2)$ *[by definition of* $\text{neg}(Q_1 \wedge Q_2)$*]*
   $\Leftrightarrow E \models \text{neg}(Q_1)$ or $E \models \text{neg}(Q_2)$ *[by semantic definition for* $\vee$*]*
   $\Leftrightarrow E \not\models Q_1$ or $E \not\models Q_2$ *[by induction hypothesis 2]*
   $\Leftrightarrow \neg(E \models Q_1$ and $E \models Q_2)$ *[by De Morgan's Law]*
   $\Leftrightarrow E \not\models Q_1 \wedge Q_2$ *[by semantic definition for* $\wedge$*]*

$\underline{P = Q_1 \vee Q_2}$:

1. $E \models \text{pos}(Q_1 \vee Q_2)$
   $\Leftrightarrow E \models \text{pos}(Q_1) \vee \text{pos}(Q_2)$ *[by definition of* $\text{pos}(Q_1 \vee Q_2)$*]*
   $\Leftrightarrow E \models \text{pos}(Q_1)$ or $E \models \text{pos}(Q_2)$ *[by semantic definition for* $\vee$*]*
   $\Leftrightarrow E \models Q_1$ or $E \models Q_2$ *[by induction hypothesis 1]*
   $\Leftrightarrow E \models Q_1 \vee Q_2$ *[by semantic definition for* $\vee$*]*

2. $E \models \text{neg}(Q_1 \vee Q_2)$
   $\Leftrightarrow E \models \text{neg}(Q_1) \wedge \text{neg}(Q_2)$ *[by definition of* $\text{neg}(Q_1 \vee Q_2)$*]*
   $\Leftrightarrow E \models \text{neg}(Q_1)$ and $E \models \text{neg}(Q_2)$ *[by semantic definition for* $\wedge$*]*
   $\Leftrightarrow E \not\models Q_1$ and $E \not\models Q_2$ *[by induction hypothesis 2]*
   $\Leftrightarrow \neg(E \models Q_1$ or $E \models Q_2)$ *[by De Morgan's Law]*
   $\Leftrightarrow E \not\models Q_1 \vee Q_2$ *[by semantic definition for* $\vee$*]*

$\underline{P = \langle a \rangle Q}$:

1. $E \models \text{pos}(\langle a \rangle Q)$

$\Leftrightarrow\ E \models \langle a \rangle \text{pos}(Q)$ *[by definition of* $\text{pos}(\langle a \rangle Q)$*]*

$\Leftrightarrow\ F \models \text{pos}(Q)$ for some F such that $E \xrightarrow{a} F$

[by semantic definition for $\langle a \rangle$*]*

$\Leftrightarrow\ F \models Q$ for some F such that $E \xrightarrow{a} F$

[by induction hypothesis 1]

$\Leftrightarrow\ E \models \langle a \rangle Q$ *[by semantic definition for* $\langle a \rangle$*]*

2. $E \models \text{neg}(\langle a \rangle Q)$

$\Leftrightarrow\ E \models [a]\text{neg}(Q)$ *[by definition of* $\text{neg}(\langle a \rangle Q)$*]*

$\Leftrightarrow\ F \models \text{neg}(Q)$ for all F such that $E \xrightarrow{a} F$

[by semantic definition for $[a]$*]*

$\Leftrightarrow\ F \not\models Q$ for all F such that $E \xrightarrow{a} F$

[by induction hypothesis 2]

$\Leftrightarrow\ E \not\models \langle a \rangle Q$ *[by semantic definition for* $\langle a \rangle$*]*

$\underline{P = [a]Q}$:

1. $E \models \text{pos}([a]Q)$

$\Leftrightarrow\ E \models [a]\text{pos}(Q)$ *[by definition of* $\text{pos}([a]Q)$*]*

$\Leftrightarrow\ F \models \text{pos}(Q)$ for all F such that $E \xrightarrow{a} F$

[by semantic definition for $[a]$*]*

$\Leftrightarrow\ F \models Q$ for all F such that $E \xrightarrow{a} F$

[by induction hypothesis 1]

$\Leftrightarrow\ E \models [a]Q$ *[by semantic definition for* $[a]$*]*

2. $E \models \text{neg}([a]Q)$

$\Leftrightarrow\ E \models \langle a \rangle \text{neg}(Q)$ *[by definition of* $\text{neg}([a]Q)$*]*

$\Leftrightarrow\ F \models \text{neg}(Q)$ for some F such that $E \xrightarrow{a} F$

[by semantic definition for $\langle a \rangle$*]*

$\Leftrightarrow\ F \not\models Q$ for some F such that $E \xrightarrow{a} F$

[by induction hypothesis 2]

$\Leftrightarrow\ E \not\models [a]Q$ *[by semantic definition for* $[a]$*]*

$\square$

Exercise 13.8 (page 343)

Fact: For all modal properties P, $neg(neg(P)) = P$.

Proof: By induction on the structure of P, arguing by cases on the structure of P.

$\underline{P = \textbf{true}}$: $neg(neg(true)) = neg(false) = true$.

$\underline{P = \textbf{false}}$: $neg(neg(false)) = neg(true) = false$.

$\underline{P = Q_1 \wedge Q_2}$: By the inductive hypothesis we assume that $neg(neg(Q_1)) = Q_1$ and $neg(neg(Q_2)) = Q_2$.

Then $neg(neg(Q_1 \wedge Q_2)) = neg(neg(Q_1) \vee neg(Q_2)) = neg(neg(Q_1)) \wedge neg(neg(Q_2)) = Q_1 \wedge Q_2$

$\underline{P = Q_1 \vee Q_2}$: By the inductive hypothesis we assume that $neg(neg(Q_1)) = Q_1$ and $neg(neg(Q_2)) = Q_2$.

Then $neg(neg(Q_1 \vee Q_2)) = neg(neg(Q_1) \wedge neg(Q_2)) = neg(neg(Q_1)) \vee neg(neg(Q_2)) = Q_1 \vee Q_2$

$\underline{P = \langle a \rangle Q}$: By the inductive hypothesis we assume that $neg(neg(Q)) = Q$.

Then $neg(neg(\langle a \rangle Q)) = neg([a]neg(Q)) = \langle a \rangle neg(neg(Q)) = \langle a \rangle Q$

$\underline{P = [a]Q}$: By the inductive hypothesis we assume that $neg(neg(Q)) = Q$.

Then $neg(neg([a]Q)) = neg(\langle a \rangle neg(Q)) = [a]neg(neg(Q)) = [a]Q$

$\square$

Exercise 13.9 (page 345)

The properties distinguishing between C and D were presented informally in the solution to Exercise 11.16(b). We need simply express these properties in the language of **HML**.

- From state C you may do an 'a' action and be in a state in which, no matter how you do a 'b' action you will either not be able to do a 'c' action or you will not be able to do a 'd' action. Formally:

$$C \models \langle a \rangle [b]([c]false \vee [d]false)$$

- On the other hand, from state D no matter how you do an 'a' action, you will be able to do a 'b' action and end up in a state in which you can both do a 'c' action as well as a 'd' action. Formally:

$$D \models [a]\langle b\rangle(\langle c\rangle\text{true} \wedge \langle d\rangle\text{true})$$

Note that these properties are, naturally, the negations of each other: $D = \text{neg}(C)$ and $C = \text{neg}(D)$.

Exercise 13.11 (page 350)

1. Consider the formula

    $$\langle a\rangle\text{true} \wedge [-a]\text{false} \wedge [-][-]\text{false}.$$

    Clearly this characterises the process $a.0$:

    - The first conjunct says that it is possible to do an a transition;
    - The second conjunction says that it is not possible to do anything other than an a transition.
    - The final conjunct says that it is not possible to do two transitions.

2. The characteristic formula for $a.(b.0 + c.0)$ is

    $$\langle a\rangle\text{true} \wedge [-a]\text{false} \wedge [a]\big(\langle b\rangle\text{true} \wedge \langle c\rangle\text{true} \wedge [-][-]\text{false}\big).$$

Exercise 13.12 (page 353)

1. $\|\langle a\rangle\text{true}\| = \{ E, E_1, F \}$
2. $\|\langle b\rangle\text{true}\| = \{ E_1, E_2 \}$
3. $\|\langle a\rangle\langle a\rangle\text{true}\| = \{ E, E_1 \}$
4. $\|\langle b\rangle\langle b\rangle\text{true}\| = \emptyset$
5. $\|\langle a\rangle[a]\text{false}\| = \{ F \}$
6. $\|[b]\langle a\rangle\text{true}\| = \{ E, E_1, E_2, F \}$

Chapter 14

Exercise 14.2 (page 360)

Let $A \stackrel{\text{def}}{=} 0$ with $\text{Sort}(A) = \emptyset$, and $B \stackrel{\text{def}}{=} 0$ with $\text{Sort}(B) = \{a\}$.

Then clearly $A \sim B$ although $\text{Sort}(A) \neq \text{Sort}(B)$.

If we let $X \stackrel{\text{def}}{=} a.0$ with $\text{Sort}(X) = \emptyset$, then $A \parallel X \sim a.0$, but $B \parallel X \sim 0$.

Thus, $A \sim B$ whereas $A \parallel X \not\sim B \parallel X$.

Exercise 14.3 (page 362)

The relevant bisimulation relation is

$$\big\{ (C_2, C \parallel C), (C_2', dec.C \parallel C), (C_2', C \parallel dec.C), (C_2'', dec.C \parallel dec.C) \big\}.$$

Exercise 14.4 (page 365)

The safety property holds: a car may cross only if the barrier is up, and a train may cross only if the signal is green; and the controller ensures that the barrier is never up at the same time that the signal is green by raising the barrier only when the signal is red and turning the signal green only when the barrier is down.

The liveness properties, however, fail to hold as given. When a car arrives, it is not necessarily the case that the barrier will eventually go up. It may be the case that an endless stream of trains arrive, and that the controller repeatedly turns the signal green to allow each of these trains to cross the intersection without ever raising the barrier to allow the waiting car to pass. Equally, the controller may allow an endless stream of cars to pass, never changing the signal to green to allow a waiting train to pass.

These liveness properties can be weakened to read:

- If a car arrives, eventually the barrier may go up.

- If a train arrives, eventually the signal may turn green.

These weakened properties do hold of the system.

In reality, a barrier typically remains up, to allow cars to cross the intersection freely, until a train arrives; the arrival of a train signals the controller, which then lowers the barrier, then turns the signal to green, then turns the signal to red again, and finally raises the barrier once again. If the components are built correctly following this protocol, then the original liveness properties will hold, along with the safety properties.

Exercise 14.5 (page 368)

The only way that the system can deadlock is if every philosopher is wanting to pick up a fork which is not available. (No philosopher would ever be hindered from eating nor from setting a fork down on the table.) No two philosophers can be wanting to pick up the same fork, as each one of them must be prevented from picking it up by the other already holding it. Since each philosopher is stopped by the absence of a different fork, every fork must be in the hand of some philosopher, and thus each philosopher must be in the state of having just picked up their first fork. But that would mean that philosophers 1 and 2 are both holding fork 2, which is impossible.

Exercise 14.6 (page 371)

We argue that if the first process reaches the state where it is ready to enter the critical section, then the second process will not be able to reach the analogous state until the first process enters and then exist the critical section. A symmetric argument shows that the second process being in the critical section prevents the first from also being so.

- When the first process becomes ready to enter the critical section (ie, enters state R_1), then the b1 process must be in the state B_1t, and either the b2 process is in the state B_2f or the k processor is in the state K_1.

- Before the first process enters and exists the critical section, if the second process is waiting to be allowed to enter the critical section (ie, is in state W_2), then the b2 process must be in state B_2t. Hence (from above) the k processor is in the state K_1. Thus this process will not be able to move to state R_2 and enter the critical section.

Exercise 14.7 (page 373)

The enhanced message-passing protocol requires no change to the SENDER, only to the RECEIVER and the MEDIUM. The acknowledgement that the SENDER is awaiting will come from the MEDIUM rather than directly from the RECEIVER, but this difference is not noticeable from the point of view of the SENDER. Thus its definition remains unchanged:

$$\text{Sender} \stackrel{\text{def}}{=} \text{in.snd.S} \qquad \text{S} \stackrel{\text{def}}{=} \text{ack.Sender} + \text{err.snd.S}$$

$$\text{Sort(Sender)} = \{\text{snd}, \text{ack}, \text{err}\}$$

Again, its transition system is depicted thus:

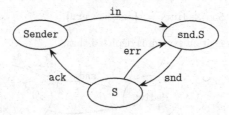

The enhanced RECEIVER must cater for the possibility of its acknowledgement being lost. After receiving a message (via the "rcv" action) and forwarding it on (via the "out" action), it will issue an auxiliary acknowledgement to the MEDIUM (via a "rack" action). At this point it will be ready to receive a new message. However, it may instead receive an auxiliary error message from the MEDIUM (modelled by a "rerr" action), indicating that the acknowledgement was lost, in which case it will retransmit this acknowledgement. The new definition is as follows:

Receiver $\stackrel{\text{def}}{=}$ rcv.out.rack.Receiver + rerr.rack.Receiver

Sort(Receiver) = {rcv, rack, rerr}

Its transition system is depicted thus:

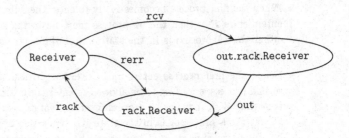

The behaviour of the MEDIUM must now interact with the RECEIVER in delivering the acknowledgement from the RECEIVER to the SENDER of the safe arrival of the message being delivered. After passing the message to the RECEIVER (via the "rcv" action), the Medium awaits the auxiliary acknowledgement from the RECEIVER (modelled by the "rack" action). It then either passes the acknowledgement along to the SENDER (via the "ack" action); or it may lose the acknowledgement (modelled by a "rerr" action), and await a new acknowledgement from the RECEIVER. The new definition is as follows:

Medium $\stackrel{\text{def}}{=}$ snd.(rcv.rack.M + err.Medium)

M $\stackrel{\text{def}}{=}$ ack.Medium + rerr.rack.M

Sort(Medium) = {snd, rcv, err}

Its transition system is depicted thus:

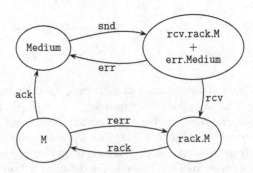

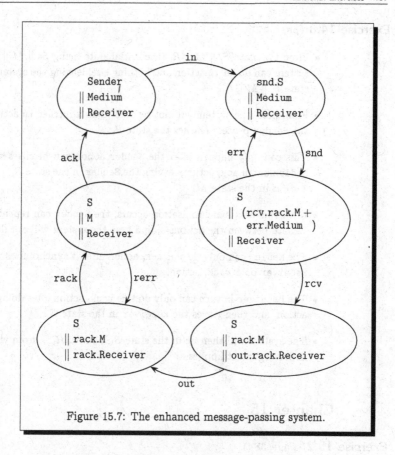

Figure 15.7: The enhanced message-passing system.

The symmetry reflected in the transition diagram makes clear the similarity in the manner that the MEDIUM treats the SENDER and RECEIVER.

The complete system is again defined to be the composition of these three components:

$$\text{System} \overset{\text{def}}{=} \text{Sender} \parallel \text{Medium} \parallel \text{Receiver}$$

but now the following configuration:

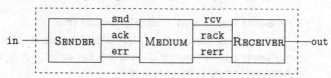

The behaviour of the complete enhanced system is thus depicted by the transition system depicted in Figure 15.7. The symmetry between the SENDER and the RECEIVER is immediately noticeable in this transition system.

Exercise 14.8 (page 377)

- From the state $S_i \parallel M \parallel R_i$ (the initial state being $S_0 \parallel M \parallel R_0$) the system can do an in action and nothing else, leaving the system in the state $S_i' \parallel M \parallel R_i$.

- From here, the system will not be able to do a further in action until the Sender process reaches the state S_{1-i}.

- This can only happen after the Sender action synchronises with the Medium on an ack_i action, leaving the Sender in the state S_{1-i} and the Medium in the state M.

- Until this ack_i synchronisation occurs, the Sender can repeatedly alternate between the actions s_i and t, so the system will not deadlock.

- The Medium can only do this ack_i action after it synchronises with the Receiver on a $rack_i$ action.

- The Receiver in turn can only do this $rack_i$ action after doing an out action, and then leaves the Receiver in the state R_{1-i}.

- The system will then be in the state $S_{1-i} \parallel M \parallel R_{1-i}$, from which the above argument applies.

Chapter 15

Exercise 15.2 (page 384)

$$
\begin{array}{lll}
\text{Deadlock-free} &= \Box\langle-\rangle\text{true} & \text{(by definition)}\\
&= \neg\Diamond\neg\langle-\rangle\text{true} & \text{(since } \Box P = \neg\Diamond\neg P)\\
&= \neg\Diamond[-]\text{false} & \text{(since } \neg\langle-\rangle P = [-]\neg P)\\
&= \neg\text{Deadlockable} & \text{(by definition)}
\end{array}
$$

Exercise 15.3 (page 385)

Ev P asserts that P must eventually become true.

This is almost the same as $Q \cup P$ which also asserts that P must eventually become true; the only difference is the added requirement that until P becomes true, Q must remain true.

However, this added requirement is vacuous if we take the property Q to be true, as of course true is always true anyways.

Hence, Ev P = true $\cup P$.

Exercise 15.4 (page 386)

Fact: $E \models_V P$ if, and only if, $E \in \|P\|_V$.

Proof: By induction on the structure of P, arguing by cases on the structure of P.

$\underline{P = \mathbf{true}}$: $E \models_V \mathbf{true}$ $\Leftrightarrow$ $E \in \text{States}$ $\Leftrightarrow$ $E \in \|\mathbf{true}\|_{V[X \mapsto S]}$

$\underline{P = \mathbf{false}}$: $E \models_V \mathbf{false}$ $\Leftrightarrow$ $E \in \emptyset$ $\Leftrightarrow$ $E \in \|\mathbf{false}\|_{V[X \mapsto S]}$

$\underline{P = X}$: $E \models_V X$ $\Leftrightarrow$ $E \in V(X)$ $\Leftrightarrow$ $E \in \|X\|_{V[X \mapsto S]}$

$\underline{P = \neg P}$: $E \models_V \neg P$ $\Leftrightarrow$ $E \not\models_V P$

$\qquad\qquad\qquad \Leftrightarrow E \notin \|P\|_{V[X \mapsto S]}$

$\qquad\qquad\qquad \Leftrightarrow E \in \overline{\|P\|_{V[X \mapsto S]}}$ $\Leftrightarrow$ $E \in \|\neg P\|_{V[X \mapsto S]}$

$\underline{P = Q_1 \wedge Q_2}$: $E \models_V Q_1 \wedge Q_2$ $\Leftrightarrow$ $E \models_V Q_1$ and $E \models_V Q_2$

$\qquad\qquad\qquad\qquad \Leftrightarrow E \in \|Q_1\|_{V[X \mapsto S]}$ and $E \in \|Q_2\|_{V[X \mapsto S]}$

$\qquad\qquad\qquad\qquad \Leftrightarrow E \in \|Q_1\|_{V[X \mapsto S]} \cap \|Q_2\|_{V[X \mapsto S]}$

$\qquad\qquad\qquad\qquad \Leftrightarrow E \in \|Q_1 \wedge Q_2\|_{V[X \mapsto S]}$

$\underline{P = Q_1 \vee Q_2}$: $E \models_V Q_1 \vee Q_2$ $\Leftrightarrow$ $E \models_V Q_1$ or $E \models_V Q_2$

$\qquad\qquad\qquad\qquad \Leftrightarrow E \in \|Q_1\|_{V[X \mapsto S]}$ or $E \in \|Q_2\|_{V[X \mapsto S]}$

$\qquad\qquad\qquad\qquad \Leftrightarrow E \in \|Q_1\|_{V[X \mapsto S]} \cup \|Q_2\|_{V[X \mapsto S]}$

$\qquad\qquad\qquad\qquad \Leftrightarrow E \in \|Q_1 \vee Q_2\|_{V[X \mapsto S]}$

$\underline{P = \langle a \rangle Q}$: $E \models_V \langle a \rangle Q$ $\Leftrightarrow$ $E \xrightarrow{a} E'$ such that $E' \models_V Q$

$\qquad\qquad\qquad\qquad \Leftrightarrow E \xrightarrow{a} E'$ such that $E' \in \|Q\|_{V[X \mapsto S]}$

$\qquad\qquad\qquad\qquad \Leftrightarrow E \in \|\langle a \rangle Q\|_{V[X \mapsto S]}$

$\underline{P = [a]Q}$: $E \models_V [a]Q$ $\Leftrightarrow$ $E \xrightarrow{a} E'$ implies $E' \models_V Q$

$\qquad\qquad\qquad \Leftrightarrow E \xrightarrow{a} E'$ implies $E' \in \|Q\|_{V[X \mapsto S]}$

$\qquad\qquad\qquad \Leftrightarrow E \in \|[a]Q\|_{V[X \mapsto S]}$ $\qquad\qquad\qquad\qquad$ $\square$

Exercise 15.5 (page 388)

$\|\langle a \rangle X\|_{V[X \mapsto \emptyset]} = \{\, E \in \text{States} : E \xrightarrow{a} E' \text{ for some } E' \in \emptyset \,\} = \emptyset.$

Exercise 15.6 (page 389)

By Exercise 15.5, the empty set $S = \emptyset$ satisfies $S = \|\langle a \rangle X\|_{V[X \mapsto S]}$ and hence must be the least fixed point of the function $f(S) = \|\langle a \rangle X\|_{V[X \mapsto S]}$.

Let $A = \{ E \in \text{States} \ : \ E \xrightarrow{a} \cdot \xrightarrow{a} \cdot \xrightarrow{a} \cdots \}$ be the set of states which we intended to capture in Example 15.4 with the recursive property $X = \langle a \rangle X$. As demonstrated in Example 15.4, this set is a fixed point; we shall demonstrate that A must in fact be the greatest fixed point.

To this end, suppose that S is any fixed point:

$$S = f(S) = \|\langle a \rangle X\|_{V[X \mapsto S]}$$
$$= \{ E \in \text{States} \ : \ E \xrightarrow{a} E' \text{ for some } E' \in S \}$$

and suppose further that $E \in S$. We need to show that $E \in A$.

- Since $E \in S$, $E \xrightarrow{a} E'$ for some $E' \in S$.
- Since $E' \in S$, $E' \xrightarrow{a} E''$ for some $E'' \in S$.
- Since $E'' \in S$, $E'' \xrightarrow{a} E'''$ for some $E''' \in S$.

Continuing in this fashion, it becomes clear that $E \in S$.

As for a fixed point of the function $f(S) = \|\langle a \rangle X\|_{V[X \mapsto S]}$ which is neither the least nor greatest fixed point, consider the process with two states A and B and two transitions $A \xrightarrow{a} A$ and $B \xrightarrow{a} A$. Then $\emptyset$, $\{A\}$ and $\{A, B\}$ are all fixed points of this function.

Exercise 15.7 (page 392)

We prove this by induction – and arguing by cases – on the structure of P. However, we only present the three cases which don't appear in the proof of the analogous result for **HML** (Theorem 13.6, page 343).

$\underline{P = X}$:
$$E \models_{\overline{V}} \text{neg}(X) \ \Leftrightarrow \ E \models_{\overline{V}} X \ \Leftrightarrow \ E \in \overline{V}(X) \ \Leftrightarrow \ E \notin V(X) \ \Leftrightarrow \ E \not\models_V X$$

$\underline{P = \mu X.Q}$:
$$E \models_{\overline{V}} \text{neg}(\mu X.Q) \ \Leftrightarrow \ E \models_{\overline{V}} \nu X.\text{neg}(Q)$$
$$\Leftrightarrow \ \exists S \subseteq \text{States} : E \in S \text{ and } \forall F \in S : F \models_{\overline{V}[X \mapsto S]} \text{neg}(Q)$$
$$\Leftrightarrow \ \exists S \subseteq \text{States} : E \notin S \text{ and } \forall F \notin S : F \models_{\overline{V[X \mapsto S]}} \text{neg}(Q)$$
$$\Leftrightarrow \ \exists S \subseteq \text{States} : E \notin S \text{ and } \forall F \notin S : F \not\models_{V[X \mapsto S]} Q$$
$$\Leftrightarrow \ E \not\models_V \mu X.Q$$

$\underline{P = \nu X.Q}$:

$$E \models_{\overline{\nabla}} \operatorname{neg}(\nu X.Q) \quad \Leftrightarrow \quad E \models_{\overline{\nabla}} \mu X.\operatorname{neg}(Q)$$

$$\Leftrightarrow \ \forall S \subseteq \text{States} : \text{if } E \notin S \text{ then } \exists F \notin S \text{ such that } F \models_{\overline{\nabla}[X \mapsto S]} \operatorname{neg}(Q)$$

$$\Leftrightarrow \ \forall S \subseteq \text{States} : \text{if } E \in S \text{ then } \exists F \in S \text{ such that } F \models_{\overline{\nabla[X \mapsto S]}} \operatorname{neg}(Q)$$

$$\Leftrightarrow \ \forall S \subseteq \text{States} : \text{if } E \in S \text{ then } \exists F \in S \text{ such that } F \not\models_{\nabla[X \mapsto S]} Q$$

$$\Leftrightarrow \ E \not\models_{\nabla} \nu X.Q \qquad\qquad\qquad\qquad \square$$

Exercise 15.12 (page 399)

1. With a least fixed point, we cannot be allowed to unroll the recursive equation infinitely often in verifying that the property P is true in every state.

   At every state we reach, the property P must hold. But we must eventually have nowhere to go; that it, the process must eventually deadlock.

   Thus this property is true as long as P is true in every state of the process and every run of the process deadlocks.

2. With a greatest fixed point, we are allowed to unroll the recursive property infinitely often in our search for a state in which P is true.

   At each state, either the property P must hold, or it must be possible to make a transition and continue the search for a state in which P holds; however, we need never complete this search.

   Thus this property is true if P is true in some state, or if there is an infinite path through the process.

3. With a greatest fixed point, we are allowed to unroll the recursive property infinitely often in our search for a state in which Q is true.

   Thus the property is true if P is true for as long as Q is not true, but until Q becomes true – if ever – it must be possible to do something.

Exercise 15.13 (page 401)

1. P almost always holds along some a^{ω} path.

   In order for this property to hold, there must be a state reachable by a sequence of a transitions from which an a^{ω} path exists along which P is always true.

   We have already seen how to express the property that P always holds along some a^{ω} path:

   $$\Phi \ = \ \nu X.P \wedge \langle a \rangle X.$$

   We need only find a state satisfying this property which can be reached by a sequence of a-transitions:

$$\mu X.\Phi \lor \langle a \rangle X.$$

Writing this out in full by substituting in the formula for Φ – whilst at the same time changing one of the variables to avoid confusion – we get the following:

$$\mu Z.\big(\nu X.P \land \langle a \rangle X\big) \lor \langle a \rangle Z.$$

2. P holds infinitely often along some a^ω path.

In order for this property to be true, we must be able to reach a state by doing a sequence of a transitions in which P holds, and then to repeat this forever.

We will, therefore, have a least fixed point construction – to allow us to look for the state in which P holds – embedded within a greatest fixed point construction – to allow us to repeat this search over and over again forever.

$$\nu Z.\mu X.(P \land \langle a \rangle Z) \lor \langle a \rangle X.$$

Index

F. Moller, G. Struth, *Modelling Computing Systems*,
Undergraduate Topics in Computer Science,
DOI 10.1007/978-1-84800-322-4, © Springer-Verlag London 2013